KEW INDEX
FOR 1989

KEW INDEX FOR 1989

NAMES OF SEED-BEARING PLANTS,
FERNS, AND FERN ALLIES
AT THE RANK OF FAMILY AND BELOW
PUBLISHED DURING 1989
WITH SOME OMISSIONS FROM
EARLIER YEARS

COMPILED BY

R. A. DAVIES
and
K. M. LLOYD

AT THE HERBARIUM OF THE ROYAL BOTANIC GARDENS
KEW

CLARENDON PRESS · OXFORD
1990

Oxford University Press, Walton Street, Oxford OX2 6DP
Oxford New York Toronto
Delhi Bombay Calcutta Madras Karachi
Petaling Jaya Singapore Hong Kong Tokyo
Nairobi Dar es Salaam Cape Town
Melbourne Auckland

and associated companies in
Berlin Ibadan

Oxford is a trade mark of Oxford University Press

Published in the United States
by Oxford University Press, New York

British Library Cataloguing in Publication Data
Kew index for 1989: names of seed-bearing plants, ferns, and fern allies at the rank of family and below published during 1989 with some omissions from earlier years.—
1989
1. Seed plants. Terminology: Lists
582'.0014
ISBN 0–19–854283–6

Library of Congress Cataloging in Publication Data
(Data available)
ISBN 0–19–854283–6

Typeset by Promenade Graphics Ltd., Cheltenham
Printed and bound in
Great Britain by Biddles Ltd.
Guildford and King's Lynn

CONTENTS

FOREWORD

In this fourth annual *Kew Index* the format is similar to *Kew Index 1988*. We have scanned books and journals received in the *Index Kewensis* office by 31 December 1989 or seen at the Natural History Museum (BM) during 1989. Some extra entries missed in previous years have been included.

Names at all taxonomic ranks from family downwards are included. Names at ranks above the genus are arranged alphabetically under families in a separate sequence at the beginning. Names at the rank of genus downwards are arranged alphabetically under genus headings. These are followed by a section containing a list of new genera in alphabetic order under family.

Entries for ferns and fern-allies are listed in Appendix 1 at the end of the book. These will form part of the basis of *Index Filicum*, Supplement 7.

At each rank there are three categories of names: new taxon, new name (nom. nov.), and new combination.

An entry can be divided up as follows:
Name/Author/Bibliographic details and either geography if a new taxon, or replaced synonym if a new name, or basionym if a new combination.

Name

The spelling of names is changed, if they are incorrect, in accordance with the appropriate Article of the International Code of Botanical Nomenclature (ICBN). Each genus is assigned to a family according to the list of families agreed by Kew, the Natural History Museum (BM) and R.B.G. Edinburgh (E).

Authorship

The author of the basionym appears in brackets before the publishing author in entries of new combinations and is quoted in the form given by the publishing author, without any attempt to standardize abbreviations. The name of the publishing author is always given in full, again irrespective of any list of standard abbreviations.

Use of 'ex'

A recurring problem in citation of authorship is whether to use 'in' or 'ex' when a name is attributed to one person in a publication of which someone else is the author. We look forward to the recommendations of the special committee set up at the 14th International Congress in Berlin (1987).

Bibliographic details

The same journal abbreviations are used as are found in the *Kew Record of Taxonomic Literature relating to Vascular Plants*. New journal title abbreviations are published regularly in *Kew Record*.

Nomenclatural status

A note is sometimes added after the bibliographic details, indicating the following:

(1) the original spelling of the epithet if it has been changed.
(2) the original status of a name, if incorrectly stated by the author, for example when an author writes 'sp. nov.' where 'nom. nov.' or 'comb. nov' is appropriate.
(3) the term 'nom. nov.', used to indicate an avowed substitute for a previously validly published name according to Arts. 7.11 and 33.2 of ICBN.
(4) details which might affect nomenclatural validity, as defined by the ICBN.
(5) the term 'isonym', used to indicate a new combination which has already been made by another author; we do not go out of our way to find this information but it is included if known.

Geography

Countries are listed for specimens cited in the article, not where they might occur. If the distribution range is wide, the country of the type collection is given first. In the future, we shall follow the standard list to be compiled for the International Working Group on Taxonomic Databases in Plant Sciences when it is published.

Geography is not stated for ranks above species level.

Basionym/Replaced Synonym

Sometimes the basionyms are names published some years ago when symbols were used instead of the name of the rank, e.g. *** or letters from the Greek alphabet. Where possible, we reproduce these as they appear in the original publication.

Corrected entries

A dagger (†) at the end of an entry indicates the corrected re-entry of a name cited in an earlier volume.

Transliteration

Cyrillic characters have been transliterated in accordance with *British Standard* 2979 (1958): *Transliteration of Cyrillic and Greek Characters*.

Thanks are again due to Dr R. K. Brummitt for nomenclatural advice and the Kew Computer Department for processing the data.

Royal Botanic Gardens
Kew
January 1990

G. T. Prance
Director

KEW INDEX FOR 1989
NAMES OF ALL RANKS ABOVE GENUS UP TO AND INCLUDING FAMILY

COMMELINACEAE
subfam. **Cartonematoideae** (Pichon) R. Faden ex G.C. Tucker in J. Arnold Arbor., 70(1): 99 (1989): Cartonemataceae.

COMPOSITAE
tribus **Plucheae** A.A. Anderberg in Canad. J. Bot., 67(8): 2293 (1989), as '(Benth.)'.

CYPERACEAE
tribus **Koyamaeae** W.W. Thomas & G. Davidse in Syst. Bot., 14(2): 189 (1989).

DROSOPHYLLACEAE
fam. **Drosophyllaceae** J. Chrtek, Z. Slavíkova & M. Studnička in Preslia, 61(2): 122 (1989).

EUPHORBIACEAE
subtribus **Cleidiocarpinae** N.N. Thin in J. Biol. (Vietnam), 10(2): 32 (1988).
subtribus **Lobanilinae** A. Radcliffe-Smith in Kew Bull., 44(2): 339 (1989).

HAMAMELIDACEAE
subtribus **Dicoryphinae** P.K. Endress in Taxon, 38(3): 374 (1989).
subtribus **Loropetalinae** P.K. Endress in Taxon, 38(3): 374 (1989).

LACANDONIACEAE
fam. **Lacandoniaceae** E. Martínez & C.H. Ramos in Ann. Missouri Bot. Gard., 76(1): 128 (1989).

LARDIZABALACEAE
subfam. **Decaisneoideae** H. Loconte & J.R. Estes in Syst. Bot., 14(4): 574 (1989).
tribus **Sinofranchetieae** H. Loconte & J.R. Estes in Syst. Bot., 14(4): 575 (1989).

LEGUMINOSAE
subtribus **Pseudarthriinae** (Hutch.) A. Pramanik & K. Thothathri in Taxon, 38(1): 109 (1989): tribus Pseudarthrieae.

MONOTACEAE
fam. **Monotaceae** A.J.G.H. Kostermans in Taxon, 38(1): 123 (1989), isonym; see A. Takhtadzhyan, Sistema Magnoliofitov: 129 (1987).

PINACEAE
subfam. **Piceoideae** M.P. Frankis in Notes Roy. Bot. Gard. Edinburgh, 45(3): 544 (1988 publ. 1989).

RANUNCULACEAE
subtribus **Asteropyrinae** (W.T. Wang & C.Y. Chang) M. Tamura & K. Kosuge in Acta Phytotax. Geobot., 40(1–4): 33 (1989): tribus Asteropyreae.
subtribus **Coptidinae** M. Tamura & K. Kosuge in Acta Phytotax. Geobot., 40(1–4): 33 (1989).
subtribus **Dichocarpinae** M. Tamura & K. Kosuge in Acta Phytotax. Geobot., 40(1–4): 33 (1989).

RUBIACEAE
tribus **Prismatomereae** Y.Z. Ruan in Acta Phytotax. Sin., 26(6): 446 (1988).

TAXODIACEAE
subfam. **Taiwanioideae** L.C. Li in Acta Bot. Yunnanica, 11(2): 123 (1989).

KEW INDEX FOR 1989
NAMES AT GENERIC RANK AND BELOW

A

ABELIOPHYLLUM (Oleac.).
distichum
f. eburneum *T.B. Lee in J. Korean Pl. Taxon.*, 7(1–2): 22 (1976), without type—Korea.
var. obtusicarpum *T.B. Lee in J. Korean Pl. Taxon.*, 7(1–2): 22 (1976)—Korea.
f. viridicalycinum *T.B. Lee in J. Korean Pl. Taxon.*, 7(1–2): 22 (1976), without type—Korea.

ABELMOSCHUS (Malvac.).
moschatus
var. lanyunatus *S.S. Ying, Colour. Illustr. Pl. Taiwan, Pt. 3; Colour. Illustr. Ligneous Pl. Taiwan,* 1: 414 (1987)—Taiwan.

ABUTILON (Malvac.).
sect. Mexabutilon *P.A. Fryxell, Malvac. of Mexico (Syst. Bot. Monog., 25):* 25 (1988).
sect. Pluriovulata *P.A. Fryxell, Malvac. of Mexico (Syst. Bot. Monog., 25):* 25 (1988).

ACACIA (Legumin.).
alvaroi *L. Cárdenas & G. De Martino in Ernstia,* 56: 10 (1989)—Venezuela.
caerulescens *B.R. Maslin & A.B. Court in Muelleria,* 7(1): 131 (1989)—Australia (Victoria).
caurina *R.C. Barneby & T.A. Zanoni in Moscosoa,* 5: 20 (1989)—Hispaniola (Haiti).
cucuyo *R.C. Barneby & T.A. Zanoni in Moscosoa,* 5: 14 (1989)—Hispaniola (Dominican Republic).
densispina *M. Thulin in Nordic J. Bot.,* 8(5): 460 (1989)—Somali Republic.
farnesiana
var. guanacastensis *H.D. Clarke, D.S. Seigler & J.E. Ebinger in Syst. Bot.,* 14(4): 562 (1989)—Costa Rica, Salvador, Guatemala, Mexico.
flagellaris *M. Thulin in Nordic J. Bot.,* 8(5): 461 (1989)—Somali Republic.
piptadenioides *G.P. Lewis in Kew Bull.,* 44(1): 171 (1989)—Brazil.
pringlei
subsp. californica *(Brandegee) Y.S. Lee, D.S. Seigler & J.E. Ebinger in Syst. Bot.,* 14(1): 99 (1989): A. californica.
veronica *B.R. Maslin in Nuytsia,* 7(1): 43 (1989)—Australia (W. Australia).

ACAENA (Rosac.).
emittens *B.H. Macmillan in New Zealand J. Bot.,* 27(1): 115 (1989)—New Zealand.
juvenca *B.H. Macmillan in New Zealand J. Bot.,* 27(1): 109 (1989)—New Zealand.

ACALYPHA (Euphorbiac.).
×cristata *A. Radcliffe-Smith in Kew Bull.,* 44(3): 443 (1989)—Kenya, Tanzania. (A. ciliata × fimbriata).
lanceolata
var. glandulosa *(Muell. Arg.) A. Radcliffe-Smith in Kew Bull.,* 44(3): 444 (1989): A. crenata var. β glandulosa.
×malawiensis *A. Radcliffe-Smith in Kew Bull.,* 44(3): 444 (1989)—Malawi. (A. chirindica × ornata).

ACANTHOLIMON (Plumbaginac.).
karatavicum *N.V. Pavlov in Vestn. Akad. Nauk Kas. SSR,* 1949(1): 35 (1949)—U.S.S.R. (Middle Asia).
minshelkense *N.V. Pavlov in Vestn. Akad. Nauk Kaz. SSR,* 1949(1): 35 (1949)—U.S.S.R. (Middle Asia).
squarrosum *N.V. Pavlov in Vestn. Akad. Nauk Kaz. SSR,* 1949(1): 36 (1949)—U.S.S.R. (Middle Asia).

ACER (Acerac.).
stenolobum
var. monochladea *S.C. Cui & J.X. Yu in Acta Phytotax. Sin.,* 27(2): 131 (1989)—China.

ACHILLEA (Compos.).
×albinea *Ž. Bjelčić & Maly in Glasn. Zem. Muz. Bosne Herceg. Prir. Nauke,* 8: 68 (1969 publ. 1971)—Jugoslavia. (A. abrotanoides × clavenae).

ACHNATHERUM (Gramin.).
duthiei *(Hook. f.) P.C. Kuo & S.L. Lu ex J.L. Yang in Fl. Sichuanica,* 5(2): 190 (1988), without basionym date: Stipa duthiei.

ACHYRANTHES (Amaranthac.).
megaphylla *Y.H. Li in Guihaia,* 9(1): 26 (1989)—China.

ACIDOSASA (Gramin.).
breviclavata *W.T. Lin in Bamboo Res.,* 1986(2): 27 (1986)—China.
brilletii *(A. Camus) C.S. Chao & S.A. Renvoize in Kew Bull.,* 44(2): 351 (1989): Arundinaria brilletii.

ACIDOSASA :–
guangxiensis *Q.H. Dai & C.F. Huang in Bamboo Res.*, 1986(3): 64 (1986), with two types cited—China.

ACILEPIDOPSIS H. Robinson in Phytologia, 67(4): 291 (1989). *COMPOSITAE*.
echitifolia *(Mart. ex DC.) H. Robinson in Phytologia,* 67(4): 291 (1989): Vernonia echitifolia.

ACONITUM (Ranunculac.).
alboviolaceum
var. albiflorum *(S.H. Li & Y.H. Huang) S.H. Li in Fl. Liaoningica, chief ed. S.X. Li,* 1: 476 (1988): A. alboviolaceum f. albiflorum.
delphinifolium
subsp. pavlovae *(Worosch.) V.N. Voroshilov in Byull. Mosk. Obshch. Ispȳt. Prir., Biol.,* 94(2): 87 (1989): A. pavlovae.
fanjingshanicum *W.T. Wang in Bull. Bot. Res. North-East. Forest. Inst.,* 9(2): 3 (1989)—China.
hopeiense *(W.T. Wang) V.N. Voroshilov in Byull. Glavn. Bot. Sada (Moscow),* 151: 43 (1988 publ. 1989): A. leucostomum var. hopeiense.
sczukinii
subsp. subalpinum *(Baran.) V.N. Voroshilov in Byull. Mosk. Obshch. Ispȳt. Prir., Biol.,* 94(2): 83 (1989): A. subalpinum.
subglandulosum *A.P. Khokhryakov, Analiz Fl. Kolȳmskogo Nagor'ya:* 42 (1989), nom. nov.: A. glandulosum *Khokhr.*
sungpanense
var. villosulum *W.T. Wang in Bull. Bot. Res. North-East. Forest. Inst.,* 9(2): 5 (1989)—China.
taigicola *V.N. Voroshilov in Byull. Glavn. Bot. Sada (Moscow),* 151: 42 (1988 publ. 1989)—U.S.S.R. (Soviet Far East).
variegatum
subsp. paniculatum *(Arcangeli) W. Greuter & Burdet in Willdenowia,* 19(1): 42 (1989): A. cammarum subsp. paniculatum.
subsp. valesiacum *(Gáyer) W. Greuter & Burdet in Willdenowia,* 19(1): 42 (1989): A. valesiacum.
wolongense *W.T. Wang in Bull. Bot. Res. North-East. Forest. Inst.,* 9(2): 1 (1989)—China.

ACONOGONON (Polygonac.).
ajanense
var. divaricatum *(Fr. Schmidt) N.N. Tsvelev in Sosud. Rast. Sovet. Dal'nego Vostoka,* 4: 96 (1989): Polygonum polymorphum var. divaricatum.
diffusum *(Willd. ex Spreng.) N.N. Tsvelev in Novosti Sist. Vȳssh. Rast.,* 25: 184 (1988): Polygonum diffusum.

ACONOGONON :–
ocreatum
var. jurtzevii *N.N. Tsvelev in Sosud. Rast. Sovet. Dal'nego Vostoka,* 4: 96 (1989)—U.S.S.R. (Soviet Far East).
var. laxmannii *(Lepech.) N.N. Tsvelev in Sosud. Rast. Sovet. Dal'nego Vostoka,* 4: 96 (1989): Polygonum laxmannii.
var. riparium *(Georgi) N.N. Tsvelev in Sosud. Rast. Sovet. Dal'nego Vostoka,* 4: 96 (1989): Polygonum riparium.
pseudoajanense *V.Yu. Barkalov & I.B. Vȳshin in Bot. Zhurn.,* 74(8): 183 (1989)—Kuril Is.
savatieri
var. iturupense *(Mischurov) N.N. Tsvelev in Sosud. Rast. Sovet. Dal'nego Vostoka,* 4: 100 (1989): Polygonum iturupense.
tzvelevii *V.Yu. Barkalov & I.B. Vȳshin in Bot. Zhurn.,* 74(8): 182 (1989)—U.S.S.R. (Soviet Far East).

ACORELLUS (Cyperac.).
laevigatus
subsp. distachyos *(All.) J. Holub in Folia Geobot. Phytotax.,* 23(4): 413 (1988): Cyperus distachyos.

ACROCERAS (Gramin.).
chaseae *F.O. Zuloaga & O. Morrone in Darwiniana,* 28(1–4): 198 (1987 publ. 1988)—Brazil.
excavatum *(Henrard) F.O. Zuloaga & O. Morrone in Darwiniana,* 28(1–4): 195 (1987 publ. 1988): Panicum excavatum.
fluminense *(Hackel) F.O. Zuloaga & O. Morrone in Darwiniana,* 28(1–4): 197 (1987 publ. 1988): Panicum fluminense.

ACROGLOCHIN (Chenopodiac.).
persicarioides
var. muliensis *T.P. Soong in Fl. Sichuanica,* 4: 3 (1988)—China.
var. multiflora *T.P. Soong in Fl. Sichuanica,* 4: 3 (1988)—China.

ACTINIDIA (Actinidiac.).
lijiangensis *C.F. Liang & Y.X. Lu in Guihaia,* 9(1): 21 (1989)—China.

ACTINODAPHNE (Laurac.).
macgregorii *(Merr.) A.J.G.H. Kostermans in Reinwardtia,* 10(5): 467 (1988): Litsea macgregorii.

ACTINOSCIRPUS (Ohwi) R.W. Haines & K.A. Lye in Bot. Notiser, 124: 481 (1971): Scirpus sect. Actinoscirpus (Cyperac.).

ADENIA (Passiflorac.).
lewallei *A. Robyns in Bull. Jard. Bot. Nation. Belg.,* 59(1–2): 227 (1989)—Burundi.

ADENOPHORA (Campanulac.).
changaica *I.A. Gubanov & R.V. Kamelin in Byull. Mosk. Obshch. Ispȳt. Prir., Biol.,* 93(5): 111 (1988)—Mongolia.

ADONIS (Ranunculac.).
multiflora *T. Nishikawa & K. Ito in J. Jap. Bot.*, 64(2): 51 (1989)—Japan, Korea.

AEGILOPS (Gramin.).
cylindrica
var. kastorianum *S.S. Karataglis in Pl. Syst. Evol.*, 163(1–2): 19 (1989)—Greece.
neglecta
subsp. contracta *(Eig) H. Scholz in Willdenowia*, 19(1): 105 (1989): A. triaristata subsp. contracta.

AEONIUM (Crassulac.).
ser. Chrysocome *H.Y. Liu, Syst. Aeonium (Crassulac.)(NMNS, Taiwan, Special Publ. 3):* 47 (1989).
ser. Patinaria *H.Y. Liu, Syst. Aeonium (Crassulac.)(NMNS, Taiwan, Special Publ. 3):* 54 (1989).
sect. Petrothamnium *(Webb ex Christ) H.Y. Liu, Syst. Aeonium (Crassulac.)(NMNS, Taiwan, Special Publ. 3):* 39 (1989): Greenovia sect. Petrothamnium.
ser. Praegeri *H.Y. Liu, Syst. Aeonium (Crassulac.)(NMNS, Taiwan, Special Publ. 3):* 53 (1989).
ser. Simsii *H.Y. Liu, Syst. Aeonium (Crassulac.)(NMNS, Taiwan, Special Publ. 3):* 46 (1989).
arboreum
var. holochrysum *H.Y. Liu, Syst. Aeonium (Crassulac.)(NMNS, Taiwan, Special Publ. 3):* 67 (1989), nom. nov.: Sempervivum urbicum *Lindley*.
var. rubrolineatum *(Svent.) H.Y. Liu, Syst. Aeonium (Crassulac.)(NMNS, Taiwan, Special Publ. 3):* 68 (1989): A. rubrolineatum.
canariense
var. palmense *(Webb ex Christ) H.Y. Liu, Syst. Aeonium (Crassulac.)(NMNS, Taiwan, Special Publ. 3):* 57 (1989): A. palmense.
var. subplanum *(Praeger) H.Y. Liu, Syst. Aeonium (Crassulac.)(NMNS, Taiwan, Special Publ. 3):* 59 (1989): A. subplanum.
var. virgineum *(Webb ex Christ) H.Y. Liu, Syst. Aeonium (Crassulac.)(NMNS, Taiwan, Special Publ. 3):* 60 (1989): A. virgineum.
davidbramwellii *H.Y. Liu, Syst. Aeonium (Crassulac.)(NMNS, Taiwan, Special Publ. 3):* 88 (1989)—Canary Is.
korneliuslemsii *H.Y. Liu, Syst. Aeonium (Crassulac.)(NMNS, Taiwan, Special Publ. 3):* 63 (1989)—Morocco.
lindleyi
var. viscatum *(Bolle) H.Y. Liu, Syst. Aeonium (Crassulac.)(NMNS, Taiwan, Special Publ. 3):* 41 (1989): A. viscatum.
stuessyi *H.Y. Liu, Syst. Aeonium (Crassulac.)(NMNS, Taiwan, Special Publ. 3):* 77 (1989)—Tanzania, Ethiopia, Kenya.

AEONIUM :–
×velutinum *(N.E. Br.) H.Y. Liu, Syst. Aeonium (Crassulac.)(NMNS, Taiwan, Special Publ. 3):* 96 (1989): Sempervivum velutinum.

AERANGIS (Orchidac.).
gracillima *(Kränzlin) J.C. Arends & J. Stewart in Lindleyana*, 4(1): 24 (1989): Angraecum gracillimum.

AERANTHES (Orchidac.).
tenella
var. borbonica *J. Bosser in Bull. Mus. Nation. Hist. Nat., B, Adansonia*, Ser. 4, 11(1): 36 (1989)—Réunion.

AETHYOPYS (Benth.) F.M. Opiz, Seznam: 11 (1852): Salvia sect. Aethyopys (Labiat.).

AFROCARPUS (Buchh. & E. Gray) C.N. Page in Notes Roy. Bot. Gard. Edinburgh, 45(2): 383 (1988 publ. 1989): Podocarpus sect. Afrocarpus (Podocarpac.).
dawei *(Stapf) C.N. Page in Notes Roy. Bot. Gard. Edinburgh*, 45(2): 384 (1988 publ. 1989): Podocarpus dawei.
falcata *(Thunb.) C.N. Page in Notes Roy. Bot. Gard. Edinburgh*, 45(2): 383 (1988 publ. 1989): Taxus falcata.
gaussenii *(Woltz) C.N. Page in Notes Roy. Bot. Gard. Edinburgh*, 45(2): 384 (1988 publ. 1989): Podocarpus gaussenii.
gracilior *(Pilger) C.N. Page in Notes Roy. Bot. Gard. Edinburgh*, 45(2): 383 (1988 publ. 1989): Podocarpus gracilior.
mannii *(Hook. f.) C.N. Page in Notes Roy. Bot. Gard. Edinburgh*, 45(2): 384 (1988 publ. 1989): Podocarpus mannii.
usambarensis *(Pilger) C.N. Page in Notes Roy. Bot. Gard. Edinburgh*, 45(2): 384 (1988 publ. 1989): Podocarpus usambarensis.

AGARISTA (Ericac.).
villarrealana *L.M. González V. in Acta Bot. Mex.*, 5: 13 (1989)—Mexico.

AGATHISANTHEMUM (Rubiac.).
bojeri
subsp. angolense *(Bremek.) B. Verdcourt in Fl. Zambes.*, 5(1): 110 (1989): A. angolense.
var. linearifolia *B. Verdcourt in Fl. Zambes.*, 5(1): 110 (1989)—Zambia.

AGAVE (Agavac.).
scabra *C. Gomez Ortega, Nov. Pl. Descr. Dec.:* 13 (1797)—Mexico.

AGERATINA (Compos.).
chazaroana *B.L. Turner in Phytologia*, 67(2): 115 (1989)—Mexico.
perezii *B.L. Turner in Phytologia*, 67(2): 116 (1989)—Mexico.

AGLOSSORHYNCHA (Orchidac.).
bilobula *P.J. Kores in Allertonia*, 5(1): 117 (1989)—Fiji.

AGRIOPHYLLUM (Chenopodiac.).
montasirii *A. El-Gazzar in Mitt. Bot. Staatssamml. München,* 27: 16 (1988), as 'montasiri'—Saudi Arabia.

AGROSTIS (Gramin.).
brevipes *Keng ex J.L. Yang in Fl. Sichuanica,* 5(2): 122 (1988)—China, Xizang.
micrandra *Keng ex J.L. Yang in Fl. Sichuanica,* 5(2): 128 (1988)—China.

AGROSTOPHYLLUM (Orchidac.).
aristatum *P.J. Kores in Allertonia,* 5(1): 107 (1989)—Fiji.

AIPHANES (Palmae).
tricuspidata *F. Borchsenius, R.G. Bernal & M. Ruiz in Brittonia,* 41(2): 156 (1989)—Ecuador, Colombia.

AJUGA (Labiat.).
davisiana *Kit Tan & B. Yildiz in Notes Roy. Bot. Gard. Edinburgh,* 45(3): 448 (1988 publ. 1989)—Turkey.

AKANIA (Akaniac.).
bidwillii *(Hogg) D.J. Mabberley, Plant-book:* 707 (1989 reprint): Lomatia bidwillii.

ALBIZIA (Legumin.).
adianthifolia
var. intermedia *(De Wildeman & T. Durand) J.F. Villiers in Fl. Gabon,* 31: 147 (1989): A. intermedia.
corymbosa *(L.C. Rich.) G.P. Lewis & P.E. Owen, Legumes of Ilha de Maracá:* 40 (1989): Mimosa corymbosa.
glabripetala *(H.S. Irwin) G.P. Lewis & P.E. Owen, Legumes of Ilha de Maracá:* 42 (1989): Pithecellobium glabripetalum.

ALCHEMILLA (Rosac.).
deylii *Plocek ex J. Soják in Sborn. Nár. Muz. Praze, B,* 39(1): 57 (1983)—U.S.S.R. (Ukraine).
grenieri *J. Guillot in Bull. Soc. Bot. France, Lett. Bot.,* 136(2): 152 (1989)—France.

ALCHORNEA (Euphorbiac.).
hunanensis *H.S. Kiu in Acta Phytotax. Sin.,* 26(6): 458 (1988)—China.
trewioides
var. sinica *H.S. Kiu in Acta Phytotax. Sin.,* 26(6): 460 (1988)—China.

ALDAMA (Compos.).
mesoamericana *N.A. Harriman in Syst. Bot.,* 14(4): 580 (1989)—Nicaragua.

ALINULA (Cyperac.).
malawica *(J. Raynal) P. Goetghebeur & P. Vorster in Bull. Jard. Bot. Nation. Belg.,* 58(3–4): 462 (1988): Mariscus malawicus.
paradoxa *(Chermezon) P. Goetghebeur & P. Vorster in Bull. Jard. Bot. Nation. Belg.,* 58(3–4): 461 (1988): Lipocarpha paradoxa.

ALINULA :–
peteri *(Kük.) P. Goetghebeur & P. Vorster in Bull. Jard. Bot. Nation. Belg.,* 58(3–4): 464 (1988): Ascolepis peteri.

ALISMA (Alismatac.).
canaliculatum
var. azuminoense *Y. Kadono & S. Hamashima in J. Jap. Bot.,* 63(12): 411 (1988)—Japan.

ALLIUM (Alliac.).
subsect. Tenuissima *(Tzag.) N.V. Frizen, Lukovýe Sibiri:* 119 (1988): A. ser. Tenuissima.
cyaneum
var. deltoides *S.O. Yu, S. Lee & W.T. Lee in J. Korean Pl. Taxon.,* 11(1–2): 29 (1981)—Korea.
deltoide-fistulosum *S.O. Yu, S. Lee & W.T. Lee in J. Korean Pl. Taxon.,* 11(1–2): 30 (1981), with two types cited—Korea.
dshambulicum *N.V. Pavlov in Vestn. Akad. Nauk Kaz. SSR,* 1949(1): 27 (1949)—U.S.S.R. (Middle Asia).
ecornutum *F.O. Khasanov & I.I. Mal'tsev in Uzb. Biol. Zhurn.,* 1988(4): 43 (1988)—U.S.S.R. (Middle Asia).
jepsonii *(Ownbey ex Traub) S.S. Denison & D.W. McNeal in Madroño,* 36(2): 127 (1989): A. sanbornii var. jepsonii.
maniaticum *S. Brullo & D. Tzanoudakis in Willdenowia,* 19(1): 111 (1989)—Greece.
omeiense *Z.Y. Zhu in Bull. Bot. Res. North-East. Forest. Inst.,* 9(4): 65 (1989)—China.
pentadactyli *S. Brullo, P. Pavone & G. Spampinato in Willdenowia,* 19(1): 115 (1989)—Italy.
schoenoprasum
subsp. gredense *(Rivas Mateos) S. Rivas-Martínez, F. Fernández González & D. Sánchez-Mata in Opusc. Bot. Pharm. Complutensis,* 2: 103 (1986): A. gredense.
subsp. latiorifolium *(Pau) S. Rivas-Martínez, F. Fernández González & D. Sánchez-Mata in Opusc. Bot. Pharm. Complutensis,* 2: 103 (1986): A. schoenoprasum var. latiorifolium.
senescens
var. minor *S.O. Yu, S. Lee & W.T. Lee in J. Korean Pl. Taxon.,* 11(1–2): 32 (1981)—Korea.
tuolumnense *(Ownbey ex Traub) S.S. Denison & D.W. McNeal in Madroño,* 36(2): 128 (1989): A. sanbornii var. tuolumnense.
victorialis
f. variegatum *(Nakai ex Mori) S.O. Yu, W.T. Lee & S. Lee in J. Korean Pl. Taxon.,* 11(1–2): 35 (1981): A. victorialis var. variegatum.
vvedenskyanum *N.V. Pavlov in Vestn. Akad. Nauk Kaz. SSR,* 1949(1): 27 (1949)—U.S.S.R. (Middle Asia).
yuchuanii *Y.Z. Zhao & J.Y. Chao in Acta Sci. Nat. Univ. Intramongolicae,* 20(2): 241 (1989)—China.

ALLOCASUARINA (Casuarinac.).
sect. Acanthopitys *(Miq.) L.A.S. Johnson in Fl. Australia,* 3: 191, 173 (1989): Casuarina sect. Acanthopitys.
sect. Amorphopitys *L.A.S. Johnson in Fl. Australia,* 3: 191, 143 (1989).
sect. Ceropitys *L.A.S. Johnson in Fl. Australia,* 3: 191, 135 (1989).
sect. Cylindropitys *L.A.S. Johnson in Fl. Australia,* 3: 191, 147 (1989).
sect. Dolichopitys *L.A.S. Johnson in Fl. Australia,* 3: 192, 127 (1989).
sect. Echinopitys *L.A.S. Johnson in Fl. Australia,* 3: 192, 132 (1989).
sect. Inopitys *L.A.S. Johnson in Fl. Australia,* 3: 192, 147 (1989).
sect. Nannopitys *L.A.S. Johnson in Fl. Australia,* 3: 192, 170 (1989).
sect. Oopitys *L.A.S. Johnson in Fl. Australia,* 3: 192, 144 (1989).
sect. Oxypitys *L.A.S. Johnson in Fl. Australia,* 3: 193, 127 (1989).
sect. Platypitys *L.A.S. Johnson in Fl. Australia,* 3: 193, 132 (1989).
sect. Trachypitys *(Benth.) L.A.S. Johnson in Fl. Australia,* 3: 193, 170 (1989): Casuarina sect. Trachypitys.
sect. Trichopitys *L.A.S. Johnson in Fl. Australia,* 3: 193, 146 (1989).
brachystachya *L.A.S. Johnson in Fl. Australia,* 3: 193, 168 (1989)—Australia (New South Wales).
crassa *L.A.S. Johnson in Fl. Australia,* 3: 194, 159 (1989)—Tasmania.
defungens *L.A.S. Johnson in Fl. Australia,* 3: 194, 157 (1989)—Australia (New South Wales).
diminuta *L.A.S. Johnson in Fl. Australia,* 3: 194, 155 (1989)—Australia (New South Wales).
subsp. annectens *L.A.S. Johnson in Fl. Australia,* 3: 194, 157 (1989)—Australia (New South Wales).
subsp. mimica *L.A.S. Johnson in Fl. Australia,* 3: 195, 155 (1989)—Australia (New South Wales).
emuina *L.A.S. Johnson in Fl. Australia,* 3: 195, 150 (1989)—Australia (Queensland).
eriochlamys *(L. Johnson) L.A.S. Johnson in Fl. Australia,* 3: 195, 137 (1989): Casuarina campestris subsp. eriochlamys.
subsp. grossa *(L. Johnson) L.A.S. Johnson in Fl. Australia,* 3: 195, 139 (1989): Casuarina campestris subsp. grossa.
filidens *L.A.S. Johnson in Fl. Australia,* 3: 195, 150 (1989)—Australia (Queensland).
glareicola *L.A.S. Johnson in Fl. Australia,* 3: 195, 157 (1989)—Australia (New South Wales).
globosa *L.A.S. Johnson in Fl. Australia,* 3: 196, 136 (1989)—Australia (W. Australia).
grampiana *L.A.S. Johnson in Fl. Australia,* 3: 196, 158 (1989)—Australia (Victoria).

ALLOCASUARINA :–
gymnanthera *L.A.S. Johnson in Fl. Australia,* 3: 196, 166 (1989)—Australia (New South Wales).
lehmanniana
subsp. ecarinata *L.A.S. Johnson in Fl. Australia,* 3: 196, 161 (1989)—Australia (W. Australia).
mackliniana *L.A.S. Johnson in Fl. Australia,* 3: 196, 162 (1989)—Australia (S. Australia, Victoria).
subsp. hirtilinea *L.A.S. Johnson in Fl. Australia,* 3: 197, 162 (1989)—Australia (Victoria).
subsp. xerophila *L.A.S. Johnson in Fl. Australia,* 3: 197, 162 (1989)—Australia (S. Australia, Victoria).
media *L.A.S. Johnson in Fl. Australia,* 3: 197, 148 (1989)—Australia (Victoria).
misera *L.A.S. Johnson in Fl. Australia,* 3: 197, 163 (1989)—Australia (Victoria).
muelleriana
subsp. alticola *L.A.S. Johnson in Fl. Australia,* 3: 197, 169 (1989)—Australia (S. Australia).
subsp. notocolpica *L.A.S. Johnson in Fl. Australia,* 3: 198, 169 (1989)—Australia (S. Australia).
ophiolitica *L.A.S. Johnson in Fl. Australia,* 3: 198, 152 (1989)—Australia (New South Wales).
portuensis *L.A.S. Johnson in Fl. Australia,* 3: 198, 152 (1989)—Australia (New South Wales).
rigida
subsp. exsul *L.A.S. Johnson in Fl. Australia,* 3: 198, 151 (1989)—Australia (Queensland).
rupicola *L.A.S. Johnson in Fl. Australia,* 3: 199, 152 (1989)—Australia (New South Wales, Queensland).
simulans *L.A.S. Johnson in Fl. Australia,* 3: 199, 154 (1989)—Australia (New South Wales).
spinosissima *(C. Gardner) L.A.S. Johnson in Fl. Australia,* 3: 199, 133 (1989): Casuarina spinosissima.
thalassoscopica *L.A.S. Johnson in Fl. Australia,* 3: 199, 150 (1989)—Australia (Queensland).
tortiramula *E.M. Bennett in Fl. Australia,* 3: 190, 137 (1989)—Australia (W. Australia).
zephyrea *L.A.S. Johnson in Fl. Australia,* 3: 199, 158 (1989)—Tasmania.

ALLOMARKGRAFIA (Apocynac.).
laxiflora *A.H. Gentry in Ann. Missouri Bot. Gard.,* 76(3): 923 (1989)—Colombia.

ALLOPHYLUS (Sapindac.).
dodsonii *A.H. Gentry in Ann. Missouri Bot. Gard.,* 75(4): 1438 (1988 publ. 1989)—Ecuador.
parimensis *J.A. Steyermark in Ann. Missouri Bot. Gard.,* 75(1): 333 (1988)—Venezuela.

ALLOSIDASTRUM (Hochreutiner) A. Krapovickas, P.A. Fryxell & D.M. Bates in P.A. Fryxell, Malvac. of Mexico (Syst. Bot. Monog., 25): 70 (1988); et in Bol. Soc. Bot. Méx., 48: 24 (1988 publ. 1989): Pseudabutilon subgen. Allosidastrum (Malvac.).
dolichophyllum *A. Krapovickas, P.A. Fryxell & D.M. Bates in Bol. Soc. Bot. Méx.*, 48: 25 (1988 publ. 1989)—Venezuela.
hilarianum *(K. Presl) A. Krapovickas, P.A. Fryxell & D.M. Bates, in P.A. Fryxell, Malvac. of Mexico (Syst. Bot. Monog., 25):* 71 (1988); *et in Bol. Soc. Bot. Méx.*, 48: 28 (1988 publ. 1989): Sida hilariana.
interruptum *(Balbis ex de Candolle) A. Krapovickas, P.A. Fryxell & D.M. Bates in P.A. Fryxell, Malvac. of Mexico (Syst. Bot. Monog., 25):* 71 (1988); *et in Bol. Soc. Bot. Méx.*, 48: 30 (1988 publ. 1989): Sida interrupta.
pyramidatum *(Cavanilles) A. Krapovickas, P.A. Fryxell & D.M. Bates in P.A. Fryxell, Malvac. of Mexico (Syst. Bot. Monog., 25):* 74 (1988); *et in Bol. Soc. Bot. Méx.*, 48: 31 (1988 publ. 1989): Sida pyramidata.

ALNUS (Betulac.).
japonica
var. villosa *L. Zhao & D. Chen in Fl. Liaoningica, chief ed. S.X. Li,* 1: 1361, 247 (1988)—China.

ALOE (Aloeac.).
molederana *J.J. Lavranos & H.F. Glen in Flow. Pl. Afr.*, 50(2): t. 1982 (1989)—Somali Republic.

ALOITIS (Gentianac.).
falconeri *(Clarke) S. Omer, M. Qaiser & S.I. Ali in Pakistan J. Bot.*, 20(2): 155 (1988): Gentiana falconeri.
holosteoides *(Pritchard) S. Omer, M. Qaiser & S.I. Ali in Pakistan J. Bot.*, 20(2): 159 (1988): Gentianella holosteoides.
maddenii *(Clarke) S. Omer, M. Qaiser & S.I. Ali in Pakistan J. Bot.*, 20(2): 155 (1988): Gentiana moorcroftiana var. maddenii.
moorcroftiana *(Wall. ex G. Don) S. Omer, M. Qaiser & S.I. Ali in Pakistan J. Bot.*, 20(2): 157 (1988): Gentiana moorcroftiana.
stoliczkae *(Clarke) S. Omer, M. Qaiser & S.I. Ali in Pakistan J. Bot.*, 20(2): 159 (1988), as 'stoliczkai': Gentiana stoliczkae.
umbellata *(M. Bieb.) S. Omer, M. Qaiser & S.I. Ali in Pakistan J. Bot.*, 20(2): 157 (1988): Gentiana umbellata.

ALOPECURUS (Gramin.).
arundinaceus
subsp. castellanus *(Boiss. & Reuter) S. Rivas-Martínez, F. Fernández González & D. Sánchez-Mata in Opusc. Bot. Pharm. Complutensis,* 2: 104 (1986): A. castellanus.

ALPINIA (Zingiberac.).
beamanii *R.M. Smith in Notes Roy. Bot. Gard. Edinburgh,* 45(2): 341 (1988 publ. 1989)—Borneo (Sabah).
emaculata *S.Q. Tong in Acta Phytotax. Sin.*, 27(4): 285 (1989)—China.
guangdongensis *S.J. Chen & Z.Y. Chen in Guihaia,* 8(4): 317 (1988)—China.
rubromaculata *S.Q. Tong in Acta Phytotax. Sin.*, 27(4): 284 (1989), as 'rubro-maculata'—China.

ALTHAEA (Malvac.).
ludwigii
var. microcalyculata *A. Dobignard in Saussurea,* 19: 103 (1988 publ. 1989)—Morocco.

ALYSICARPUS (Legumin.).
heyneanus
var. meeboldii *(Schindl.) A. Pramanik & K. Thothathri in J. Econ. Taxon. Bot.*, 12(2): 364 (1988 publ. 1989): A. meeboldii.
vaginalis
var. venosa *(Blatt. & Hall.) A. Pramanik & K. Thothathri in J. Econ. Taxon. Bot.*, 12(2): 363 (1988 publ. 1989): A. monilifer var. venosa.

ALYSSOPSIS (Crucif.).
trinervis *V.P. Bochantsev & É.M. Seĭfulin in Novosti Sist. Vÿssh. Rast.*, 25: 89 (1988)—U.S.S.R. (Middle Asia), Iran.

ALYSSUM (Crucif.).
diffusum
subsp. corymbosum *(Pau) C. Morales Torres, A.B. Robles Cruz & A.T. Romero García in Lazaroa,* 9: 148 (1986 publ. 1988): A. diffusum var. corymbosum.
samium *T.R. Dudley & D. Christodoulakis in Notes Roy. Bot. Gard. Edinburgh,* 45(3): 433 (1988 publ. 1989)—East Aegean Is.

AMOMUM (Zingiberac.).
borneense *(K. Schum.) R.M. Smith in Notes Roy. Bot. Gard. Edinburgh,* 45(2): 337 (1988 publ. 1989): Zingiber borneense.
capsiciforme *S.Q. Tong in Acta Phytotax. Sin.*, 27(4): 282 (1989)—China.
epiphyticum *R.M. Smith in Notes Roy. Bot. Gard. Edinburgh,* 45(2): 338 (1988 publ. 1989)—Borneo (Sarawak).
fragile *S.Q. Tong in Acta Phytotax. Sin.*, 27(4): 277 (1989)—China.
ghaticum *K.G. Bhat in Indian J. Forest.*, 11(4): 322 (1988 publ. 1989)—India.
glabrum *S.Q. Tong in Acta Phytotax. Sin.*, 27(4): 282 (1989)—China.
quadratolaminare *S.Q. Tong in Acta Phytotax. Sin.*, 27(4): 279 (1989), as 'quadrato-laminare'—China.
verrucosum *S.Q. Tong in Acta Phytotax. Sin.*, 27(4): 280 (1989)—China.

AMORPHOPHALLUS (Arac.).
eburneus *J. Bogner in Willdenowia,* 18(2): 441 (1989)—Borneo (Sarawak).
nanus *H. Li & C.L. Long in Aroideana,* 11(1): 8 (1988 publ. 1989)—China.
pingbianensis *H. Li & C.L. Long in Aroideana,* 11(1): 4 (1988 publ. 1989)—China.
smithsonianus *M. Sivadasan in Willdenowia,* 18(2): 435 (1989)—India.

AMPELOCISSUS (Vitac.).
wightiana *B.V. Shetty & P. Singh in Kew Bull.,* 44(3): 469 (1989)—India, Sri Lanka.

AMPHICARPAEA (Legumin.).
edgeworthii
f. alba *M.H. Lee in Korean J. Pl. Taxon.,* 13(2): 109 (1983)—Korea.

AMYGDALUS (Rosac.).
×iranshahrii *M. Khatamsaz in Iranian J. Bot.,* 4(1): 114 (1988)—Iran. (A. eburnea × scoparia).
×kamiaranensis *M. Khatamsaz & Assadi in Iranian J. Bot.,* 4(1): 113 (1988)—Iran. (A. haussknechtii × scoparia).
×mozaffarianii *M. Khatamsaz in Iranian J. Bot.,* 4(1): 114 (1988)—Iran. (A. brahuica × scoparia).
×yasujensis *M. Khatamsaz in Iranian J. Bot.,* 4(1): 116 (1988), contrary to Art. H.5.2 ICBN (1988)—Iran. (A. elaeagnifolia subsp. elaeagnifolia × scoparia).

ANABASIS (Chenopodiac.).
sect. Ebracteolatae *U. Pratov in Uzb. Biol. Zhurn.,* 1976(3): 49 (1976).

ANACAMPSEROS (Portulacac.).
bayeriana *S. Hammer in Cact. Succ. J. (U.S.A.),* 60(5): 215 (1988)—South Africa (Cape Province).

ANAPHYLLOPSIS A. Hay in Aroideana, 11(1): 25 (1988 publ. 1989). *ARACEAE.*
americana *(Engler) A. Hay in Aroideana,* 11(1): 26 (1988 publ. 1989): Cyrtosperma americanum.
cururuana *A. Hay in Aroideana,* 11(1): 29 (1988 publ. 1989)—Brazil.
pinnata *A. Hay in Aroideana,* 11(1): 26 (1988 publ. 1989)—Venezuela.

ANARRHINUM (Scrophulariac.).
forskaohlii
subsp. abyssinicum *(Jaub. & Spach) D.A. Sutton, Revis. tribe Antirrhineae:* 260 (1988): A. abyssinicum.
subsp. pubescens *(Fresen.) D.A. Sutton, Revis. tribe Antirrhineae:* 258 (1988): A. pubescens.
fruticosum
subsp. brevifolium *(Cosson & Kralik) D.A. Sutton, Revis. tribe Antirrhineae:* 257 (1988): A. brevifolium.

ANAXETON (Compos.).
lundgrenii *B. Nordenstam in S. Afr. J. Bot.,* 54(6): 631 (1988)—South Africa (Cape Province).

ANCANA (Annonac.).
hirsuta *L.W. Jessup in Austrobaileya,* 3(1): 65 (1989)—Australia (Queensland).

ANCISTRORHYNCHUS (Orchidac.).
crystalensis *P.J. Cribb & F.M. Van der Laan in Kew Bull.,* 44(3): 481 (1989)—Gabon.

ANCYLOSTEMON (Gesneriac.).
aureus
var. angustifolius *K.Y. Pan in Acta Phytotax. Sin.,* 26(6): 434 (1988)—China.
gamosepalus *K.Y. Pan in Acta Phytotax. Sin.,* 26(6): 434 (1988)—China.
mairei
var. emeiensis *K.Y. Pan in Acta Phytotax. Sin.,* 26(6): 434 (1988)—China.
rhombifolius *K.Y. Pan in Acta Phytotax. Sin.,* 26(6): 435 (1988)—China.
ronganensis *K.Y. Pan in Acta Phytotax. Sin.,* 26(6): 436 (1988)—China.

ANDRACHNE (Euphorbiac.).
schweinfurthii
var. papillosa *A. Radcliffe-Smith in Kew Bull.,* 44(3): 447 (1989)—Socotra.

ANDROGRAPHIS (Acanthac.).
monglunensis *H.T. Chang & H. Chu in Acta Sci. Nat. Univ. Sunyatseni,* 28(2): 64 (1989)—China.

ANDROPOGON (Gramin.).
sanguineus
var. brevipedicellatus *(Beal) Y. Herrera A. in Bol. Soc. Bot. Méx.,* 48: 22 (1988 publ. 1989): A. hirtiflorus var. brevipedicellatus.
var. oligostachyus *(Chapm.) Y. Herrera A. in Bol. Soc. Bot. Méx.,* 48: 22 (1988 publ. 1989): A. oligostachyus.

ANEMONASTRUM (Ranunculac.).
sect. Elongata *V.N. Starodubtsev in Bot. Zhurn.,* 74(9): 1345 (1989).
sect. Imbricata *V.N. Starodubtsev in Bot. Zhurn.,* 74(9): 1346 (1989).

ANEMONE (Ranunculac.).
subsect. Carolinianae *V.N. Starodubtsev in Bot. Zhurn.,* 74(9): 1345 (1989).
subsect. Sylvestres *V.N. Starodubtsev in Bot. Zhurn.,* 74(9): 1345 (1989).
subsect. Vitifoliae *V.N. Starodubtsev in Bot. Zhurn.,* 74(9): 1345 (1989).
exigua
var. shanxiensis *B.L. Li & X.Y. Yu in Acta Phytotax. Sin.,* 27(2): 152 (1989)—China.
narcissifolia
subsp. willdenowii *(Boiss.) W. Greuter & Burdet in Willdenowia,* 19(1): 43 (1989): A. narcissiflora var. willdenowii.

ANEMONE :–
trifolia
subsp. brevidentata *D. Ubaldi & G. Puppi in Candollea,* 44(1): 145 (1989)—Italy.

ANEMONIDIUM (Ranunculac.).
subsect. Helleborifolia *V.N. Starodubtsev in Bot. Zhurn.,* 74(9): 1345 (1989).
subgen. Meridium *V.N. Starodubtsev in Bot. Zhurn.,* 74(9): 1345 (1989).
sect. Meridium *V.N. Starodubtsev in Bot. Zhurn.,* 74(9): 1345 (1989).
sect. Mexicana *V.N. Starodubtsev in Bot. Zhurn.,* 74(9): 1345 (1989), without latin descr.
helleborifolium *(DC.) V.N. Starodubtsev in Bot. Zhurn.,* 74(9): 1345 (1989): Anemone helleborifolia.
mexicanum *(Humb., Bonpl. & Kunth) V.N. Starodubtsev in Bot. Zhurn.,* 74(9): 1345 (1989): Anemone mexicana.

ANEMONOIDES (Ranunculac.).
subsect. Altaicae *V.N. Starodubtsev in Bot. Zhurn.,* 74(9): 1346 (1989).
subsect. Ranunculoides *V.N. Starodubtsev in Bot. Zhurn.,* 74(9): 1346 (1989).
sect. Umbrosa *V.N. Starodubtsev in Bot. Zhurn.,* 74(9): 1346 (1989).

ANGELICA (Umbellif.).
sachalinensis
var. kawakamii *(Koidz.) T. Yamazaki in J. Jap. Bot.,* 64(3): 90 (1989): A. kawakamii.
f. saninensis *T. Yamazaki in J. Jap. Bot.,* 64(3): 90 (1989)—Japan.

ANGOSTURA (Rutac.).
kunorum *G. McPherson in Ann. Missouri Bot. Gard.,* 75(1): 375 (1988)—Panama.

ANGRAECUM (Orchidac.).
peyrotii *J. Bosser in Bull. Mus. Nation. Hist. Nat., B, Adansonia,* Ser. 4, 11(1): 32 (1989)—Madagascar.

ANOECTOCHILUS (Orchidac.).
zhejiangensis *Z. Wei & Y.B. Chang in Bull. Bot. Res. North-East. Forest. Inst.,* 9(2): 39 (1989)—China.

ANTENNARIA (Compos.).
rosea
subsp arida *(E. Nels.) R.J. Bayer in Brittonia,* 41(1): 57 (1989): Antennaria arida.
subsp. confinis *(Greene) R.J. Bayer in Brittonia,* 41(1): 57 (1989): A. confinis.
subsp. pulvinata *(Greene) R.J. Bayer in Brittonia,* 41(1): 59 (1989): A. pulvinata.

ANTERIORCHIS E. Klein & D. Strack in Phytochemistry, 28(8): 2136 (1989). *ORCHIDACEAE.*
coriophora *(L.) E. Klein & D. Strack in Phytochemistry,* 28(8): 2137 (1989): Orchis coriophora.

ANTERIORCHIS :–
sancta *(L.) E. Klein & D. Strack in Phytochemistry,* 28(8): 2137 (1989): Orchis sancta.

ANTHERICUM (Anthericac.).
collinum *P. Ravenna in Onira, Bot. Leafl.,* 1(3): 26 (1988)—Peru.
glareosum *P. Ravenna in Onira, Bot. Leafl.,* 1(3): 29 (1988)—Peru.
machupiccense *P. Ravenna in Onira, Bot. Leafl.,* 1(3): 27 (1988)—Peru.
pratense *P. Ravenna in Onira, Bot. Leafl.,* 1(3): 25 (1988)—Bolivia.
ticsanianum *P. Ravenna in Onira, Bot. Leafl.,* 1(3): 28 (1988)—Peru.
tunarianum *P. Ravenna in Onira, Bot. Leafl.,* 1(3): 24 (1988)—Bolivia.

ANTHOSPERMUM (Rubiac.).
rigidum
subsp. pumilum *(Sond.) C. Puff in Fl. Zambes.,* 5(1): 158 (1989): A. pumilum.

ANTHOTIUM (Goodeniac.).
junciforme *(Vriese) D.A. Morrison in Nuytsia,* 7(1): 55 (1989): Goodenia junciformis.

ANTHURIUM (Arac.).
iramireziae *G.S. Bunting in Ann. Missouri Bot. Gard.,* 76(3): 917 (1989), as 'iramirezae'—Venezuela.

ANTHYLLIS (Legumin.).
vulneraria
var. busambarensis *(Lojac.) J.R. Akeroyd in Notes Roy. Bot. Gard. Edinburgh,* 45(2): 372 (1988 publ. 1989): A. heterophylla var. busambarensis.

ANTIMIMA (Aizoac.).
subgen. Caespitosa *M. Dehn in Mitt. Inst. Allg. Bot. Hamburg,* 22: 211 (1988).
subgen. Clavipes *M. Dehn in Mitt. Inst. Allg. Bot. Hamburg,* 22: 212 (1988).
subgen. Microphylla *M. Dehn in Mitt. Inst. Allg. Bot. Hamburg,* 22: 213 (1988).
subgen. Virgata *M. Dehn in Mitt. Inst. Allg. Bot. Hamburg,* 22: 213 (1988).
alborubra *(L. Bolus) M. Dehn in Mitt. Inst. Allg. Bot. Hamburg,* 22: 212 (1988): Ruschia alborubra.
longipes *(L. Bolus) M. Dehn in Mitt. Inst. Allg. Bot. Hamburg,* 22: 211 (1988): Ruschia longipes.
microphylla *(Haw.) M. Dehn in Mitt. Inst. Allg. Bot. Hamburg,* 22: 213 (1988): Mesembryanthemum microphyllum.
virgata *(Haw.) M. Dehn in Mitt. Inst. Allg. Bot. Hamburg,* 22: 213 (1988): Mesembryanthemum virgatum.

ANTIRRHINUM (Scrophulariac.).
graniticum
subsp. brachycalyx *D.A. Sutton, Revis. tribe Antirrhineae:* 83 (1988)—Spain.

APHELANDRA (Acanthac.).
antioquiensis *D.C. Wasshausen in Rev. Acad. Colomb. Cienc. Exact. Fis. Nat.,* 65: 253 (1989)—Colombia.
tetroicia *D.C. Wasshausen in Rev. Acad. Colomb. Cienc. Exact. Fis. Nat.,* 65: 251 (1989)—Colombia.

APONOGETON (Aponogetonac.).
afroviolaceus
var. angustifolius *K.A. Lye in Lidia,* 2(1): 8 (1989)—Kenya.

×APOROHELIOCEREUS (Cactac.).
mallisonii *(Otto & Dietrich) P.V. Heath in Taxon,* 38(2): 281 (1989): Cereus mallisonii.
f. smithii *(Pfeiffer) P.V. Heath in Taxon,* 38(2): 281 (1989): Cereus smithii.

APPENDICULA (Orchidac.).
sect. Stenodesme *P.J. Kores in Allertonia,* 5(1): 120 (1989), nom. nov.: A. sect. Eu-Appendicula *Schlechter.*

AQUILEGIA (Ranunculac.).
blecicii *A. Podobnik in Biosistematika,* 12(1): 16 (1986 publ. 1987)—Jugoslavia.

ARACHNOTHRYX (Rubiac.).
brenesii *(Standl.) A. Borhidi in Acta Bot. Hung.,* 33(3–4): 301 (1987): Rondeletia brenesii.
buddleioides
var. aspera *(Kirkbr.) A. Borhidi in Acta Bot. Hung.,* 33(3–4): 301 (1987): Rondeletia buddleioides var. aspera.
chinajensis *(Standl. & Steyerm.) A. Borhidi in Acta Bot. Hung.,* 33(3–4): 301 (1987): Rondeletia chinajensis.
cordovana *(Standl. & Steyerm.) A. Borhidi in Acta Bot. Hung.,* 33(3–4): 301 (1987): Rondeletia cordovana.
edwardsii *(Standl.) A. Borhidi in Acta Bot. Hung.,* 33(3–4): 301 (1987): Rondeletia edwardsii.
gonzalaeoides *(Standl.) A. Borhidi in Acta Bot. Hung.,* 33(3–4): 302 (1987): Rondeletia gonzalaeoides.
gracilispica *(Standl.) A. Borhidi in Acta Bot. Hung.,* 33(3–4): 302 (1987): Rondeletia gracilispica.
izabalensis *(Standl. & Steyerm.) A. Borhidi in Acta Bot. Hung.,* 33(3–4): 302 (1987): Rondeletia izabalensis.
liebmanii *(Standl.) A. Borhidi in Acta Bot. Hung.,* 33(3–4): 302 (1987): Rondeletia liebmanii.
lundelliana *(Standl.) A. Borhidi in Acta Bot. Hung.,* 33(3–4): 302 (1987): Rondeletia lundelliana.
macrocalyx *(Standl. & Steyerm.) A. Borhidi in Acta Bot. Hung.,* 33(3–4): 302 (1987): Rondeletia macrocalyx.
mexiae *(Standl.) A. Borhidi in Acta Bot. Hung.,* 33(3–4): 302 (1987): Rondeletia mexiae.

ARACHNOTHRYX :–
myriantha *(Standl. & Steyerm.) A. Borhidi in Acta Bot. Hung.,* 33(3–4): 302 (1987): Rondeletia myriantha.
nebulosa *(Standl.) A. Borhidi in Acta Bot. Hung.,* 33(3–4): 302 (1987): Rondeletia nebulosa.
oaxacana *(Standl.) A. Borhidi in Acta Bot. Hung.,* 33(3–4): 302 (1987): Rondeletia oaxacana.
phyllocephala *(Standl. & L.O. Wms.) A. Borhidi in Acta Bot. Hung.,* 33(3–4): 302 (1987): Rondeletia phyllocephala.
skutchii *(Standl. & Steyerm.) A. Borhidi in Acta Bot. Hung.,* 33(3–4): 302 (1987): Rondeletia skutchii.
standleyana *(A. Molina) A. Borhidi in Acta Bot. Hung.,* 33(3–4): 302 (1987): Rondeletia standleyana.
stenostachya *(Standl.) A. Borhidi in Acta Bot. Hung.,* 33(3–4): 303 (1987): Rondeletia stenostachya.
tacanensis *(Lundell) A. Borhidi in Acta Bot. Hung.,* 33(3–4): 303 (1987): Rondeletia tacanensis.
torresii *(Standl.) A. Borhidi in Acta Bot. Hung.,* 33(3–4): 303 (1987): Rondeletia torresii.
vulcanicola *(Standley & Steyerm.) A. Borhidi in Acta Bot. Hung.,* 33(3–4): 303 (1987): Rondeletia vulcanicola.
zolleriana *(Standl. & Steyerm.) A. Borhidi in Acta Bot. Hung.,* 33(3–4): 303 (1987): Rondeletia zolleriana.

ARAFOE M.G. Pimenov & T.V. Lavrova in Bot. Zhurn., 74(1): 102 (1989). *UMBELLIFERAE.*
aromatica *M.G. Pimenov & T.V. Lavrova in Bot. Zhurn.,* 74(1): 102 (1989), nom. nov.: Ligusticum arafoe *Albov.*

ARCEUTHOBIUM (Viscac.).
durangense *(Hawksworth & Wiens) F.G. Hawksworth & D. Wiens in Phytologia,* 66(1): 7 (1989): A. vaginatum subsp. durangense.
nigrum *(Hawksworth & Wiens) F.G. Hawksworth & D. Wiens in Phytologia,* 66(1): 9 (1989): A. gillii subsp. nigrum.
oaxacanum *F.G. Hawksworth & D. Wiens in Phytologia,* 66(1): 7 (1989)—Mexico.
yecorense *F.G. Hawksworth & D. Wiens in Phytologia,* 66(1): 6 (1989)—Mexico.

ARCTOSTAPHYLOS (Ericac.).
bakeri
subsp. sublaevis *P.V. Wells in Four Seasons,* 8(2): 63 (1988)—U.S.A. (California).
canescens
f. candidissima *(Eastw.) P.V. Wells in Madroño,* 35(4): 332 (1988 publ. 1989): A. candidissima.
subsp. sonomensis *(Eastw.) P.V. Wells in Madroño,* 35(4): 333 (1988 publ. 1989): A. sonomensis.

ARCTOSTAPHYLOS :–
columbiana
f. setosissima *(Eastw.) P.V. Wells in Madroño,* 35(4): 333 (1988 publ. 1989): A. setosissima.
f. tracyi *(Eastw.) P.V. Wells in Madroño,* 35(4): 334 (1988 publ. 1989): A. tracyi.
edmundsii
f. parvifolia *(Roof) P.V. Wells in Madroño,* 35(4): 334 (1988 publ. 1989): A. edmundsii var. parvifolia.
glauca
f. eremicola *(Jeps.) P.V. Wells in Madroño,* 35(4): 334 (1988 publ. 1989): A. glauca var. eremicola.
f. puberula *(J.T. Howell) P.V. Wells in Madroño,* 35(4): 334 (1988 publ. 1989): A. glauca var. puberula.
imbricata
subsp. montaraensis *(Roof) P.V. Wells in Madroño,* 35(4): 334 (1988 publ. 1989), without exact basionym page: A. montaraensis.
nevadensis
subsp. knightii *(Gankin & Hildreth) P.V. Wells in Madroño,* 35(4): 335 (1988 publ. 1989): A. knightii.
patula
f. acutifolia *(Eastw.) P.V. Wells in Madroño,* 35(4): 337 (1988 publ. 1989): A. acutifolia.
f. coalescens *(W. Knight) P.V. Wells in Madroño,* 35(4): 337 (1988 publ. 1989): A. patula var. coalescens.
f. platyphylla *(A. Gray) P.V. Wells in Madroño,* 35(4): 337 (1988 publ. 1989): A. pungens var. platyphylla.
stanfordiana
f. decumbens *P.V. Wells in Madroño,* 35(4): 338 (1988 publ. 1989), nom. nov.: A. stanfordiana var. repens *Roof.*
uva-ursi
f. adenotricha *(Fern. & Macbr.) P.V. Wells in Madroño,* 35(4): 339 (1988 publ. 1989): A. uva-ursi var. adenotricha.
f. coactilis *(Fern. & Macbr.) P.V. Wells in Madroño,* 35(4): 339 (1988 publ. 1989): A. uva-ursi var. coactilis.
f. leobreweri *(Roof) P.V. Wells in Madroño,* 35(4): 339 (1988 publ. 1989): A. uva-ursi var. leobreweri.
f. longipilosa *(Packer & Denford) P.V. Wells in Madroño,* 35(4): 340 (1988 publ. 1989): A. uva-ursi subsp. longipilosa.
f. marinensis *(Roof) P.V. Wells in Madroño,* 35(4): 340 (1988 publ. 1989): A. uva-ursi var. marinensis.
f. stipitata *(Packer & Denford) P.V. Wells in Madroño,* 35(4): 340 (1988 publ. 1989): A. uva-ursi subsp. stipitata.
f. suborbiculata *(W. Knight) P.V. Wells in Madroño,* 35(4): 340 (1988 publ. 1989): A. uva-ursi var. suborbiculata.

ARDISIA (Myrsinac.).
montana
var. selangorensis *B.C. Stone in Tree Fl. Malaya,* 4: 274 (1989)—Malaya.

ARENARIA (Caryophyllac.).
commagenae *T. Çelebioğlu & C. Favarger in Candollea,* 44(1): 329 (1989)—Turkey.
souliei *A.M. Romo in Soc. Ech. Pl. Vasc. Eur. Bassin Médit. Bull.,* 22: 79 (1986–87 publ. 1988), nom. nov.: A. hispida var. hispanica *Coste & Soulié.*

ARENGA (Palmae).
micrantha *C.F. Wei in Acta Phytotax. Sin.,* 26(5): 404 (1988)—Xizang.

ARGOSTEMMA (Rubiac.).
apiculatum *B. Bremer in Ann. Missouri Bot. Gard.,* 76(1): 21 (1989)—Borneo (Sarawak).
brookei *B. Bremer in Ann. Missouri Bot. Gard.,* 76(1): 38 (1989)—Borneo (Sarawak).
burttii *B. Bremer in Ann. Missouri Bot. Gard.,* 76(1): 22 (1989)—Borneo (Sarawak).
calcicolum *B. Bremer in Ann. Missouri Bot. Gard.,* 76(1): 40 (1989)—Borneo (Sarawak).
chaii *B. Bremer in Ann. Missouri Bot. Gard.,* 76(1): 34 (1989), nom. nov.: A. enerve *Ridley (1927).*
gaharuense *B. Bremer in Ann. Missouri Bot. Gard.,* 76(1): 26 (1989)—Borneo (Sarawak).
geesinkii *B. Bremer in Ann. Missouri Bot. Gard.,* 76(1): 25 (1989)—Borneo (Kalimantan, Sabah, Sarawak).

ARGUSIA (Boraginac.).
argentea *(L.f.) G. Murata in Acta Phytotax. Geobot.,* 39(4–6): 164 (1988): Tournefortia argentea.

ARISAEMA (Arac.).
amurense
f. violaceum *(Engler) Y.S. Kim & S.C. Ko in Korean J. Pl. Taxon.,* 15(2): 76 (1985): A. amurense var. violaceum.
daochengense *P.C. Kao in Acta Bot. Yunnanica,* 11(3): 309 (1989)—China.
omeiense *P.C. Kao in Acta Bot. Yunnanica,* 11(3): 308 (1989)—China.
peninsulae
f. convolutum *(Nakai) Y.S. Kim & S.C. Ko in Korean J. Pl. Taxon.,* 15(2): 80 (1985): A. convolutum.

ARISTOCAPSA J.L. Reveal & C.B. Hardham in Phytologia, 66(2): 84 (1989). *POLYGONACEAE.*
insignis *(Curran) J.L. Reveal & C.B. Hardham in Phytologia,* 66(2): 84 (1989): Chorizanthe insignis.

ARISTOLOCHIA (Aristolochiac.).
ser. Hexodon *(Duch.) J.S. Ma in Acta Phytotax. Sin.,* 27(5): 350 (1989): A. sect. Hexodon.

ARISTOLOCHIA :-
sect. Leptosiphisia *J.S. Ma in Acta Phytotax. Sin.*, 27(5): 346 (1989).
sect. Obliquosiphisia *J.S. Ma in Acta Phytotax. Sin.*, 27(5): 350 (1989).
sect. Odontosiphisia *J.S. Ma in Acta Phytotax. Sin.*, 27(5): 346 (1989).
amara *(Aublet) O. Poncy in Bull. Mus. Nation. Hist. Nat., B, Adansonia*, Ser. 4, 10(4): 344 (1988 publ. 1989): Abuta amara.
austrochinensis *C.Y. Cheng & J.S. Ma in Acta Phytotax. Sin.*, 27(4): 293 (1989)—China, Hainan.
bukuti *O. Poncy in Bull. Mus. Nation. Hist. Nat., B, Adansonia*, Ser. 4, 10(4): 337 (1988 publ. 1989)—French Guiana.
caulialata *C.Y. Wu ex C.Y. Cheng & J.S. Ma in Acta Phytotax. Sin.*, 27(4): 294 (1989)—China.
cremersii *O. Poncy in Bull. Mus. Nation. Hist. Nat., B, Adansonia*, Ser. 4, 10(4): 339 (1988 publ. 1989)—French Guiana.
flava *O. Poncy in Bull. Mus. Nation. Hist. Nat., B, Adansonia*, Ser. 4, 10(4): 341 (1988 publ. 1989)—French Guiana.
forrestiana *J.S. Ma in Acta Bot. Yunnanica*, 11(3): 321 (1989)—China.
guianensis *O. Poncy in Bull. Mus. Nation. Hist. Nat., B, Adansonia*, Ser. 4, 10(4): 343 (1988 publ. 1989)—French Guiana, Surinam.
krisagathra *V.V. Sivarajan & A.K. Pradeep in Pl. Syst. Evol.*, 163(1–2): 31 (1989)—India.
kunmingensis *C.Y. Cheng & J.S. Ma in Acta Phytotax. Sin.*, 27(4): 296 (1989)—China.
salweenensis *C.Y. Cheng & J.S. Ma in Acta Phytotax. Sin.*, 27(4): 295 (1989)—China.
wardiana *J.S. Ma in Acta Phytotax. Sin.*, 27(5): 348 (1989)—Burma, India.
zhongdianensis *J.S. Ma in Acta Phytotax. Sin.*, 27(5): 339 (1989)—China.

ARMATOCEREUS (Cactac.).
brevispinus *J.E. Madsen in Fl. Ecuador*, 35(45): 9 (1989)—Ecuador.

ARMENIACA (Rosac.).
holosericea
var. xupuensis *T.Z. Li in Bull. Bot. Res. North-East. Forest. Inst.*, 9(4): 67 (1989)—China.
mandshurica
f. major *T.Z. Li in Bull. Bot. Res. North-East. Forest. Inst.*, 9(4): 67 (1989)—China.
sibirica
var. multipetala *G.S. Liu & L.B. Zhang in Acta Phytotax. Sin.*, 27(5): 394 (1989)—China.
var. pleniflora *J.Y. Zhang, T.Z. Li, X.J. Li & Y. He in Bull. Bot. Res. North-East. Forest. Inst.*, 9(3): 65 (1989)—China.
vulgaris
var. meixianensis *J.Y. Zhang, T.Z. Li, X.J. Li & Y. He in Bull. Bot. Res. North-East. Forest. Inst.*, 9(3): 66 (1989)—China.

ARMENIACA :-
var. xiongyueensis *T.Z. Li, J.Y. Zhang, X.J. Li & Y. He in Bull. Bot. Res. North-East. Forest. Inst.*, 9(3): 65 (1989)—China.

ARMERIA (Plumbaginac.).
arenaria
subsp. bupleuroides *(Godron & Gren.) W. Greuter & Burdet in Willdenowia*, 19(1): 39 (1989): A. bupleuroides.
segoviensis
subsp. lacaitae *(Huguet del Villar) S. Rivas-Martínez, F. Fernández González & D. Sánchez-Mata in Opusc. Bot. Pharm. Complutensis*, 2: 104 (1986): Statice lacaitae.

ARNICA (Compos.).
griscomii
subsp. frigida *(Iljin) S.J. Wolf in Taxon*, 38(1): 142 (1989): A. frigida.

ARRHENATHERUM (Gramin.).
tuberosum
subsp. baeticum *(Romero Zarco) S. Rivas-Martínez, F. Fernández González & D. Sánchez-Mata in Opusc. Bot. Pharm. Complutensis*, 2: 105 (1986): A. elatius subsp. baeticum.

ARROJADOA (Cactac.).
aureispina
var. anguinea *P.J. Braun & E. Esteves Pereira in Cact. Succ. J. (U.S.A.)*, 60(4): 174 (1988)—Brazil.
beateae *P.J. Braun & E. Esteves Pereira in Kakt. And. Sukk.*, 40(10): 254 (1989)—Brazil.
rhodantha
subsp. canudosensis *(Buining & Brederoo) P.J. Braun in Bradleya*, 6: 90 (1988): A. canudosensis.
var. theunisseniana *(Buining & Brederoo) P.J. Braun in Bradleya*, 6: 90 (1988): A. theunisseniana.

ARSENJEVIA V.N. Starodubtsev in Bot. Zhurn., 74(9): 1344 (1989). *RANUNCULACEAE*.
baikalensis *(Fisch. ex Turcz.) V.N. Starodubtsev in Bot. Zhurn.*, 74(9): 1345 (1989): Anemone baikalensis.
flaccida *(Fr. Schmidt.) V.N. Starodubtsev in Bot. Zhurn.*, 74(9): 1345 (1989): Anemone flaccida.
glabrata *(Maxim.) V.N. Starodubtsev in Bot. Zhurn.*, 74(9): 1345 (1989): Anemone baikalensis var. glabrata.
prattii *(Huth ex Ulbr.) V.N. Starodubtsev in Bot. Zhurn.*, 74(9): 1345 (1989): Anemone prattii.
rossii *(S. Moore) V.N. Starodubtsev in Bot. Zhurn.*, 74(9): 1345 (1989): Anemone rossii.

ARTEMISIA (Compos.).
sect. Salsoloides *T.G. Leonova in Novosti Sist. Vyssh. Rast.*, 25: 144 (1988).

ARTEMISIA :–
condensata *(Korobk.) A.P. Khokhryakov, Analiz Fl. Kolÿmskogo Nagor'ya:* 76 (1989): A. kruhsiana subsp. condensata.
jarovoi *(Korobk.) A.P. Khokhryakov, Analiz Fl. Kolÿmskogo Nagor'ya:* 66 (1989): A. lagopus subsp. jarovoi.
mendozana
var. paramilloensis *F.A. Roig & J.A. Ambrosetti in Parodiana,* 5(2): 364 (1989)—Argentina.

ARTOCARPUS (Morac.).
gongshanensis *S.K. Wu ex C.Y. Wu & S.S. Chang in Acta Bot. Yunnanica,* 11(1): 29 (1989)—China.
nanchuanensis *S.S. Chang, S.C. Tan & Z.Y. Liu in Acta Bot. Yunnanica,* 11(1): 29 (1989)—China.
nigrifolius *C.Y. Wu in Acta Bot. Yunnanica,* 11(1): 28 (1989)—China.
pithecogallus *C.Y. Wu in Acta Bot. Yunnanica,* 11(1): 26 (1989)—China.

ARUM (Arac.).
subsect. Alpina *P. Boyce in Kew Bull.,* 44(3): 391 (1989).
subsect. Cretica *(Engler) P. Boyce in Kew Bull.,* 44(3): 392 (1989): A. § Cretica.
sect. Dioscoridea *(Engler) P. Boyce in Kew Bull.,* 44(3): 390 (1989): A. § Dioscoridea.
subsect. Hygrophila *P. Boyce in Kew Bull.,* 44(3): 391 (1989).
subsect. Tenuifila *(Engler) P. Boyce in Kew Bull.,* 44(3): 391 (1989): A. § Tenuifila.

ARUNDINARIA (Gramin.).
chino
f. angustifolia *(Mitf.) C.S. Chao & S.A. Renvoize in Kew Bull.,* 44(2): 368 (1989): Bambusa angustifolia.
pygmaea
var. disticha *(Mitf.) C.S. Chao & S.A. Renvoize in Kew Bull.,* 44(2): 368 (1989): Bambusa disticha.
scandens *T.R. Soderstrom & R.P. Ellis in Smithsonian Contrib. Bot.,* 72: 20 (1988)—Sri Lanka.

ASARUM (Aristolochiac.).
celsum *F. Maek. ex S. Hatusima & E. Yamahata in J. Phytogeogr. Taxon.,* 36(1): 3 (1988)—Ryukyu Is.
dissitum *F. Maek. ex S. Hatusima & E. Yamahata in J. Phytogeogr. Taxon.,* 36(1): 1 (1988)—Ryukyu Is.
gelasinum *S. Hatusima & E. Yamahata in J. Phytogeogr. Taxon.,* 36(1): 1 (1988)—Ryukyu Is.
gusk *S. Hatusima & E. Yamahata in J. Phytogeogr. Taxon.,* 36(1): 5 (1988)—Ryukyu Is.
hatsushimae *F. Maek. ex S. Hatusima & E. Yamahata in J. Phytogeogr. Taxon.,* 36(1): 3 (1988)—Ryukyu Is.

ASARUM :–
pellucidum *S. Hatusima & E. Yamahata in J. Phytogeogr. Taxon.,* 36(1): 5 (1988)—Ryukyu Is.
simile *S. Hatusima & E. Yamahata in J. Phytogeogr. Taxon.,* 36(1): 3 (1988)—Japan.
tokarense *Hatusima ex S. Hatusima & E. Yamahata in J. Phytogeogr. Taxon.,* 36(1): 5 (1988)—Ryukyu Is.
trinacriforme *S. Hatusima & E. Yamahata in J. Phytogeogr. Taxon.,* 36(1): 4 (1988)—Ryukyu Is.

ASCLEPIAS (Asclepiadac.).
sanjuanensis *K.D. Heil, J.M. Porter & S.L. Welsh in Great Basin Nat.,* 49(1): 100 (1989)—U.S.A. (New Mexico).

ASPASIA (Orchidac.).
silvana *F. de Barros in Hoehnea,* 15: 94 (1988 publ. 1989)—Brazil.

ASPERULA (Rubiac.).
badachschanica *M.G. Pakhomova in Opred. Rast. Sred. Azii,* 9: 367 (1987)—U.S.S.R. (Middle Asia).
botschantzevii *M.G. Pakhomova in Opred. Rast. Sred. Azii,* 9: 366 (1987)—U.S.S.R. (Middle Asia).
ciliatula *M.G. Pakhomova in Opred. Rast. Sred. Azii,* 9: 363 (1987)—U.S.S.R. (Middle Asia).
cynanchica
subsp. occidentalis *(Rouy) C.A. Stace in Watsonia,* 17(4): 444 (1989): A. occidentalis.
czukavinae *M.G. Pakhomova & Karim. in Opred. Rast. Sred. Azii,* 9: 365 (1987)—U.S.S.R. (Middle Asia).
gypsacea *M.G. Pakhomova in Opred. Rast. Sred. Azii,* 9: 367 (1987)—U.S.S.R. (Middle Asia).
insolita *M.G. Pakhomova in Opred. Rast. Sred. Azii,* 9: 366 (1987)—U.S.S.R. (Middle Asia).
karategini *M.G. Pakhomova & Karim. in Opred. Rast. Sred. Azii,* 9: 364 (1987)—U.S.S.R. (Middle Asia).
kovalevskiana *M.G. Pakhomova in Opred. Rast. Sred. Azii,* 9: 365 (1987)—U.S.S.R. (Middle Asia).
nuratensis *M.G. Pakhomova in Opred. Rast. Sred. Azii,* 9: 366 (1987)—U.S.S.R. (Middle Asia).
oppositifolia
var. albidula *M.G. Pakhomova in Opred. Rast. Sred. Azii,* 9: 290 (1987)—U.S.S.R. (Middle Asia).
pseudoglomerata *M.G. Pakhomova in Opred. Rast. Sred. Azii,* 9: 364 (1987)—U.S.S.R. (Middle Asia).
strishovae *M.G. Pakhomova & Karim. in Opred. Rast. Sred. Azii,* 9: 364 (1987)—U.S.S.R. (Middle Asia).

ASPIDISTRA (Convallariac.).
carinata *Y. Wan & X.H. Lu in Bull. Bot. Res. North-East. Forest. Inst.*, 9(2): 97 (1989)—China.
cyathiflora *Y. Wan & C.C. Huang in Bull. Bot. Res. North-East. Forest. Inst.*, 9(2): 100 (1989)—China.
patentiloba *Y. Wan & X.H. Lu in Bull. Bot. Res. North-East. Forest. Inst.*, 9(2): 99 (1989)—China.
yingjiangensis *L.J. Peng in Acta Bot. Yunnanica*, 11(2): 173 (1989)—China.

ASPIDOGLOSSUM (Asclepiadac.).
difficile *O.M. Hilliard in Notes Roy. Bot. Gard. Edinburgh*, 45(2): 179 (1988 publ. 1989)—South Africa (Natal).
xanthosphaerum *O.M. Hilliard in Notes Roy. Bot. Gard. Edinburgh*, 45(2): 181 (1988 publ. 1989)—South Africa (Natal).

ASPIDOGYNE (Orchidac.).
steyermarkii *G. Carnevali & E. Foldats in Ann. Missouri Bot. Gard.*, 76(2): 596 (1989)—Venezuela.

ASPILIA (Compos.).
malaissei *S. Lisowski in Bull. Jard. Bot. Nation. Belg.*, 58(3–4): 557 (1988)—Zaïre.

ASTER (Compos.).
hintonii *G.L. Nesom in Phytologia*, 67(4): 342 (1989)—Mexico.
saxicastellii *J.J.N. Campbell & M.E. Medley in Sida*, 13(3): 277 (1989)—U.S.A. (Kentucky).

ASTEROMYRTUS (Myrtac.).
angustifolia *(Gaertner) L.A. Craven in Austral. Syst. Bot.*, 1(4): 377 (1988 publ. 1989): Melaleuca angustifolia.
arnhemica *(Byrnes) L.A. Craven in Austral. Syst. Bot.*, 1(4): 378 (1988 publ. 1989): Melaleuca arnhemica.
brassii *(Byrnes) L.A. Craven in Austral. Syst. Bot.*, 1(4): 379 (1988 publ. 1989): Melaleuca brassii.
lysicephala *(F. Muell. & F.M. Bail.) L.A. Craven in Austral. Syst. Bot.*, 1(4): 380 (1988 publ. 1989): Melaleuca lysicephala.
magnifica *(Specht) L.A. Craven in Austral. Syst. Bot.*, 1(4): 381 (1988 publ. 1989): Melaleuca magnifica.
symphyocarpa *(F. Muell.) L.A. Craven in Austral. Syst. Bot.*, 1(4): 383 (1988 publ. 1989): Melaleuca symphyocarpa.
tranganensis *L.A. Craven in Austral. Syst. Bot.*, 1(4): 384 (1988 publ. 1989)—New Guinea (W. Irian).

ASTRACANTHA (Legumin.).
ambigua *(O. Kuntze) W. Greuter & Burdet in Willdenowia*, 19(1): 30 (1989): Astragalus ambiguus.

ASTRACANTHA :-
baytopiana *(Chamberlain & Mathews) W. Greuter & Burdet in Willdenowia*, 19(1): 30 (1989): Astragalus baytopianus.
elazigensis *(Ekim) W. Greuter & Burdet in Willdenowia*, 19(1): 30 (1989): Astragalus elazigensis.

ASTRAGALUS (Legumin.).
sect. Cercidophaca *D. Podlech & A.R. Maassoumi in Davis & Hedge Festschrift (ed. Kit Tan)*: 72 (1989).
sect. Plagiophaca *A.A. Ramak Maassoumi & D. Podlech in Iranian J. Bot.*, 4(1): 83 (1988).
absentivus *A.A. Ramak Maassoumi in Iranian J. Bot.*, 4(1): 128 (1988)—Iran.
adulterinus *D. Podlech in Mitt. Bot. Staatssamml. München*, 27: 52 (1988)—Iran.
anisacanthus
subsp. schurabicus *(Bunge) S. Tietz in Mitt. Bot. Staatssamml. München*, 27: 244 (1988): A. schurabicus.
aridicola *P.C. Li in Acta Bot. Yunnanica*, 11(3): 295 (1989)—China, Xizang.
aridovallicola *P.C. Li in Acta Bot. Yunnanica*, 11(3): 299 (1989)—China.
armatus
subsp. numidicus *(Murb.) S. Tietz in Mitt. Bot. Staatssamml. München*, 27: 251 (1988): A. numidicus.
austrosachalinensis *N.S. Pavlova in Bot. Zhurn.*, 74(6): 906 (1989)—Sakhalin Is. (Moneron Is.).
baba-alliar
subsp. nudicarpus *(Širj. & Rech. f.) S. Tietz in Mitt. Bot. Staatssamml. München*, 27: 259 (1988): A. bruguieri var. nudicarpus.
babakhanloui *A.A. Ramak Maassoumi & D. Podlech in Iranian J. Bot.*, 4(1): 72 (1988)—Iran.
bazmanicus *D. Podlech in Mitt. Bot. Staatssamml. München*, 27: 63 (1988)—Iran.
brotherusii *D. Podlech in Mitt. Bot. Staatssamml. München*, 27: 54 (1988)—U.S.S.R. (Middle Asia).
calliphysa
subsp. angustifolius *S. Tietz in Mitt. Bot. Staatssamml. München*, 27: 268 (1988)—Iran.
callistachys
subsp. porphyrobaphis *(Fischer) S. Tietz in Mitt. Bot. Staatssamml. München*, 27: 190 (1988): A. porphyrobaphis.
cercidophacos *D. Podlech & A.R. Maassoumi in Davis & Hedge Festschrift (ed. Kit Tan)*: 77 (1989)—Iran.
chingianus *P.C. Li in Acta Bot. Yunnanica*, 11(3): 290 (1989)—China.
davidii
var. acutidentatus *P.C. Li in Acta Bot. Yunnanica*, 11(3): 294 (1989)—China.

ASTRAGALUS :–

demavendicolus
subsp. microphysopsis *S. Tietz in Mitt. Bot. Staatssamml. München,* 27: 204 (1988)—Iran.
doshman-ziariensis *A.A. Ramak Maassoumi & D. Podlech in Iranian J. Bot.,* 4(1): 74 (1988)—Iran.
dulungkiangensis *P.C. Li in Acta Bot. Yunnanica,* 11(3): 298 (1989)—China.
enneaphyllus *P.C. Li in Acta Bot. Yunnanica,* 11(3): 289 (1989)—China.
facetus *A.A. Ramak Maassoumi & D. Podlech in Iranian J. Bot.,* 4(1): 74 (1988)—Iran.
fasciculifolius
subsp. arbusculinus *(Bornm. & Gauba) S. Tietz in Mitt. Bot. Staatssamml. München,* 27: 276 (1988): A. arbusculinus.
galbineus *A.A. Ramak Maassoumi in Iranian J. Bot.,* 4(1): 130 (1988)—Iran.
ghahremanii *A.A. Ramak Maassoumi & D. Podlech in Iranian J. Bot.,* 4(1): 76 (1988)—Iran.
gonggamontis *P.C. Li in Acta Bot. Yunnanica,* 11(3): 296 (1989)—China.
grubovii *Cheng f. ex P.C. Li in Acta Bot. Yunnanica,* 11(3): 293 (1989)—China.
herbertii *A.A.R. Maassoumi in Mitt. Bot. Staatssamml. München,* 27: 129 (1988)—Iran.
hypoglottis
subsp. gremlii *(Burnat) W. Greuter & Burdet in Willdenowia,* 19(1): 30 (1989): A. gremlii.
inexpectatus *A.A. Ramak Maassoumi & D. Podlech in Iranian J. Bot.,* 4(1): 77 (1988)—Iran.
inopinatus
subsp. pseudoadsurgens *(Jurtz.) N.S. Pavlova in Sosud. Rast. Sovet. Dal'nego Vostoka,* 4: 234 (1989): A. pseudoadsurgens.
inquilinus *A.A. Ramak Maassoumi in Iranian J. Bot.,* 4(1): 132 (1988)—Iran.
iranshahrii *A.A. Ramak Maassoumi & D. Podlech in Iranian J. Bot.,* 4(1): 79 (1988)—Iran.
kashafensis *D. Podlech in Mitt. Bot. Staatssamml. München,* 27: 61 (1988)—Iran.
kunarensis *D. Podlech in Mitt. Bot. Staatssamml. München,* 27: 55 (1988)—Afghanistan.
lentiginosus
var. pohlii *S.L. Welsh & R.C. Barneby in Iselya,* 2(1): 1 (1981)—U.S.A. (Utah).
leonardii *A.A. Ramak Maassoumi in Iranian J. Bot.,* 4(1): 132 (1988)—Iran.
lepidus *D. Podlech in Mitt. Bot. Staatssamml. München,* 27: 57 (1988)—Iran.
lineariaurifer *P.C. Li in Acta Bot. Yunnanica,* 11(3): 292 (1989)—China.
magistratus *A.A. Ramak Maassoumi, A. Ghahreman & V. Mozaffarian in Iranian J. Bot.,* 4(1): 110 (1988)—Iran.

ASTRAGALUS :–

marinus
subsp. boreomarinus *(Khokhr.) N.S. Pavlova in Sosud. Rast. Sovet. Dal'nego Vostoka,* 4: 235 (1989): A. boreomarinus.
matiensis *P.C. Li in Acta Bot. Yunnanica,* 11(3): 294 (1989)—China.
memnonius *A.A. Ramak Maassoumi & D. Podlech in Iranian J. Bot.,* 4(1): 81 (1988)—Iran.
meuselii *A. Romo in Flora (Germany),* 182(3–4): 265 (1989)—Morocco.
miyalomontis *P.C. Li in Acta Bot. Yunnanica,* 11(3): 288 (1989)—China.
montigenus *(Hand.-Mazz.) Cheng f. ex P.C. Li in Acta Bot. Yunnanica,* 11(3): 289 (1989): A. bhotanensis var. montigenus.
neomobayenii *A.A. Ramak Maassoumi in Iranian J. Bot.,* 4(1): 135 (1988), as 'neo-mobayenii'—Iran.
nigrocarpus *F.O. Khasanov & I.I. Mal'tsev in Dokl. Akad. Nauk Uzb. SSR,* 1989(3): 54 (1989)—U.S.S.R. (Middle Asia).
nutriosensis *M.J. Sanderson in Madroño,* 35(4): 325 (1988 publ. 1989)—U.S.A. (Arizona).
papillosus *D. Podlech in Mitt. Bot. Staatssamml. München,* 27: 60 (1988)—Afghanistan.
patrius *A.A. Ramak Maassoumi in Iranian J. Bot.,* 4(1): 137 (1988)—Iran.
perpexus *A.A.R. Maassoumi in Mitt. Bot. Staatssamml. München,* 27: 131 (1988)—Iran.
pish-chakensis *A.A.R. Maassoumi in Mitt. Bot. Staatssamml. München,* 27: 132 (1988)—Iran.
plagiophacos *A.A. Ramak Maassoumi & D. Podlech in Iranian J. Bot.,* 4(1): 83 (1988)—Iran.
pseudofragiferus *S. Tietz in Mitt. Bot. Staatssamml. München,* 27: 213 (1988)—Iran.
pseudomultijugus *D. Podlech in Mitt. Bot. Staatssamml. München,* 27: 58 (1988)—Iran.
rivashensis *A.A. Ramak Maassoumi in Iranian J. Bot.,* 4(1): 139 (1988)—Iran.
rubrocalycinus *A.A. Ramak Maassoumi & D. Podlech in Iranian J. Bot.,* 4(1): 85 (1988)—Iran.
sanandajianus *S. Tietz in Mitt. Bot. Staatssamml. München,* 27: 222 (1988)—Iran.
semicircularis *P.C. Li in Acta Bot. Yunnanica,* 11(3): 297 (1989), as 'semi-circularis'—China.
siccaneus *P.C. Li in Acta Bot. Yunnanica,* 11(3): 291 (1989)—China.
simonii *A. Huber-Morath in Bauhinia,* 9(1): 38 (1988)—Turkey.
stepporum *D. Podlech in Mitt. Bot. Staatssamml. München,* 27: 51 (1988)—Iran.

ASTRAGALUS :–
stevenianus
subsp. glochideus *(Boriss.) A.K. Sytin in Novosti Sist. Vyssh. Rast.*, 25: 107 (1988): A. glochideus.
subsp. meskheticus *(Manden.) A.K. Sytin in Novosti Sist. Vyssh. Rast.*, 25: 106 (1988): A. meskheticus.
susianus
subsp. sericeus *S. Tietz in Mitt. Bot. Staatssamml. München*, 27: 174 (1988)—Iran.
termeanus *A.A. Ramak Maassoumi & D. Podlech in Iranian J. Bot.*, 4(1): 87 (1988)—Iran.
typhaeformis *A.A. Ramak Maassoumi in Iranian J. Bot.*, 4(1): 141 (1988)—Iran.
venustus *A.R. Maassoumi & D. Podlech in Davis & Hedge Festschrift (ed. Kit Tan)*: 78 (1989)—Iran.
vessaliae *A.A. Ramak Maassoumi & D. Podlech in Iranian J. Bot.*, 4(1): 88 (1988), as 'vessalae'—Iran.
wolungensis *P.C. Li in Acta Bot. Yunnanica*, 11(3): 295 (1989)—China.

ASTRANTIA (Umbellif.).
tenorei *M.G. Mariotti in Webbia*, 43(1): 11 (1989)—Italy.

ASTROCODON (Campanulac.).
auritus *(Greene) A.P. Khokhryakov, Analiz Fl. Kolymskogo Nagor'ya*: 122 (1989): Campanula aurita.
piperi *(Howell) A.P. Khokhryakov, Analiz Fl. Kolymskogo Nagor'ya*: 122 (1989): Campanula piperi.

ASTROTRICHILIA (Meliac.).
asterotricha *(Radlk.) M. Cheek in Kew Bull.*, 44(2): 369 (1989): Trichilia asterotricha.
elliotii *(Harms) M. Cheek in Kew Bull.*, 44(2): 369 (1989): Trichilia elliotii.

ASYNEUMA (Campanulac.).
anhuiense *B.A. Shen in Acta Phytotax. Sin.*, 26(6): 463 (1988)—China.

ATRACTYLIS (Compos.).
auranitica *K.H. Rechinger in Willdenowia*, 19(1): 156 (1989)—Jordan.
sojakii *K.H. Rechinger in Willdenowia*, 19(1): 155 (1989)—Iran.

ATRIPLEX (Chenopodiac.).
acanthocarpa
subsp. coahuilensis *J. Henrickson in Southwest. Nat.*, 33(4): 458 (1988)—Mexico, U.S.A. (Texas).
subsp. pringlei *(Standley) J. Henrickson in Southwest. Nat.*, 33(4): 461 (1988): A. pringlei.
subsp. stewartii *(I.M. Johnston) J. Henrickson in Southwest. Nat.*, 33(4): 457 (1988): A. stewartii.

ATRIPLEX :–
×gustafssoniana *P.M. Taschereau in Watsonia*, 17(3): 256 (1989)—Great Britain, Sweden, Norway, Denmark. (A. longipes × prostrata).
var. kattegatensis *(Turesson) P.M. Taschereau in Watsonia*, 17(3): 258 (1989): A. longipes subsp. kattegatense.
×hulmeana *P.M. Taschereau in Watsonia*, 17(3): 261 (1989)—Great Britain. (A. littoralis × prostrata).

ATUNA (Chrysobalanac.).
racemosa
subsp. excelsa *(Jack) G.T. Prance in Fl. Males.*, 10(4): 670 (1989): Petrocarya excelsa.

AUSTRALOPYRUM (Gramin.).
uncinatum *J.F. Veldkamp in Blumea*, 34(1): 67 (1989)—New Guinea (Papua New Guinea).

AUSTROCEPHALOCEREUS (Cactac.).
estevesii
subsp. grandiflora *L. Diers & E. Esteves Pereira in Kakt. And. Sukk.*, 40(4): 81 (1989)—Brazil.
subsp. insigniflorus *L. Diers & E. Esteves Pereira in Kakt. And. Sukk.*, 39(11): 268 (1988)—Brazil.

AUSTROMYRTUS (Myrtac.).
floribunda *(A.J. Scott) G.P. Guymer in Austrobaileya*, 2(5): 569 (1988): Backhousia floribunda.

AUTUMNALIA M.G. Pimenov in Bot. Zhurn., 74(10): 1492 (1989). *UMBELLIFERAE.*
botschantzevii *M.G. Pimenov in Bot. Zhurn.*, 74(10): 1493 (1989)—U.S.S.R. (Middle Asia).
innopinata *M.G. Pimenov in Bot. Zhurn.*, 74(10): 1495 (1989)—U.S.S.R. (Middle Asia).

AVENELLA (Gramin.).
flexuosa
subsp. corsica *(Tausch) J. Holub in Folia Geobot. Phytotax.*, 23(4): 413 (1988): Aira corsica.

AXYRIS (Chenopodiac.).
prostrata
f. ovatifolia *T.P. Soong in Fl. Sichuanica*, 4: 14 (1988)—China.

AZORELLA (Umbellif.).
macquariensis *A.E. Orchard in Muelleria*, 7(1): 16 (1989)—Macquarie Is.

B

BACCHARIS (Compos.).
sect. Baccharidastrum *(Cabrera) G.L. Nesom in Phytologia,* 65(3): 170 (1988): gen. Baccharidastrum.
breviseta
var. colombianum *(Cuatr.) G.L. Nesom in Phytologia,* 65(3): 171 (1988): Baccharidastrum argutum var. colombianum.
schomburkkii
subsp. occidentalis *J. Steyermark & B. Maguire in Acta Bot. Venezuel.,* 14(3): 25 (1984)—Venezuela.

BACKHOUSIA (Myrtac.).
kingii *G.P. Guymer in Austrobaileya,* 2(5): 567 (1988)—Australia (Queensland).

BACTRIS (Palmae).
dardanoi *J.T. de Medeiros-Costa in Prog. Resum. Soc. Bot. Brasil, XXXVI Congr. Nac. Bot., Curitiba:* 217 (1985), without latin descr. or type—Brazil.
glassmanii *J.T. de Medeiros-Costa in Prog. Resum. Soc. Bot. Brasil, XXXVI Congr. Nac. Bot., Curitiba:* 217 (1985), without latin descr. or type—Brazil.

BALIOSPERMUM (Euphorbiac.).
angustifolium *Y.T. Chang in Acta Phytotax. Sin.,* 27(2): 148 (1989)—Xizang.
malayanum *J.D. Hooker, Fl. Brit. India,* 5: 463 (1888), as 'Baliospereum analayanum' in Index Kewensis Supp. 1—Borneo.†
yui *Y.T. Chang in Acta Bot. Yunnanica,* 11(4): 413 (1989), with two types cited—China, Burma.

BAMBUSA (Gramin.).
albofolia *T.H. Wen & Hua in J. Bamboo Res.,* 8(1): 20 (1989)—China.
angustiaurita *W.T. Lin in Bamboo Res.,* 1983(2): 52 (1983)—China.
dissimilis *W.T. Lin in Bamboo Res.,* 1983(2): 50 (1983)—China, Hainan.
flavonoda *W.T. Lin in Bamboo Res.,* 1986(2): 23 (1986), as 'flavo-noda'—China.
parvifolia *W.T. Lin in Bamboo Res.,* 1986(2): 22 (1986)—China.
textilis
var. purpurascens *N.H. Xia in Bamboo Res.,* 1985(1): 38 (1985)—China.
viridivittata *W.T. Lin in Bamboo Res.,* 1983(2): 54 (1983), as 'viridi-vittata'—China.

BANKSIA (Proteac.).
ser. Bauerinae *A.S. George in Nuytsia,* 6(3): 309 (1988).
epica *A.S. George in Nuytsia,* 6(3): 310 (1988)—Australia (W. Australia).

BANKSIA :–
leptophylla
var. melletica *A.S. George in Nuytsia,* 6(3): 314 (1988), as 'mellitica'—Australia (W. Australia).
occidentalis
subsp. formosa *S.D. Hopper in Nuytsia,* 7(1): 21 (1989)—Australia (W. Australia).
oligantha *A.S. George in Nuytsia,* 6(3): 312 (1988)—Australia (W. Australia).
seminuda
subsp. remanens *S.D. Hopper in Nuytsia,* 7(1): 16 (1989)—Australia (W. Australia).
spinulosa
var. neoanglica *A.S. George in Nuytsia,* 6(3): 315 (1988)—Australia (New South Wales, Queensland).

BAPHICACANTHUS (Acanthac.).
multibracteolata *H.T. Chang & H. Chu in Acta Sci. Nat. Univ. Sunyatseni,* 28(2): 65 (1989)—China.

BARKERIA (Orchidac.).
obovata *(Presl) E.A. Christenson in Lindleyana,* 3(4): 221 (1988 publ. 1989): Oncidium obovatum.

BASANANTHE (Passiflorac.).
aristolochioides *A. Robyns in Bull. Jard. Bot. Nation. Belg.,* 59(1–2): 228 (1989)—Zaïre.
botryoidea *A. Robyns in Bull. Jard. Bot. Nation. Belg.,* 59(1–2): 231 (1989)—Zaïre.
cupricola *A. Robyns in Bull. Jard. Bot. Nation. Belg.,* 59(1–2): 234 (1989)—Zaïre.
kundelunguensis *A. Robyns in Bull. Jard. Bot. Nation. Belg.,* 59(1–2): 236 (1989)—Zaïre.
malaissei *A. Robyns in Bull. Jard. Bot. Nation. Belg.,* 59(1–2): 238 (1989)—Zaïre.
scabrida *A. Robyns in Bull. Jard. Bot. Nation. Belg.,* 59(1–2): 239 (1989)—Zaïre.

BASHANIA (Gramin.).
spanostachya *T.P. Yi in Acta Bot. Yunnanica,* 11(1): 35 (1989)—China.

BASTARDIASTRUM (Malvac.).
tarasoides *P.A. Fryxell, Malvac. of Mexico (Syst. Bot. Monog., 25):* 126 (1988)—Mexico.

BAUHINIA (Legumin.).
chrysophylla *K. Larsen & S.S. Larsen in Nordic J. Bot.,* 9(3): 253 (1989)—Thailand.
leucantha *M. Thulin in Nordic J. Bot.,* 9(1): 51 (1989)—Somali Republic.
sessilifolia *(Mutis ex DC.) L.M. Quiñones in Ann. Missouri Bot. Gard.,* 75(3): 1156 (1988): Amaria sessilifolia.

BECCARINDA (Gesneriac.).
minima *K.Y. Pan in Acta Phytotax. Sin.,* 26(6): 437 (1988)—China.

BECIUM (Labiat.).
subgen. Monobecium *O. Sebald in Stuttgarter Beitr. Naturk., A,* 419: 34 (1988).

BECIUM :–
subgen. Orthobecium *O. Sebald in Stuttgarter Beitr. Naturk., A,* 405: 10 (1987).
sect. Serpyllifolium *O. Sebald in Stuttgarter Beitr. Naturk., A,* 405: 10 (1987).
centrali-africanum *(R.E. Fries) O. Sebald in Stuttgarter Beitr. Naturk., A,* 405: 10 (1987): Ocimum centrali-africanum.
dhofarense *O. Sebald in Stuttgarter Beitr. Naturk., A,* 405: 7 (1987), nom. nov.: Orthosiphon comosus *Baker.*
fimbriatum *(Briq.) O. Sebald in Stuttgarter Beitr. Naturk., A,* 405: 10 (1987): Ocimum fimbriatum.
var. angustilanceolatum *(De Wild.) O. Sebald in Stuttgarter Beitr. Naturk., A,* 405: 10 (1987): Ocimum angustilanceolatum.
var. bequaertii *(De Wild.) O. Sebald in Stuttgarter Beitr. Naturk., A,* 419: 49 (1988): Ocimum bequaertii.
var. ctenodon *(Gilli) O. Sebald in Stuttgarter Beitr. Naturk., A,* 405: 10 (1987): B. ctenodon.
var. microphyllum *O. Sebald in Stuttgarter Beitr. Naturk., A,* 419: 51 (1988)—Zambia.
irvinei *(Morton) O. Sebald in Stuttgarter Beitr. Naturk., A,* 419: 34 (1988): Ocimum irvinei.
minutiflorum *O. Sebald in Stuttgarter Beitr. Naturk., A,* 405: 11 (1987)—Burundi, Rwanda, Uganda, Tanzania, Zambia, Zimbabwe.

BEGONIA (Begoniac.).
fenchihuensis *S.S. Ying, Coloured Illustr. Fl. Taiwan,* 3: 618 (1988)—Taiwan.
taiwaniana
var. albomaculata *S.S. Ying, Coloured Illustr. Fl. Taiwan,* 3: 623 (1988), as 'albo-maculata'—Taiwan.

BELLUCIA (Melastomatac.).
beckii *S.S. Renner in Mem. New York Bot. Gard.,* 50: 28 (1989)—Bolivia, Brazil.
huberi *(Wurdack) S.S. Renner in Mem. New York Bot. Gard.,* 50: 28 (1989): Loreya huberi.

BENNETTIODENDRON (Flacourtiac.).
leprosipes
var. pilosum *G.S. Fan & Y.C. Hsu in Guihaia,* 9(1): 32 (1989)—China.

BERGIA (Elatinac.).
auriculata *G.J. Leach in J. Adelaide Bot. Gard.,* 11(2): 91 (1989)—Australia (W. Australia).
barklyana *G.J. Leach in J. Adelaide Bot. Gard.,* 11(2): 89 (1989)—Australia (N. Terr.).
diacheiron *Verdon ex G.J. Leach in J. Adelaide Bot. Gard.,* 11(2): 97 (1989)—Australia (N. Terr., Queensland, S. Australia).
henshallii *G.J. Leach in J. Adelaide Bot. Gard.,* 11(2): 83 (1989)—Australia (N. Terr., W. Australia).

BERGIA :–
occultipetala *G.J. Leach in J. Adelaide Bot. Gard.,* 11(2): 95 (1989)—Australia (N. Terr., S. Australia).
perennis
subsp. exigua *G.J. Leach in J. Adelaide Bot. Gard.,* 11(2): 89 (1989)—Australia (S. Australia, W. Australia, Queensland).
subsp. obtusifolia *G.J. Leach in J. Adelaide Bot. Gard.,* 11(2): 87 (1989)—Australia (N. Terr., W. Australia, S. Australia).

BERKHEYA (Compos.).
leucaugeta *O.M. Hilliard in Notes Roy. Bot. Gard. Edinburgh,* 45(2): 183 (1988 publ. 1989)—South Africa (Natal).

BERTYA (Euphorbiac.).
glabrescens *(C. White) G.P. Guymer in Austrobaileya,* 2(5): 429 (1988): B. oleifolia var. glabrescens.
sharpeana *G.P. Guymer in Austrobaileya,* 2(5): 427 (1988)—Australia (Queensland).

BESCHORNERIA (Agavac.).
septentrionalis *A. García Mendoza in Cact. Suc. Mex.,* 33(1): 3 (1988)—Mexico.

BETULA (Betulac.).
ceratoptera *G.H. Liu & Y.C. Ma in Bull. Bot. Res. North-East. Forest. Inst.,* 9(4): 55 (1989)—China.
pendula
f. bircalensis *(Mela) L. Hämet-Ahti in Mem. Soc. Fauna Flora Fenn.,* 65(1): 9 (1989): B. verrucosa bircalensis.
f. crispa *(Reichenb.) L. Hämet-Ahti in Mem. Soc. Fauna Flora Fenn.,* 65(1): 9 (1989): Betula laciniata var. crispa.
var. lapponica *(Lindquist) L. Hämet-Ahti in Mem. Soc. Fauna Flora Fenn.,* 65(1): 9 (1989): B. verrucosa var. lapponica.
pubescens
var. czerepanovii *(Orlova) L. Hämet-Ahti in Mem. Soc. Fauna Flora Fenn.,* 65(1): 9 (1989): B. czerepanovii.
f. hibernifolia *T. Ulvinen in Mem. Soc. Fauna Flora Fenn.,* 65(4): 133 (1989)—Finland.
f. rubra *T. Ulvinen in Mem. Soc. Fauna Flora Fenn.,* 65(4): 135 (1989)—Finland.

BHIDEA (Gramin.).
borii *U.R. Deshpande, V. Prakash & N.P. Singh in Current Sci.,* 58(19): 1094 (1989)—India.

BIARUM (Arac.).
ditschianum *J. Bogner & P. Boyce in Willdenowia,* 18(2): 409 (1989)—Turkey.

BIDARIA (Asclepiadac.).
indica *M.A. Rahman & C.C. Wilcock in Blumea,* 34(1): 99 (1989), nom. nov.: Gymnema montanum var. beddomei *Hook. f.*
lacei *(Craib) M.A. Rahman & C.C. Wilcock in Blumea,* 34(1): 100 (1989): Gymnema lacei.

BIDARIA :–
montana *(Roxb.) M.A. Rahman & C.C. Wilcock in Blumea,* 34(1): 99 (1989): Asclepias montana.

×**BIFRENIELLA** hort. in Orchid Rev., 97(1154): centre page pull-out p. 8 (1989). *ORCHIDACEAE.* (Bifrenaria × Rudolfiella).

BIKKIOPSIS (Rubiac.).
comptonii *(S. Moore) M.G. Baumann-Bodenheim, Syst. Fl. Neu-Caledonien,* 4: 102 (1988 publ. Jan. 1989), without basionym ref.; *et Syst. Fl. Neu-Caledonien,* 5: 97 (May 1989), with basionym ref.: Bikkia comptonii.

BILLBERGIA (Bromeliac.).
domingosmartinsis *E. Gross in Abh. Akad. Wiss. Lit. Mainz, Math.-Naturwiss. Klasse, Trop. Subtrop. Pflanzenwelt,* 65: 65 (1988)—Brazil.

BIOPHYTUM (Oxalidac.).
hermanni *J.F. Veldkamp in Taxon,* 38(1): 114 (1989)—Sri Lanka.

BISTORTA (Polygonac.).
alopecuroides
var. zeaensis *(Kom.) N.N. Tsvelev in Sosud. Rast. Sovet. Dal'nego Vostoka,* 4: 59 (1989): Polygonum zeaense.
ensigera *(Juz.) N.N. Tsvelev in Novosti Sist. Vyssh. Rast.,* 25: 185 (1988): Polygonum ensigerum.
pacifica
var. tomentella *(Kom.) N.N. Tsvelev in Sosud. Rast. Sovet. Dal'nego Vostoka,* 4: 57 (1989): Polygonum ussuriense f. tomentellum.

BLAKEA (Melastomatac.).
fuchsioides *F. Almeda in Proc. Calif. Acad. Sci.,* Ser. 4, 46(5): 137 (1989)—Panama.

BLEEKRODEA (Morac.).
malayana *(Corner) C.C. Berg in Proc. Kon. Nederl. Akad. Wetensch., C,* 91(4): 359 (1988): Streblus malayanus.

BOEHMERIA (Urticac.).
chingshuishaniana *S.S. Ying, Coloured Illustr. Fl. Taiwan,* 3: 468 (1988)—Taiwan.
nivea
var. viridula *(Yamamoto) S.S. Ying, Coloured Illustr. Fl. Taiwan,* 3: 470 (1988): B. nivea var. viridula.
taiwaniana
var. pilushanensis *(Liu & Lu) S.S. Ying, Coloured Illustr. Fl. Taiwan,* 3: 466 (1988): B. pilushanensis.

BOEICA (Gesneriac.).
multinervia *K.Y. Pan in Acta Phytotax. Sin.,* 26(6): 439 (1988)—China.
stolonifera *K.Y. Pan in Acta Phytotax. Sin.,* 26(6): 439 (1988)—China.

BOENNINGHAUSENIA (Rutac.).
albiflora
var. pilosa *Z.M. Tan in Bull. Bot. Res. North-East. Forest. Inst.,* 9(2): 47 (1989)—China.

BOLBOSCHOENUS (Cyperac.).
maritimus
subsp. paludosus *(A. Nels.) T. Koyama in Bishop Mus. Occas. Pap.,* 29: 128 (1989): Scirpus paludosus.

×**BOLLOPETALUM** hort. in Orchid Rev., 97(1146): centre page pull-out p. 8 (1989). *ORCHIDACEAE.* (Bollea × Zygopetalum).

BOMAREA (Alstroemeriac.).
bracteolata *R.E. Gereau in Ann. Missouri Bot. Gard.,* 76(2): 601 (1989)—Panama.
caudatisepala *R.E. Gereau in Ann. Missouri Bot. Gard.,* 76(2): 599 (1989)—Panama.
suberecta *R.E. Gereau in Ann. Missouri Bot. Gard.,* 76(2): 599 (1989)—Panama, Costa Rica.

BONAMIA (Convolvulac.).
leonii *A.H. Gentry & Austin in Ann. Missouri Bot. Gard.,* 75(4): 1429 (1988 publ. 1989)—Colombia.

BONTIA (Myoporac.).
bontioides *(A. Gray) L.V. Aver'yanov in L.V. Aver'yanov, N.T. Ban & E.P. Kudryavtseva, Mater. Fl. Rast. Ostrov. V'etnama:* 73 (1988): Myoporum bontioides.

BOTHRIOCHLOA (Gramin.).
bladhii
subsp. glabra *(Roxb.) B.K. Simon in Austrobaileya,* 3(1): 79 (1989): Andropogon glaber.

BOTHRIOCLINE (Compos.).
kundelungensis *S. Lisowski in Bull. Jard. Bot. Nation. Belg.,* 58(3–4): 558 (1988)—Zaïre.

BRACHYCHITON (Sterculiac.).
sect. Delabechea *(Mitchell ex Lindley) G.P. Guymer in Austral. Syst. Bot.,* 1(3): 239 (1988): gen. Delabechea.
sect. Oxystele *G.P. Guymer in Austral. Syst. Bot.,* 1(3): 222 (1988).
acuminatus *G.P. Guymer in Austral. Syst. Bot.,* 1(3): 251 (1988)—Australia (W. Australia).
albidus *G.P. Guymer in Austral. Syst. Bot.,* 1(3): 271 (1988)—Australia (Queensland).
×allochrous *G.P. Guymer in Austral. Syst. Bot.,* 1(3): 316 (1988)—Australia (Queensland). (B. grandiflorus × muellerianus).
×carneus *G.P. Guymer in Austral. Syst. Bot.,* 1(3): 315 (1988)—Australia (Queensland). (B. garawayae × grandiflorus).
chillagoensis *G.P. Guymer in Austral. Syst. Bot.,* 1(3): 296 (1988)—Australia (Queensland).

BRACHYCHITON :-
collinus *G.P. Guymer in Austral. Syst. Bot.,* 1(3): 249 (1988)—Australia (Queensland, N. Terr.).
compactus *G.P. Guymer in Austral. Syst. Bot.,* 1(3): 240 (1988)—Australia (Queensland).
diversifolius
subsp. orientalis *G.P. Guymer in Austral. Syst. Bot.,* 1(3): 231 (1988)—Australia (Queensland).
×excellens *G.P. Guymer in Austral. Syst. Bot.,* 1(3): 314 (1988)—Australia (New South Wales). (B. bidwillii × discolor).
fitzgeraldianus *G.P. Guymer in Austral. Syst. Bot.,* 1(3): 282 (1988)—Australia (W. Australia).
garrawayae *(Bailey) G.P. Guymer in Austral. Syst. Bot.,* 1(3): 274 (1988): Sterculia garrawayae.
grandiflorus *G.P. Guymer in Austral. Syst. Bot.,* 1(3): 263 (1988)—Australia (Queensland).
×hirtellus *G.P. Guymer in Austral. Syst. Bot.,* 1(3): 317 (1988)—Australia (N. Terr.). (B. megaphyllus × multicaulis).
×incarnatus *G.P. Guymer in Austral. Syst. Bot.,* 1(3): 313 (1988)—Australia (Queensland). (B. discolor × populneus).
megaphyllus *G.P. Guymer in Austral. Syst. Bot.,* 1(3): 291 (1988)—Australia (N. Terr.).
muellerianus *G.P. Guymer in Austral. Syst. Bot.,* 1(3): 302 (1988)—Australia (Queensland).
multicaulis *G.P. Guymer in Austral. Syst. Bot.,* 1(3): 305 (1988)—Australia (N. Terr.).
obtusilobus *G.P. Guymer in Austral. Syst. Bot.,* 1(3): 254 (1988)—Australia (W. Australia).
populneus
subsp. trilobus *G.P. Guymer in Austral. Syst. Bot.,* 1(3): 235 (1988)—Australia (New South Wales, Queensland).
×roseus *G.P. Guymer in Austral. Syst. Bot.,* 1(3): 308 (1988)—Cultivation. (B. acerifolius × populneus).
spectabilis *G.P. Guymer in Austral. Syst. Bot.,* 1(3): 280 (1988)—Australia (N. Terr.).
tridentatus *G.P. Guymer in Austral. Syst. Bot.,* 1(3): 288 (1988)—Australia (W. Australia).
tuberculatus *(W. Fitzg.) G.P. Guymer in Austral. Syst. Bot.,* 1(3): 303 (1988): Sterculia tuberculata.
×turgidulus *G.P. Guymer in Austral. Syst. Bot.,* 1(3): 309 (1988)—Australia (Queensland). (B. populneus × rupestris).
×vinicolor *G.P. Guymer in Austral. Syst. Bot.,* 1(3): 311 (1988)—Australia (New South Wales, Queensland). (B. acerifolius × discolor).
viridiflorus *(W. Fitzg.) G.P. Guymer in Austral. Syst. Bot.,* 1(3): 290 (1988): Sterculia viridiflora.
viscidulus *(W. Fitzg.) G.P. Guymer in Austral. Syst. Bot.,* 1(3): 277 (1988): Sterculia viscidula.

BRACHYCHITON :-
vitifolius *(Bailey) G.P. Guymer in Austral. Syst. Bot.,* 1(3): 269 (1988): Sterculia vitifolia.
xanthophyllus *G.P. Guymer in Austral. Syst. Bot.,* 1(3): 295 (1988)—Australia (W. Australia).

BRACHYRAMPHUS (Compos.).
obtusus *A.P. de Candolle, Prodromus,* 7: 177 (1838)—India.†

BRACHYSTELMA (Asclepiadac.).
inconspicuum *S. Venter in Excelsa,* 13: 106 (1988)—South Africa (Transvaal).
praelongum
subsp. thunbergii *(N.E. Br.) C. Boele in Asklepios,* 47: 3 (1989): B. thunbergii.
simplex
subsp. banforae *J.P. Lebrun & A.L. Stork in Bull. Mus. Nation. Hist. Nat., B, Adansonia,* Ser. 4, 11(1): 71 (1989)—Nigeria, Ivory Coast, Upper Volta.

×BRASSIODA hort. in Orchid Rev., 97(1149): centre page pull-out p. 8 (1989). *ORCHIDACEAE.* (Brassia × Cochlioda).

BRAYOPSIS (Crucif.).
alpaminae
subsp. smithii *I.A. Al-Shehbaz in Nordic J. Bot.,* 8(6): 623 (1989)—Peru.

BREDIA (Melastomatac.).
amoena
var. eglandulata *B.Y. Ding in Guihaia,* 8(4): 318 (1988)—China.

BREWCARIA L.B. Smith, J. Steyermark & Robinson in Acta Bot. Venezuel., 14(3): 10 (1984). *BROMELIACEAE.*
duidensis *L.B. Smith, J. Steyermark & Robinson in Acta Bot. Venezuel.,* 14(3): 10 (1984)—Venezuela.

BRICKELLIA (Compos.).
eupatorioides
var. chlorolepis *(Woot. & Standl.) B.L. Turner in Phytologia,* 67(2): 129 (1989): Kuhnia chlorolepis.
var. floridana *(R.W. Long) B.L. Turner in Phytologia,* 67(2): 130 (1989): Kuhnia eupatorioides var. floridana.
var. gracillima *(A. Gray) B.L. Turner in Phytologia,* 67(2): 130 (1989): Kuhnia eupatorioides var. gracillima.
secundiflora
var. parryi *(A. Gray) B.L. Turner in Phytologia,* 67(4): 310 (1989): B. parryi.

BRIGGSIA (Gesneriac.).
acutiloba *K.Y. Pan in Acta Phytotax. Sin.,* 26(6): 455 (1988)—China.
dongxingensis *Chun ex K.Y. Pan in Acta Phytotax. Sin.,* 26(6): 451 (1988)—China.
humilis *K.Y. Pan in Acta Phytotax. Sin.,* 26(6): 453 (1988)—China.

BRIGGSIA :–
latisepala *Chun ex K.Y. Pan in Acta Phytotax. Sin.*, 26(6): 454 (1988)—China.
longicaulis *W.T. Wang & K.Y. Pan in Acta Phytotax. Sin.*, 26(6): 450 (1988)—China.
longifolia
var. multiflora *S.Y. Chen ex K.Y. Pan in Acta Phytotax. Sin.*, 26(6): 453 (1988)—China.
parvifolia *K.Y. Pan in Acta Phytotax. Sin.*, 26(6): 457 (1988)—China.
rosthornii
var. wenshanensis *K.Y. Pan in Acta Phytotax. Sin.*, 26(6): 456 (1988)—China.
var. xingrenensis *K.Y. Pan in Acta Phytotax. Sin.*, 26(6): 456 (1988)—China.

BROMUS (Gramin.).
madritensis
subsp. haussknechtii *(Boiss.) H. Scholz in Willdenowia,* 19(1): 107 (1989): B. haussknechtii.
sericeus
subsp. fallax *H. Scholz in Willdenowia,* 19(1): 134 (1989)—Sinai Peninsula.

BRONGNIARTIA (Legumin.).
guerrerensis *J. Jiménez Ramírez & J.L. Contreras Jiménez in An. Inst. Biol. Univ. Nac. Autón. Méx., Bot.*, 58: 63 (1987 publ. 1989)—Mexico.
vazquezii *O. Dorado in Syst. Bot.*, 14(1): 20 (1989)—Mexico.

BROTERA (Sterculiac.).
phoenicea *(L.) A.J. Cavanilles in Anales Hist. Nat.*, 1(1): 34 (1799): Pentapetes phoenicea.

BRUCEA (Simaroubac.).
bruceadelpha *(Nooteboom) A.J.G.H. Kostermans in Reinwardtia,* 10(5): 471 (1988): Laumoniera bruceadelpha.

BRUNELLIA (Brunelliac.).
subsect. Foreroae *J. Cuatrecasas in Fl. Neotrop.*, 2 Suppl.: 80 (1985).
subsect. Stuebeliae *J. Cuatrecasas in Fl. Neotrop.*, 2 Suppl.: 85 (1985).
antioquensis *(Cuatrecasas) J. Cuatrecasas in Fl. Neotrop.*, 2 Suppl.: 48 (1985): B. sibundoya subsp. antioquensis.
boliviana
var. brittonii *(Rusby) J. Cuatrecasas in Fl. Neotrop.*, 2 Suppl.: 79 (1985): B. brittonii.
carpishensis *J. Cuatrecasas in Fl. Neotrop.*, 2 Suppl.: 69 (1985), nom. nov., as 'sp. nov.': B. inermis var. petiolulata *Cuatr.*
cuatrecasana *C.I. Orozco in Fl. Neotrop.*, 2 Suppl.: 20 (1985)—Colombia.
cutervensis *J. Cuatrecasas in Fl. Neotrop.*, 2 Suppl.: 64 (1985)—Peru.
diversifolia *C.I. Orozco in Fl. Neotrop.*, 2 Suppl.: 22 (1985)—Colombia.
espinalii *J. Cuatrecasas in Fl. Neotrop.*, 2 Suppl.: 95 (1985)—Colombia.
gentryi *J. Cuatrecasas in Fl. Neotrop.*, 2 Suppl.: 48 (1985)—Colombia.

BRUNELLIA :–
hiltyana *J. Cuatrecasas in Fl. Neotrop.*, 2 Suppl.: 55 (1985)—Colombia.
littlei
subsp. caucana *J. Cuatrecasas in Fl. Neotrop.*, 2 Suppl.: 74 (1985)—Colombia.
morii *J. Cuatrecasas in Fl. Neotrop.*, 2 Suppl.: 59 (1985)—Panama.
pitayensis *J. Cuatrecasas in Fl. Neotrop.*, 2 Suppl.: 89 (1985)—Colombia.
propinqua
subsp. susaconensis *J. Cuatrecasas in Fl. Neotrop.*, 2 Suppl.: 72 (1985)—Colombia.

BUGLOSSOIDES (Boraginac.).
splitgerberi *(Guss.) S. Brullo in Colloqu. Phytosciolog.*, 12: 79 (1983); *see Pubbl. Ist. Bot. Univ. Catania* (1985): Lithospermum splitgerberi.

BULBOPHYLLUM (Orchidac.).
amplistigmaticum *P.J. Kores in Allertonia,* 5(1): 150 (1989)—Fiji.
aristopetalum *P.J. Kores in Allertonia,* 5(1): 147 (1989)—Fiji.
derchianum *S.S. Ying in J. Jap. Bot.*, 64(5): 148 (1989)—Taiwan.
grandimesense *B. Gray & D.L. Jones in Austrobaileya,* 3(1): 141 (1989)—Australia (Queensland).
hassallii *P.J. Kores in Allertonia,* 5(1): 148 (1989)—Fiji.
incommodum *P.J. Kores in Allertonia,* 5(1): 156 (1989)—Fiji.
kestron *J.J. Vermeulen & A. Lamb in Gard. Bull. Singapore,* 41(2): 30 (1988 publ. 1989)—Borneo (Sabah).
lecouflei *J. Bosser in Bull. Mus. Nation. Hist. Nat., B, Adansonia,* Ser. 4, 11(1): 30 (1989)—Madagascar.
leproglossum *J.J. Vermeulen & A. Lamb in Gard. Bull. Singapore,* 41(2): 31 (1988 publ. 1989)—Borneo (Sabah).
lewisense *B. Gray & D.L. Jones in Austrobaileya,* 3(1): 143 (1989)—Australia (Queensland).
moratii *J. Bosser in Bull. Mus. Nation. Hist. Nat., B, Adansonia,* Ser. 4, 11(1): 29 (1989)—Madagascar.
nubinatum *J.J. Vermeulen in Gard. Bull. Singapore,* 41(2): 34 (1988 publ. 1989)—Borneo (Sabah).
pelicanopsis *J.J. Vermeulen & A. Lamb in Gard. Bull. Singapore,* 41(2): 34 (1988 publ. 1989)—Borneo (Sabah).
quadricarinum *P.J. Kores in Allertonia,* 5(1): 154 (1989)—Fiji.
scabrum *J.J. Vermeulen & A. Lamb in Gard. Bull. Singapore,* 41(2): 37 (1988 publ. 1989)—Borneo (Sabah).
simmondsii *P.J. Kores in Allertonia,* 5(1): 146 (1989)—Fiji.
tengchongense *Z.H. Tsi in Bull. Bot. Res. North-East. Forest. Inst.*, 9(2): 29 (1989)—China.

BULBOPHYLLUM :-
thymophorum *J.J. Vermeulen & A. Lamb in Gard. Bull. Singapore,* 41(2): 40 (1988 publ. 1989)—Borneo (Sabah).
windsorense *B. Gray & D.L. Jones in Austrobaileya,* 3(1): 143 (1989)—Australia (Queensland).

BULBOSTYLIS (Cyperac.).
hispidula
subsp. senegalensis *(Cherm.) C. Vanden Berghen in Fl. Illustr. Sénégal,* 9: 159 (1988): Fimbristylis exilis var. senegalensis.

BUPLEURUM (Umbellif.).
greuteri *S. Snogerup in Davis & Hedge Festschrift (ed. Kit Tan):* 23 (1989)—Greece.
isphairamicum *M.G. Pimenov in Byull. Mosk. Obshch. Ispȳt. Prir., Biol.,* 94(2): 80 (1989)—U.S.S.R. (Middle Asia).

BURMANNIA (Burmanniac.).
cryptopetala
var. baxikangensis *Y.B. Chang & Z. Wei in Bull. Bot. Res. North-East. Forest. Inst.,* 9(2): 37 (1989)—China.
foliosa
subsp. marahuacensis *B. Maguire & J. Steyermark in Acta Bot. Venezuel.,* 14(3): 19 (1984)—Venezuela.

BURMEISTERA (Campanulac.).
intii *J. Gómez-Laurito & L.D. Gómez in Brenesia,* 25–26: 311 (1986 publ. 1988)—Costa Rica.

BUXUS (Buxac.).
sect. Notobuxus *(Oliv.) I. Friis in Kew Bull.,* 44(2): 297 (1989): gen. Notobuxus.
acutata *I. Friis in Kew Bull.,* 44(2): 297 (1989), nom. nov.: Macropodandra acuminata *Gilg.*
cordata *(Radcliffe-Smith) I. Friis in Kew Bull.,* 44(2): 297 (1989): Notobuxus cordata.
koreana
f. elongata *(Nakai) Y.S. Kim & J.H. Kim in Korean J. Pl. Taxon.,* 18(3): 214 (1988): B. koreana var. elongata.
f. insularis *(Nakai) Y.S. Kim & J.H. Kim in Korean J. Pl. Taxon.,* 18(3): 214 (1988): B. microphylla var. insularis.
microphylla
var. liukiuensis *(Makino) S.S. Ying, Colour. Illustr. Pl. Taiwan, Pt. 3; Colour. Illustr. Ligneous Pl. Taiwan,* 1: 375 (1987), without basionym ref.: Basionym not stated.
f. pubescens *S.S. Ying, Coloured Illustr. Fl. Taiwan,* 3: 114 (1988)—Taiwan.
sinica
var. koreana *(Nakai ex Rehd.) Q.L. Wang in Fl. Liaoningica, chief ed. S.X. Li,* 1: 1113 (1988): B. microphylla var. koreana.

BYTTNERIA (Sterculiac.).
caripensis *C.L. Cristóbal in Bonplandia,* 6(2): 125 (1989)—Venezuela.

C

CACALIA (Compos.).
chola *(W.W. Smith) R. Mathur in R.R. Rao et al., Fl. Ind. Enumerat. - Asterac.:* 20 (1988): Senecio chola.
levingii *(Clarke) R. Mathur in R.R. Rao et al., Fl. Ind. Enumerat. - Asterac.:* 20 (1988): Senecio levingii.

CADETIA (Orchidac.).
quadrangularis *P.J. Cribb & B. Lewis in Orchid Rev.,* 97(1150): 251 (1989)—Vanuatu.

CAESALPINIA (Legumin.).
epifanioi *J.L. Contreras Jiménez in An. Inst. Biol. Univ. Nac. Autón. Méx., Bot.,* 58: 55 (1987 publ. 1989)—Mexico.

CALADENIA (Orchidac.).
atroclavia *D.L. Jones & M.A. Clements in Austrobaileya,* 2(5): 552 (1988)—Australia (Queensland).

CALAMAGROSTIS (Gramin.).
subsect. Stylagrostis *(Mez) F.D. Escalona in Phytologia,* 65(5): 338 (1988): gen. Stylagrostis.
arundinacea
var. laxiflora *(Rendle) P.C. Kuo & S.L. Lu ex J.L. Yang in Fl. Sichuanica,* 5(2): 101 (1988): Deyeuxia sylvatica var. laxiflora.
var. ligulata *(Rendle) P.C. Kuo & S.L. Lu ex J.L. Yang in Fl. Sichuanica,* 5(2): 101 (1988), isonym; *see M. Kitagawa, Neo-Lineam. Fl. Manshur.:* 73 (1979): Deyeuxia sylvatica var. ligulata.
cleefii *F.D. Escalona in Phytologia,* 65(5): 339 (1988)—Colombia.
czerepanovii *Sh.A. Guseĭnov in Bot. Zhurn.,* 73(12): 1743 (1988)—U.S.S.R. (Caucasus).
diffusa *(Keng) P.C. Kuo & S.L. Lu ex J.L. Yang in Fl. Sichuanica,* 5(2): 108 (1988), isonym; *see P.C. Keng in Bull. Bot. Res. North-East. Forest. Inst.,* 4(3): 195 (1984): Deyeuxia diffusa.
effusiflora *(Rendle) P.C. Kuo & S.L. Lu ex J.L. Yang in Fl. Sichuanica,* 5(2): 88 (1988): Deyeuxia effusiflora.
filipes *(Keng) P.C. Kuo & S.L. Lu ex J.L. Yang in Fl. Sichuanica,* 5(2): 92 (1988): Deyeuxia filipes.
guamanensis *F.D. Escalona in Phytologia,* 65(5): 340 (1988)—Ecuador.

CALAMAGROSTIS :–
henryi *(Rendle) P.C. Kuo & S.L. Lu ex J.L. Yang in Fl. Sichuanica,* 5(2): 90 (1988): Deyeuxia henryi.
hupehensis *(Rendle) P.C. Kuo & S.L. Lu ex J.L. Yang in Fl. Sichuanica,* 5(2): 95 (1988), isonym; *see Tsoong in Contrib. Inst. Bot. Nat. Acad. Peiping,* 4: 151 (1936): Deyeuxia hupehensis.
levipes *(Keng) P.C. Kuo & S.L. Lu ex J.L. Yang in Fl. Sichuanica,* 5(2): 95 (1988): Deyeuxia levipes.
minarovii *Sh.A. Guseĭnov in Bot. Zhurn.,* 73(12): 1741 (1988)—U.S.S.R. (Caucasus).
ramonae *F.D. Escalona in Phytologia,* 65(5): 342 (1988)—Venezuela.
sikangensis *(Keng) P.C. Kuo & S.L. Lu ex J.L. Yang in Fl. Sichuanica,* 5(2): 97 (1988): Deyeuxia sikangensis.
sinelatior *(Keng) P.C. Kuo & S.L. Lu ex J.L. Yang in Fl. Sichuanica,* 5(2): 90 (1988): Deyeuxia sinelatior.
tzvelevii *Sh.A. Guseĭnov in Bot. Zhurn.,* 73(12): 1742 (1988)—U.S.S.R. (Caucasus).

CALAMUS (Palmae).
aidae *E.S. Fernando in Gard. Bull. Singapore,* 41(2): 49 (1988 publ. 1989)—Philippine Is.
austroguangxiensis *S.J. Pei & S.Y. Chen in Acta Phytotax. Sin.,* 27(2): 144 (1989), as 'austro-guangxiensis'—China.
balansaeanus
var. castaneolepis *(Wei) S.J. Pei & S.Y. Chen in Acta Phytotax. Sin.,* 27(2): 134 (1989): C. henryanus var. castaneolepis.
balerensis *E.S. Fernando in Gard. Bull. Singapore,* 41(2): 51 (1988 publ. 1989)—Philippine Is.
distichus
var. shangsiensis *S.J. Pei & S.Y. Chen in Acta Phytotax. Sin.,* 27(2): 140 (1989)—China.
faberi
var. brevispicatus *(Wei) S.J. Pei & S.Y. Chen in Acta Phytotax. Sin.,* 27(2): 133 (1989): C. tonkinensis var. brevispicatus.
feanus
var. medogensis *S.J. Pei & S.Y. Chen in Acta Phytotax. Sin.,* 27(2): 137 (1989)—Xizang.
flagellum
var. furvifurfuraceus *S.J. Pei & S.Y. Chen in Acta Phytotax. Sin.,* 27(2): 133 (1989)—China.
giganteus
var. robustus *S.J. Pei & S.Y. Chen in Acta Phytotax. Sin.,* 27(2): 143 (1989)—China.
karinensis *(Becc.) S.J. Pei & S.Y. Chen in Acta Phytotax. Sin.,* 27(2): 133 (1989): C. flagellum var. karinensis.
nambariensis
var. alpinus *S.J. Pei & S.Y. Chen in Acta Phytotax. Sin.,* 27(2): 141 (1989)—China.
var. furfuraceus *S.J. Pei & S.Y. Chen in Acta Phytotax. Sin.,* 27(2): 142 (1989)—China.

CALAMUS :–
var. menglongensis *S.J. Pei & S.Y. Chen in Acta Phytotax. Sin.,* 27(2): 141 (1989)—China.
var. xishuangbannaensis *S.J. Pei & S.Y. Chen in Acta Phytotax. Sin.,* 27(2): 141 (1989)—China.
var. yingjiangensis *S.J. Pei & S.Y. Chen in Acta Phytotax. Sin.,* 27(2): 140 (1989)—China.
obovoideus *S.J. Pei & S.Y. Chen in Acta Phytotax. Sin.,* 27(2): 142 (1989)—China.
ornatus
var. pulverulentus *E.S. Fernando in Gard. Bull. Singapore,* 41(2): 53 (1988 publ. 1989)—Philippine Is.
palustris
var. longistachys *S.J. Pei & S.Y. Chen in Acta Phytotax. Sin.,* 27(2): 138 (1989)—China.
platyacanthus
var. mediostachys *S.J. Pei & S.Y. Chen in Acta Phytotax. Sin.,* 27(2): 143 (1989)—China.
rhabdocladus
var. globulosus *S.J. Pei & S.Y. Chen in Acta Phytotax. Sin.,* 27(2): 137 (1989)—China.
wailong *S.J. Pei & S.Y. Chen in Acta Phytotax. Sin.,* 27(2): 138 (1989)—China.
yunnanensis *S.J. Pei & S.Y. Chen in Acta Phytotax. Sin.,* 27(2): 134 (1989), with two types cited—China.
var. densiflorus *S.J. Pei & S.Y. Chen in Acta Phytotax. Sin.,* 27(2): 135 (1989)—China.
var. intermedius *S.J. Pei & S.Y. Chen in Acta Phytotax. Sin.,* 27(2): 137 (1989)—China.

CALANTHE (Orchidac.).
candida *J. Bosser in Bull. Mus. Nation. Hist. Nat., B, Adansonia,* Ser. 4, 11(1): 34 (1989)—Mauritius, Réunion.
imthurnii *P.J. Kores in Allertonia,* 5(1): 126 (1989)—Fiji.

CALATHEA (Marantac.).
steyermarkii *H. Kennedy & K. Nagata in Brittonia,* 41(2): 164 (1989)—Venezuela.

CALEA (Compos.).
abbreviata *J.F. Pruski & L.E. Urbatsch in Brittonia,* 40(4): 348 (1988)—Brazil.
heteropappa *J.F. Pruski & L.E. Urbatsch in Brittonia,* 40(4): 346 (1988)—Brazil.
intermedia *J.F. Pruski & L.E. Urbatsch in Brittonia,* 40(4): 351 (1988)—Brazil.
lutea *J.F. Pruski & L.E. Urbatsch in Brittonia,* 40(4): 353 (1988)—Brazil.
oaxacana *(B. Turner) B.L. Turner in Phytologia,* 65(3): 241 (1988): Tetrachyron oaxacanum.
quadrifolia *J.F. Pruski & L.E. Urbatsch in Brittonia,* 40(4): 341 (1988)—Brazil.
saddiana *G.M. Barroso in Arq. Jard. Bot. Rio de Janeiro,* 28: 195 (1986–87 publ. 1988)—Brazil.

CALENDULA (Compos.).
alata *K.H. Rechinger in Fl. Iran.*, 164: 104 (1989)—Iran, Iraq, U.S.S.R. (Middle Asia), Pakistan, India.

×**CALICHARIS** A.W. Meerow in Ann. Missouri Bot. Gard., 76(1): 208 (1989). *AMARYLLIDACEAE.* (Caliphruria × Eucharis).
butcheri *(Traub.) A.W. Meerow in Ann. Missouri Bot. Gard.*, 76(1): 208 (1989): Eucharis butcheri.

CALIPHRURIA (Amaryllidac.).
korsakoffii *(Traub) A.W. Meerow in Ann. Missouri Bot. Gard.*, 76(1): 217 (1989): Eucharis korsakoffii.

CALLIANTHEMUM (Ranunculac.).
sachalinense
subsp. kirigishiense *K. Sato & K. Ito in J. Jap. Bot.*, 64(9): 268 (1989)—Japan.

CALLICARPA (Verbenac.).
chejuensis *Y.H. Chung & H. Kim in Korean J. Pl. Taxon.*, 19(1): 21 (1989)—Korea.

CALLISIA (Commelinac.).
graminea *(Small) G.C. Tucker in J. Arnold Arbor.*, 70(1): 118 (1989): Cuthbertia graminea.
f. leucantha *(Lakela) G.C. Tucker in J. Arnold Arbor.*, 70(1): 118 (1989): Cuthbertia graminea f. leucantha.
ornata *(Small) G.C. Tucker in J. Arnold Arbor.*, 70(1): 118 (1989): Cuthbertia ornata.

CALLISTHENE (Vochysiac.).
castellanosii *H.F. Martins in Bradea*, 5(13): 150 (1988)—Brazil.
kuhlmannii *H.F. Martins in Bradea*, 5(13): 149 (1988)—Brazil.

CALOCHORTUS (Liliac.).
coxii *M.R. Godfrey & F.T. Callahan in Phytologia*, 65(3): 216 (1988)—U.S.A. (Oregon).
umpquaensis *N.A. Fredricks in Syst. Bot.*, 14(1): 12 (1989)—U.S.A. (Oregon).

CALOGYNE (Goodeniac.).
pilosa
subsp. chinensis *(Benth.) H.S. Kiu in Guihaia*, 9(3): 194 (1989): C. chinensis.

CALOPOGON (Orchidac.).
tuberosus
var. simpsonii *(Small) L.K. Magrath in Sida*, 13(3): 371 (1989): Limodorum simpsonii.

CALYCERA (Calycerac.).
sympaganthera *(R. & P.) C. Marticorena & M. Quezada in Gayana, Bot.*, 44(1–4): 40 (1987 publ. 1988): Scabiosa sympaganthera.

CALYCOLPUS (Myrtac.).
alternifolius *(Gleason) L.R. Landrum in Madroño*, 36(1): 10 (1989): Myrtus alternifolia.
bolivarensis *L.R. Landrum in Ann. Missouri Bot. Gard.*, 76(3): 930 (1989)—Venezuela.

CALYCORECTES (Myrtac.).
beruttii *J.R. Mattos in Loefgrenia*, 94: 9 (1989)—Brazil.
cucullatus *J.R. Mattos in Loefgrenia*, 94: 10 (1989)—Brazil.
striatulus *J.R. Mattos in Loefgrenia*, 94: 9 (1989)—Brazil.

CALYPTRANTHES (Myrtac.).
quinoensis *J.R. Mattos in Loefgrenia*, 95: 2 (1989)—Brazil.

CALYPTROCARYA (Cyperac.).
delascioi *G. Davidse & R. Kral in Ann. Missouri Bot. Gard.*, 75(3): 855 (1988)—Venezuela.
montesii *G. Davidse & R. Kral in Ann. Missouri Bot. Gard.*, 75(3): 853 (1988)—Venezuela.

CALYPTRONOMA (Palmae).
plumeriana *(Martius) A. Lourteig in Phytologia*, 65(6): 484 (1989): Geonoma ? plumeriana.

CALYSTEGIA (Convolvulac.).
amurensis *N.S. Probatova in Sosud. Rast. Sovet. Dal'nego Vostoka*, 4: 359 (1989)—U.S.S.R. (Soviet Far East).

CAMELLIA (Theac.).
albosericea *H.T. Chang in Acta Sci. Nat. Univ. Sunyatseni*, 28(3): 51 (1989), as 'albo-sericea'—China.
brevigyna *H.T. Chang in Acta Sci. Nat. Univ. Sunyatseni*, 28(3): 51 (1989)—China.
brevipetiolata *H.T. Chang in Acta Sci. Nat. Univ. Sunyatseni*, 28(3): 52 (1989)—China.
chunii *H.T. Chang in Acta Sci. Nat. Univ. Sunyatseni*, 28(3): 52 (1989)—China.
danzaiensis *Chang & K.M. Lan in Bull. Bot. Res. North-East. Forest. Inst.*, 9(4): 59 (1989)—China.
flavida
f. polypetala *R.G. Li & S.Q. He in Guihaia*, 9(3): 200 (1989)—China.
huiliensis *H.T. Chang in Acta Sci. Nat. Univ. Sunyatseni*, 28(3): 53 (1989)—China.
jinshajiangica *H.T. Chang & S.L. Lee in Acta Sci. Nat. Univ. Sunyatseni*, 28(3): 50 (1989)—China.
lanosituba *H.T. Chang in Acta Sci. Nat. Univ. Sunyatseni*, 28(3): 51 (1989)—China.
minor *H.T. Chang in Acta Sci. Nat. Univ. Sunyatseni*, 28(3): 53 (1989)—China.
oligophlebia *H.T. Chang in Acta Sci. Nat. Univ. Sunyatseni*, 28(3): 54 (1989)—China.
pentapetala *H.T. Chang in Acta Sci. Nat. Univ. Sunyatseni*, 28(3): 54 (1989)—China.

CAMELLIA :–
pentaphylacoides *H.T. Chang in Acta Sci. Nat. Univ. Sunyatseni,* 28(3): 55 (1989)—China.
pentaphylax *H.T. Chang in Acta Sci. Nat. Univ. Sunyatseni,* 28(3): 55 (1989)—China.
stichoclada *H.T. Chang in Acta Sci. Nat. Univ. Sunyatseni,* 28(3): 56 (1989)—China.
subliberopetala *H.T. Chang in Acta Sci. Nat. Univ. Sunyatseni,* 28(3): 56 (1989)—China.
tenuivalvis *H.T. Chang in Acta Sci. Nat. Univ. Sunyatseni,* 28(3): 57 (1989)—China.
xichangensis *H.T. Chang in Acta Sci. Nat. Univ. Sunyatseni,* 28(3): 57 (1989)—China.

CAMPANULA (Campanulac.).
bluemelii *J.J. Halda in Preslia,* 61(4): 321 (1989)—Turkey.
hypocrateriformis *A. Dobignard in Saussurea,* 19: 112 (1988 publ. 1989), nom. nov.: C. dichotoma f. pseudodichotoma *Maire ex Quézel.*
jordanovii *M. Ančev & M. Kovanda in Preslia,* 61(3): 196 (1989)—Bulgaria.
praealta *A.I. Galushko, Fl. Severn. Kavkaza,* 3: 147 (1980), nom. nov.: C. elatior *(Fomin.) Grossh. ex Fed.*
sivasica *Kit Tan & B. Yildiz in Notes Roy. Bot. Gard. Edinburgh,* 45(3): 447 (1988 publ. 1989)—Turkey.
trojanensis *M. Kovanda & M. Ančev in Preslia,* 61(3): 201 (1989)—Bulgaria.

CAMPOMANESIA (Myrtac.).
acrandroides *J.R. Mattos in Loefgrenia,* 90: 4 (1986), nom. nov.: Acrandra sellowiana *Berg.*
cerasoides *(Camb.) J.R. Mattos in Loefgrenia,* 90: 4 (1986): Psidium cerasoides.
phaea *(Berg) J.R. Mattos in Loefgrenia,* 90: 4 (1986): Abbevillea phaea.
schultziana *J.R. Mattos in Loefgrenia,* 90: 4 (1986), nom. nov.: Abbevillea gardneriana *Berg.*
sellowiana *(Berg) J.R. Mattos in Loefgrenia,* 90: 4 (1986): Abbevillea sellowiana.
sparsiflora *(DC.) J.R. Mattos in Loefgrenia,* 90: 4 (1986): Eugenia sparsiflora.

CAMPYLOTROPIS (Legumin.).
macrocarpa
subsp. hengduanshanensis *C.J. Chen in Acta Bot. Yunnanica,* 10(4): 435 (1988)—China, Xizang.
pauciflora *C.J. Chen in Acta Bot. Yunnanica,* 10(4): 433 (1988)—China.
teretiracemosa *P.C. Li & C.J. Chen in Acta Bot. Yunnanica,* 10(4): 434 (1988)—China.

CANNABIS (Cannabac.).
sativa
var. ruderalis *(Janisch.) S.Z. Liou in Fl. Liaoningica, chief ed. S.X. Li,* 1: 289 (1988): C. ruderalis.

CAPPARIS (Capparac.).
grandipetala *B. Maguire & J.A. Steyermark in Mem. New York Bot. Gard.,* 51: 115 (1989)—Brazil.
ovata
subsp. sicula *(Veillard) F.I. Pugnaire de Iraola in Blancoana,* 7: 121 (1989), as '(Duham)': C. sicula.

CARAGANA (Legumin.).
opulens
var. perforata *Merrgen & Y.Q. Ma in Acta Sci. Nat. Univ. Intramongolicae,* 20(4): 554 (1989)—China.
var. trichophylla *Z.H. Gao & S.C. Zhang in Bull. Bot. Res. North-East. Forest. Inst.,* 9(3): 63 (1989)—China.
pygmaea
subsp. austrotuvinica *N.A. Bondareva in Bot. Zhurn.,* 74(6): 900 (1989)—U.S.S.R. (Siberia).

CARAPA (Meliac.).
megistocarpa *A.H. Gentry & Dodson in Ann. Missouri Bot. Gard.,* 75(4): 1434 (1988 publ. 1989)—Ecuador.

CARDAMINE (Crucif.).
nepalensis *N. Kurosaki & H. Ohba in J. Jap. Bot.,* 64(5): 135 (1989)—Nepal.

CARDUNCELLUS (Compos.).
monspelliensium
subsp. araneosus *(Boiss. & Reut.) O. de Bolòs & J. Vigo in Folia Bot. Misc.,* 6: 85 (1989): C. araneosus.

CAREX (Cyperac.).
sect. Aciculares *(Kük.) G.A. Wheeler in Syst. Bot.,* 14(2): 174 (1989): C. subsect. Aciculares.
acuta
subsp. reuteriana *(Boiss.) S. Rivas Martínez & J. Pizarro in Fontqueria,* 24: 13 (1989): C. reuteriana.
anhuiensis *S.W. Su & S.M. Xu in Bull. Bot. Res. North-East. Forest. Inst.,* 9(1): 47 (1989)—China.
dayuongensis *Z.P. Wang in Acta Bot. Yunnanica,* 11(2): 166 (1989)—China.
edwardsiana *E.L. Bridges & S.L. Orzell in Phytologia,* 67(2): 148 (1989)—U.S.A. (Texas).
×elisabeth *J. Andres, R. Carbo, F. Llamas & M. Perez in Stud. Bot. (Salamanca),* 7: 243 (1988 publ. 1989), contrary to Art. H.5.2 ICBN (1988)—Spain. (C. disticha × muricata subsp. lamprocarpa).
flacca
subsp. erythrostachys *(Hoppe) J. Holub in Folia Geobot. Phytotax.,* 23(4): 413 (1988): C. erythrostachys.
humahuacaensis *G.A. Wheeler in Syst. Bot.,* 14(1): 37 (1989)—Argentina.

CAREX :–
macloviana
subsp. subfusca *(W. Boott) T. Koyama in Bishop Mus. Occas. Pap.*, 29: 118 (1989): C. subfusca.
minquinensis *Z.P. Wang in Acta Bot. Yunnanica,* 11(2): 169 (1989)—China.
nigra
subsp. iberica *S. Rivas Martínez in Fontqueria,* 24: 14 (1989)—Spain.
qimenensis *S.W. Su & S.M. Xu in Bull. Bot. Res. North-East. Forest. Inst.*, 9(1): 49 (1989)—China.
qiyunensis *S.W. Su & S.M. Xu in Bull. Bot. Res. North-East. Forest. Inst.*, 9(1): 50 (1989)—China.
taihuensis *S.W. Su & S.M. Xu in Bull. Bot. Res. North-East. Forest. Inst.*, 9(1): 51 (1989)—China.
toroensis *G.A. Wheeler in Syst. Bot.*, 14(2): 182 (1989)—Chile.
truncatirostris *S.W. Su & S.M. Xu in Bull. Bot. Res. North-East. Forest. Inst.*, 9(1): 45 (1989) China.
f. erostris *S.W. Su & S.M. Xu in Bull. Bot. Res. North-East. Forest. Inst.*, 9(1): 47 (1989)—China.
vallis-pulchrae
var. barrosiana *G.A. Wheeler in Syst. Bot.*, 14(2): 178 (1989)—Tierra del Fuego, Argentina, Falkland Is.
viridula
var. elatior *(Schlechtendal) W.J. Crins in Canad. J. Bot.*, 67(4): 1058 (1989): C. flava var. elatior.
var. nevadensis *(Boissier & Reuter) W.J. Crins in Canad. J. Bot.*, 67(4): 1060 (1989): C. nevadensis.
var. saxilittoralis *(Robertson) W.J. Crins in Canad. J. Bot.*, 67(4): 1061 (1989): C. saxilittoralis.
wahuensis
subsp. herbstii *T. Koyama in Bishop Mus. Occas. Pap.*, 29: 118 (1989)—Hawaiian Is.
subsp. rubiginosa *(R. Krauss) T. Koyama in Bishop Mus. Occas. Pap.*, 29: 119 (1989): C. wahuensis var. rubiginosa.
xiangxiensis *Z.P. Wang in Acta Bot. Yunnanica,* 11(2): 168 (1989)—China.
xuanchengensis *S.W. Su & S.M. Xu in Bull. Bot. Res. North-East. Forest. Inst.*, 9(1): 48 (1989)—China.

CARMINATIA (Compos.).
anomala *B.L. Turner in Pl. Syst. Evol.*, 160(3–4): 173 (1988)—Mexico.

CARNEGIEA (Cactac.).
polylopha *(DC.) D. Hunt in Bradleya,* 6: 100 (1988): Cereus polylophus.

CARPESIUM (Compos.).
nepalense
var. ciliatum *(Hook. f.) S. Kumar in R.R. Rao et al., Fl. Ind. Enumerat. - Asterac.*: 21 (1988): C. cernuum var. ciliatum.

CARPESIUM :–
var. griffithii *(Hook. f.) S. Kumar in R.R. Rao et al., Fl. Ind. Enumerat. - Asterac.*: 21 (1988): C. cernuum var. griffithii.
var. nilagiricum *(Clarke) S. Kumar in R.R. Rao et al., Fl. Ind. Enumerat. - Asterac.*: 22 (1988): C. cernuum var. nilagiricum.
var. pedunculosa *(Clarke) S. Kumar in R.R. Rao et al., Fl. Ind. Enumerat. - Asterac.*: 22 (1988): C. cernuum var. pedunculosa.

CARPHALEA (Rubiac.).
kirondron
subsp. geayi *(Homolle) C. Puff in Bull. Jard. Bot. Nation. Belg.*, 58(3–4): 302 (1988): C. geayi.
somaliensis *C. Puff in Bull. Jard. Bot. Nation. Belg.*, 58(3–4): 297 (1988)—Somali Republic.

CARPINUS (Corylac.).
kawakamii
var. minutiserrata *(Hayata) S.S. Ying, Coloured Illustr. Fl. Taiwan,* 3: 271 (1988): C. minutiserrata.

CARPOBROTUS (Aizoac.).
edulis
var. chrysophthalmus *C.D. Preston & P.D. Sell in Watsonia,* 17(3): 238 (1989)—Great Britain.

CARYODAPHNOPSIS (Laurac.).
burgeri *N. Zamora V. & L.J. Poveda A. in Ann. Missouri Bot. Gard.*, 75(3): 1160 (1988)—Costa Rica.

CASIMIROA (Rutac.).
greggii *(S. Watson) F. Chiang in Taxon,* 38(1): 116 (1989): Sargentia greggii.

CASTANOPSIS (Fagac.).
cuspidata
var. longicaudata *(Hayata) S.S. Ying, Coloured Illustr. Fl. Taiwan,* 3: 222 (1988): Quercus longicaudata.
eyrei
var. brachyacantha *(Hayata) Shen ex S.S. Ying, Coloured Illustr. Fl. Taiwan,* 3: 229 (1988): C. brachyacantha.
wenchangensis *G.A. Fu & C.C. Huang in Acta Phytotax. Sin.*, 27(2): 151 (1989)—Hainan.

CASUARINA (Casuarinac.).
cunninghamiana
subsp. miodon *L.A.S. Johnson in Fl. Australia,* 3: 200, 107 (1989)—Australia (N. Terr., Queensland).
pauper *F. Muell. ex L.A.S. Johnson in Fl. Australia,* 3: 202, 110 (1989)—Australia (S. Australia, W. Australia, New South Wales, Victoria).

CATABROSELLA (Gramin.).
sect. Himalaica *(Tzvel.) E.B. Alekseev in Byull. Mosk. Obshch. Ispȳt. Prir., Biol.*, 93(5): 108 (1988): Colpodium sect. Himalaica.

CATABROSELLA :–
subgen. Nevskia *(Tzvel.) E.B. Alekseev in Byull. Mosk. Obshch. Ispўt. Prir., Biol.*, 93(5): 98 (1988): Colpodium subgen. Nevskia.
sect. Variegata *(Nevski) E.B. Alekseev in Byull. Mosk. Obshch. Ispўt. Prir., Biol.*, 93(5): 99 (1988): Colpodium ser. Variegata.
parviflora *(Boiss. & Buhse) Czop. ex E.B. Alekseev in Byull. Mosk. Obshch. Ipsўt. Prir., Biol.*, 93(5): 107 (1988), isonym; *see R.R. Mill in Fl. Turkey & E. Aegean Is.*, 9: 496 (1985): Colpodium parviflorum.
subsp. calvertii *(Boiss.) E.B. Alekseev in Byull. Mosk. Obshch. Ispўt. Prir., Biol.*, 93(5): 107 (1988), isonym; *see R.R. Mill in Fl. Turkey & E. Aegean Is.*, 9: 496 (1985): Colpodium calvertii.
subornata *E.B. Alekseev in Byull. Mosk. Obshch. Ispўt. Prir., Biol.*, 93(5): 105 (1988)—U.S.S.R. (Caucasus).
variegata
var. leiantha *(Hack.) E.B. Alekseev in Byull. Mosk. Obshch. Ispўt. Prir., Biol.*, 93(5): 103 (1988), isonym; *see R.R. Mill in Fl. Turkey & E. Aegean Is.*, 9: 496 (1985): Colpodium leianthum.

CATANANCHE (Compos.).
arenaria
var. dimorpha *A. Dobignard in Saussurea*, 19: 118 (1988 publ. 1989)—Morocco.

CATASETUM (Orchidac.).
×dunstervillei *G.A. Romero & G. Carnevali in Ann. Missouri Bot. Gard.*, 76(2): 458 (1989)—Venezuela. (C. discolor × pileatum).
gomezii *G.A. Romero & G. Carnevali in Ann. Missouri Bot. Gard.*, 76(2): 455 (1989)—Venezuela.
parguazense *G.A. Romero & G. Carnevali in Ann. Missouri Bot. Gard.*, 76(2): 457 (1989)—Venezuela.

CELSIA (Scrophulariac.).
parviflora *J. Decaisne in Ann. Sci. Nat., Ser. 2*, 2: 254 (1834)—Sinai.†

CENTAUREA (Compos.).
araxina *E.Ts. Gabriélyan in Novosti Sist. Vўssh. Rast.*, 25: 169 (1988)—U.S.S.R. (Caucasus).
bombycina
subsp. xeranthemoides *(Lange) G. Blanca, M. Cueto & M.C. Quesada in Saussurea*, 19: 29 (1988 publ. 1989): C. funkii var. xeranthemoides.
erivanensis
subsp. holargyrea *(Bornm. & Woronow) E.Ts. Gabriélyan in Novosti Sist. Vўssh. Rast.*, 25: 163 (1988): Psephellus holargyreus.
fajvuschii *E.Ts. Gabriélyan in Novosti Sist. Vўssh. Rast.*, 25: 164 (1988)—U.S.S.R. (Caucasus).

CENTAUREA :–
kartschiana
var. coronata *A.Ž. Lovrić in Acta Bot. Croat.*, 30: 136 (1971), as 'comb. nov.'—Jugoslavia.
subsp. curictana *A.Ž. Lovrić in Acta Bot. Croat.*, 30: 138 (1971), as 'comb. nov.'—Jugoslavia.
f. troglodytes *A.Ž. Lovrić in Acta Bot. Croat.*, 30: 136 (1971)—Jugoslavia.
saguntina *G. Mateo Sanz & M.B. Crespo Villalba in Bol. Soc. Brot.*, 61: 262 (1988)—Spain.
×sanctae-barbarae *G. Mateo Sanz & M.B. Crespo Villalba in Bol. Soc. Brot.*, 61: 264 (1988)—Spain. (C. paui × pinae).
×ternilli *G. Mateo Sanz & M.B. Crespo Villalba in Bol. Soc. Brot.*, 61: 265 (1988), contrary to Art. H.5.2 ICBN (1988)—Spain. (C. aspera subsp. stenophylla × spachii).
xanthocephala
subsp. xanthocephaloides *(Tzvel.) É.Ts. Gabriélyan in Novosti Sist. Vўssh. Rast.*, 25: 168 (1988): C. xanthocepaloides.

CENTELLA (Umbellif.).
macrodus *(Spreng.) B.L. Burtt in Notes Roy. Bot. Gard. Edinburgh*, 45(2): 197 (1988 publ. 1989): Hydrocotyle macrodus.

CENTROLOBIUM (Legumin.).
microchaete *(Mart. ex Benth.) H.C. de Lima in Arq. Jard. Bot. Rio de Janeiro*, 27: 180 (1983–85 publ. 1985): C. robustum var. microchaete.
sclerophyllum *H.C. de Lima in Arq. Jard. Bot. Rio de Janeiro*, 27: 182 (1983–85 publ. 1985)—Brazil.

CEPHALANTHEROPSIS (Orchidac.).
halconensis *(Ames) S.S. Ying, Coloured Illustr. Fl. Taiwan*, 3: 622 (1988): Phaius halconensis.

CEPHALARIA (Dipsacac.).
transcaucasica *(Bobr.) A.I. Galushko, Fl. Severn. Kavkaza*, 3: 130 (1980): C. syriaca subsp. transcaucasica.
uralensis
subsp. multifida *(Roman) N. Roman & A. Beldie in Fl. Republ. Social. Român.*, 13: 50 (1976): C. uralensis var. multifida.

CEPHALOPHYLLUM (Aizoac.).
numeesense *H.E.K. Hartmann in Mitt. Inst. Allg. Bot. Hamburg*, 22: 172 (1988)—South Africa (Cape Province).
parvibracteatum *(L. Bolus) H.E.K. Hartmann in Mitt. Inst. Allg. Bot. Hamburg*, 22: 172 (1988): Cheiridopsis parvibracteata.
parvulum *(Schlechter) H.E.K. Hartmann in Mitt. Inst. Allg. Bot. Hamburg*, 22: 173 (1988): Mesembryanthemum parvulum.
spissum *H.E.K. Hartmann in Mitt. Inst. Allg. Bot. Hamburg*, 22: 176 (1988)—South Africa (Cape Province).

CEPHALOPHYLLUM :–
tetrastichum *H.E.K. Hartmann in Mitt. Inst. Allg. Bot. Hamburg,* 22: 177 (1988)—South Africa (Cape Province).

CEPHALOSTEMON (Rapateac.).
flavus *(Link) J.A. Steyermark in Ann. Missouri Bot. Gard.,* 75(4): 1565 (1988 publ. 1989), without basionym ref.: Basionym not stated.
junciformis *(Mag.) J.A. Steyermark in Ann. Missouri Bot. Gard.,* 75(4): 1565 (1988 publ. 1989): Duckea junciformis.

CERASTIUM (Caryophyllac.).
utriense *G. Barberis in Webbia,* 42(2): 154 (1988)—Italy.

CERATOCENTRON K. Senghas in Orchidee, 40(3): 89 (1989). *ORCHIDACEAE.*
fesselii *K. Senghas in Orchidee,* 40(3): 90 (1989)—Philippine Is.

CERATOPHYLLUM (Ceratophyllac.).
sect. Muricatum *D.H. Les in Syst. Bot.,* 14(2): 259 (1989).
sect. Submersum *D.H. Les in Syst. Bot.,* 14(2): 259 (1989).

CEREUS (Cactac.).
subgen. Brasilicereus *(Backeb.) P.J. Braun in Bradleya,* 6: 87 (1988): gen. Brasilicereus.
subgen. Ebneria *(Backeb.) D. Hunt in Bradleya,* 6: 100 (1988): Monvillea subgen. Ebneria.
adelmarii *(Rizzini & Mattos) P.J. Braun in Bradleya,* 6: 86 (1988): Monvillea adelmari.
alacriportanus
var. bageanus *(Ritter) P.J. Braun in Bradleya,* 6: 86 (1988): Piptanthocereus bageanus.
alticostatus *(Ritter) P.J. Braun in Bradleya,* 6: 86 (1988): Monvillea alticostata.
calcirupicola
var. cabralensis *(Ritter) P.J. Braun in Bradleya,* 6: 86 (1988): Piptanthocereus cabralensis.
var. cipoensis *(Ritter) P.J. Braun in Bradleya,* 6: 86 (1988): Piptanthocereus cipoensis.
campinensis *(Backeberg & Voll) P.J. Braun in Bradleya,* 6: 86 (1988): Pilocereus campinensis.
var. piedadensis *(Ritter) P.J. Braun in Bradleya,* 6: 86 (1988): Monvillea piedadensis.
goiasensis *(Ritter) P.J. Braun in Bradleya,* 6: 86 (1988): Piptanthocereus goiasensis.
lanosus *(Ritter) P.J. Braun in Bradleya,* 6: 86 (1988): Piptanthocereus lanosus.
markgrafii *(Backeb. & Voll) P.J. Braun in Bradleya,* 6: 87 (1988): Brasilicereus markgrafii.
neonesioticus *(Ritter) P.J. Braun in Bradleya,* 6: 86 (1988): Piptanthocereus neonesioticus.
var. interior *(Ritter) P.J. Braun in Bradleya,* 6: 86 (1988): Piptanthocereus neonesioticus var. interior.

CEREUS :–
phaeacanthus
var. breviflorus *(Ritter) P.J. Braun in Bradleya,* 6: 87 (1988): Brasilicereus breviflorus.
saddianus *(Rizzini & Mattos) P.J. Braun in Bradleya,* 6: 86 (1988): Monvillea saddiana.
sericifer *(Ritter) P.J. Braun in Bradleya,* 6: 86 (1988): Piptanthocereus sericifer.

CERIOPS (Rhizophorac.).
australis *(C.T. White) E.R. Ballment, T.J. Smith & J.A. Stoddart in Austral. Syst. Bot.,* 1(4): 395 (1988 publ. 1989): C. tagal var. australis.

CEROPEGIA (Asclepiadac.).
arabica
var. abbreviata *P.V. Bruyns in Notes Roy. Bot. Gard. Edinburgh,* 45(2): 318 (1988 publ. 1989)—Yemen.
var. superba *(Field & Collenette) P.V. Bruyns in Notes Roy. Bot. Gard. Edinburgh,* 45(2): 317 (1988 publ. 1989): C. superba.
aristolochioides
subsp. deflersiana *P.V. Bruyns in Notes Roy. Bot. Gard. Edinburgh,* 45(2): 296 (1988 publ. 1989)—Yemen.
bosseri
var. razafindratsirana *W. Rauh & G. Buchloh in Kakt. And. Sukk.,* 40(5): 124 (1989)—Madagascar.
candelabrum
var. tuberosa *(Roxb.) N.P. Singh, Fl. East. Karnataka,* 1: 410 (1988): C. tuberosa.
foliosa *P.V. Bruyns in Notes Roy. Bot. Gard. Edinburgh,* 45(2): 293 (1988 publ. 1989)—Yemen.

CESTRUM (Solanac.).
jimenezii *Alain H. Liogier in Phytologia,* 65(6): 421 (1989)—Hispaniola (Dominican Republic).

CHAENORHINUM (Scrophulariac.).
sect. Hueblia *(Speta) D.A. Sutton, Revis. tribe Antirrhineae:* 127 (1988): gen. Hueblia.
sect. Microrrhinum *(Endl.) D.A. Sutton, Revis. tribe Antirrhineae:* 116 (1988): Linaria sect. Microrrhinum.
rubrifolium
subsp. gerense *(Stapf) D.A. Sutton, Revis. tribe Antirrhineae:* 114 (1988): Linaria gerensis.

CHAETOTHYLAX (Acanthac.).
rzedowskii *S. Acosta Castellanos in Acta Bot. Mex.,* 5: 5 (1989)—Mexico.

CHAMAECRISTA (Legumin.).
dunensis *M. Thulin in Nordic J. Bot.,* 8(5): 457 (1989)—Somali Republic.
nomame *(Sieb.) H. Ohashi in J. Jap. Bot.,* 64(7): 215 (1989): Sooya nomame.

CHAMAECYTISUS (Legumin.).
mollis *(Cav.) W. Greuter & Burdet in Willdenowia,* 19(1): 31 (1989): Spartium molle.

CHAMAESYCE (Euphorbiac.).
erythroclada *(Bois.) K. Seshagiri Rao & M.N.V. Prasad in Pl. Syst. Evol.,* 160(3–4): 192 (1988): Euphorbia erythroclada.

CHASALIA (Rubiac.).
longifolia *(Dalz.) K.M. Wong in Tree Fl. Malaya,* 4: 344 (1989): Psychotria longifolia.

CHASCANUM (Verbenac.).
obovatum *Sebsebe Demissew in Nordic J. Bot.,* 8(6): 629 (1989)—Ethiopia, Kenya.
sulcatum *Sebsebe Demissew in Nordic J. Bot.,* 8(6): 627 (1989)—Ethiopia, Somali Republic.

CHAUBARDIELLA (Orchidac.).
pacuarensis *R. Jenny in Orchidee,* 40(3): 91 (1989)—Costa Rica.

CHEILOTHECA (Ericac.).
humilis
var. pubescens *(K.F. Wu) C. Ling in Fl. Zhejiang,* 5: 5 (1989): Monotropastrum pubescens.

CHELONANTHUS (Gentianac.).
fruticosus *B. Maguire & B.M. Boom in Mem. New York Bot. Gard.,* 51: 42 (1989), nom. nov.: Lisianthus arboreus *Spruce ex Progel.*

CHENOPODIUM (Chenopodiac.).
album
subf. corymbosodentatum *K. Arlt & B. Jüttersonke in Feddes Repert.,* 100(1–2): 21 (1989), as 'corymboso-dentatum'—Cultivation.
subf. corymbosolanceolatum *K. Arlt & B. Jüttersonke in Feddes Repert.,* 100(1–2): 22 (1989), as 'corymboso-lanceolatum'—Cultivation.
subvar. cymigerum *(Koch) K. Arlt & B. Jüttersonke in Feddes Repert.,* 100(1–2): 20 (1989), as subvar. nov.: C. album var. cymigerum.
f. cymigerum *(Koch) K. Arlt & B. Jüttersonke in Feddes Repert.,* 100(1–2): 20 (1989), as '(A. Ludwig)': C. album var. cymigerum.
subf. cymigerum *(Koch) K. Arlt & B. Jüttersonke in Feddes Repert.,* 100(1–2): 20 (1989), as subf. nov.: C. album var. cymigerum.
f. dubium *K. Arlt & B. Jüttersonke in Feddes Repert.,* 100(1–2): 22 (1989)—Cultivation.
subf. dubium *K. Arlt & B. Jüttersonke in Feddes Repert.,* 100(1–2): 23 (1989)—Cultivation.

CHENOPODIUM :-
f. glomerulosum *(Rchb.) K. Arlt & B. Jüttersonke in Feddes Repert.,* 100(1–2): 18 (1989), as 'glomeruloseum (A. Ludwig)': C. glomerulosum.
subf. glomerulosum *(Rchb.) K. Arlt & B. Jüttersonke in Feddes Repert.,* 100(1–2): 19 (1989), as subf. nov.: C. glomerulosum.
subf. medium *K. Arlt & B. Jüttersonke in Feddes Repert.,* 100(1–2): 17 (1989)—Cultivation.
subf. ovalifolium *K. Arlt & B. Jüttersonke in Feddes Repert.,* 100(1–2): 23 (1989)—Cultivation.
subf. paucidens *K. Arlt & B. Jüttersonke in Feddes Repert.,* 100(1–2): 19 (1989)—Cultivation.
subf. thyrsoideum *K. Arlt & B. Jüttersonke in Feddes Repert.,* 100(1–2): 18 (1989)—Cultivation.
dunosum *L.E. Simón in Bull. Mus. Nation. Hist. Nat., B, Adansonia,* Ser. 4, 10(4): 407 (1988 publ. 1989)—Brazil, Argentina.

CHEVALIERA Gaudichaud ex J.G. Beer, Fam. Bromel.: 22, 150, 257 (1856). *BROMELIACEAE.*
cariocae *(L.B. Smith) L.B. Smith & W.J. Kress in Phytologia,* 66(1): 77 (1989): Aechmea cariocae.
castanea *(L.B. Smith) L.B. Smith & W.J. Kress in Phytologia,* 66(1): 77 (1989): Aechmea castanea.
conifera *(L.B. Smith) L.B. Smith & W.J. Kress in Phytologia,* 66(1): 77 (1989): Aechmea conifera.
depressa *(L.B. Smith) L.B. Smith & W.J. Kress in Phytologia,* 66(1): 78 (1989): Aechmea depressa.
digitata *(L.B. Smith & R.W. Read) L.B. Smith & W.J. Kress in Phytologia,* 66(1): 78 (1989): Aechmea digitata.
fernandae *(E. Morren) L.B. Smith & W.J. Kress in Phytologia,* 66(1): 78 (1989): Bromelia fernandae.
hostilis *(E. Pereira) L.B. Smith & W.J. Kress in Phytologia,* 66(1): 78 (1989): Aechmea hostilis.
itapoana *(Till & Morawetz) L.B. Smith & W.J. Kress in Phytologia,* 66(1): 78 (1989): Aechmea itapoana.
lateralis *(L.B. Smith) L.B. Smith & W.J. Kress in Phytologia,* 66(1): 78 (1989): Aechmea lateralis.
leucolepis *(L.B. Smith) L.B. Smith & W.J. Kress in Phytologia,* 66(1): 78 (1989): Aechmea leucolepis.
magdalenae
var. quadricolor *(M.B. Foster) L.B. Smith & W.J. Kress in Phytologia,* 66(1): 78 (1989): Aechmea magdalenae var. quadricolor.
microcephala *(E. Pereira & Leme) L.B. Smith & W.J. Kress in Phytologia,* 66(1): 78 (1989): Aechmea microcephala.

CHEVALIERA :-
multiflora *(L.B. Smith) L.B. Smith & W.J. Kress in Phytologia,* 66(1): 78 (1989): Aechmea multiflora.
muricata *(Arruda da Camara) L.B. Smith & W.J. Kress in Phytologia,* 66(1): 78 (1989): Bromelia muricata.
ornata *Gaudichaud ex J.G. Beer, Fam. Bromel.:* 150 (1856)—Locality not stated.
pallida *(L.B. Smith) L.B. Smith & W.J. Kress in Phytologia,* 66(1): 78 (1989): Aechmea pallida.
perforata *(L.B. Smith) L.B. Smith & W.J. Kress in Phytologia,* 66(1): 78 (1989): Aechmea perforata.
rodriguesiana *(L.B. Smith) L.B. Smith & W.J. Kress in Phytologia,* 66(1): 78 (1989): Gravisia rodriguesiana.
rubiginosa *(Mez) L.B. Smith & W.J. Kress in Phytologia,* 66(1): 78 (1989): Aechmea rubiginosa.
saxicola *(L.B. Smith) L.B. Smith & W.J. Kress in Phytologia,* 66(1): 78 (1989): Aechmea saxicola.
sphaerocephala *Gaudichaud ex J.G. Beer, Fam. Bromel.:* 150 (1856)—Locality not stated.
sphaerocephala *(Baker) L.B. Smith & W.J. Kress in Phytologia,* 66(1): 78 (1989), later homonym of Gaudichaud ex J.G. Beer, *Fam. Bromel.:* 150 (1856): Aechmea sphaerocephala.
strobilacea *(L.B. Smith) L.B. Smith & W.J. Kress in Phytologia,* 66(1): 78 (1989): Aechmea strobilacea.
tayoensis *(Gilmartin) L.B. Smith & W.J. Kress in Phytologia,* 66(1): 79 (1989): Aechmea tayoensis.

CHIMONANTHUS (Calycanthac.).
caespitosus *T.B. Chao, Z.X. Chen & Z.Q. Li in Bull. Bot. Res. North-East. Forest. Inst.,* 9(4): 47 (1989)—China.

CHIMONOBAMBUSA (Gramin.).
subgen. Quadrangulares *C.J. Hsueh & W.P. Zhang in Bamboo Res.,* 1988(3): 8 (1988).
damingshanensis *C.J. Hsueh & W.P. Zhang in Bamboo Res.,* 1988(3): 5 (1988)—China.
neopurpurea *T.P. Yi in J. Bamboo Res.,* 8(3): 22 (1989), as 'nom. nov.'—China.
quadrangularis
f. purpureiculma *T.H. Wen in J. Bamboo Res.,* 8(1): 24 (1989)—China.
rivularis *T.P. Yi in J. Bamboo Res.,* 8(3): 18 (1989)—China.

CHIONOLAENA (Compos.).
breweri *J. Steyermark & B. Maguire in Acta Bot. Venezuel.,* 14(3): 26 (1984)—Venezuela.

CHIRITA (Gesneriac.).
crassituba *W.T. Wang in Bull. Bot. Res. North-East. Forest. Inst.,* 9(4): 24 (1989)—China.

CHIRITA :-
dibangensis *B.L. Burtt, S.K. Srivastava & B.N. Mehrotra in Notes Roy. Bot. Gard. Edinburgh,* 45(3): 469 (1988 publ. 1989)—India.
latinervis *W.T. Wang in Bull. Bot. Res. North-East. Forest. Inst.,* 9(4): 21 (1989)—China.
xinningensis *W.T. Wang in Bull. Bot. Res. North-East. Forest. Inst.,* 9(4): 22 (1989)—China.

CHIRITOPSIS (Gesneriac.).
xiuningensis *X.L. Liu & X.H. Guo in Bull. Bot. Res. North-East. Forest. Inst.,* 9(3): 51 (1989)—China.

CHLOROCHORION C. Puff & E. Robbrecht in Bot. Jahrb., 110(4): 547 (1989). *RUBIACEAE.*
foetidum *(Verdcourt) C. Puff & E. Robbrecht in Bot. Jahrb.,* 110(4): 547 (1989): Pentanisia foetida.
monticola *(K. Krause) C. Puff & E. Robbrecht in Bot. Jahrb.,* 110(4): 547 (1989): Otomeria monticola.

CHLOROPHYTUM (Anthericac.).
dolichocarpum *M.N. Tamura in Acta Phytotax. Geobot.,* 40(1–4): 2 (1989)—Thailand.

CHONDRORHYNCHA (Orchidac.).
andreettae *R. Jenny in Orchidee,* 40(3): 92 (1989)—Ecuador.
viridisepala *K. Senghas in Orchidee,* 40(5): 181 (1989)—Ecuador.
vollesii *G. Gerlach, T. Neudecker & H.G. Seeger in Orchidee,* 40(4): 131 (1989)—Colombia.

CHORIGYNE R. Eriksson in Nordic J. Bot., 9(1): 33 (1989). *CYCLANTHACEAE.*
cylindrica *R. Eriksson in Nordic J. Bot.,* 9(1): 44 (1989)—Panama.
densiflora *R. Eriksson in Nordic J. Bot.,* 9(1): 35 (1989)—Panama.
ensiformis *(Hook. f.) R. Eriksson in Nordic J. Bot.,* 9(1): 37 (1989): Carludovica ensiformis.
paucinervis *R. Eriksson in Nordic J. Bot.,* 9(1): 40 (1989)—Panama.
pendula *(Hammel) R. Eriksson in Nordic J. Bot.,* 9(1): 39 (1989): Sphaeradenia pendula.
pterophylla *R. Eriksson in Nordic J. Bot.,* 9(1): 40 (1989)—Costa Rica, Panama.
tumescens *R. Eriksson in Nordic J. Bot.,* 9(1): 44 (1989)—Panama.

CHORIZANTHE (Polygonac.).
subsect. Acanthogonum *(Torr.) J.L. Reveal & C.B. Hardham in Phytologia,* 66(2): 179 (1989): gen. Acanthogonum.
subgen. Amphietes *J.L. Reveal & C.B. Hardham in Phytologia,* 66(2): 113 (1989).
subsect. Ansata *J.L. Reveal & C.B. Hardham in Phytologia,* 66(2): 183 (1989).

CHORIZANTHE :–
subsect. Chorizanthella *(C. Parry) J.L. Reveal & C.B. Hardham in Phytologia,* 66(2): 186 (1989): C. sect. Chorizanthella.
sect. Clastoscapa *J.L. Reveal & C.B. Hardham in Phytologia,* 66(2): 194 (1989).
subgen. Eriogonella *(Goodman) J.L. Reveal & C.B. Hardham in Phytologia,* 66(2): 110 (1989): gen. Eriogonella.
subsect. Exolophyta *J.L. Reveal & C.B. Hardham in Phytologia,* 66(2): 182 (1989).
sect. Fragile *J.L. Reveal & C.B. Hardham in Phytologia,* 66(2): 188 (1989).
subsect. Legnota *J.L. Reveal & C.B. Hardham in Phytologia,* 66(2): 116 (1989).
subsect. Parvatia *J.L. Reveal & C.B. Hardham in Phytologia,* 66(2): 176 (1989).
subsect. Petasata *J.L. Reveal & C.B. Hardham in Phytologia,* 66(2): 134 (1989).
subgen. Quintaria *J.L. Reveal & C.B. Hardham in Phytologia,* 66(2): 108 (1989).
biloba
var. immemora *J.L. Reveal & C.B. Hardham in Phytologia,* 66(2): 138 (1989)—U.S.A. (California).
pungens
var. hartwegiana *J.L. Reveal & C.B. Hardham in Phytologia,* 66(2): 125 (1989)—U.S.A. (California).
rosulenta *J.L. Reveal in Phytologia,* 66(2): 95 (1989)—Mexico.

CHRYSOBALANUS (Chrysobalanac.).
venezuelanus *G.T. Prance in Fl. Neotrop.,* 9S: 4 (1989)—Venezuela.

CHRYSOCYCNIS (Orchidac.).
tigrinum *(C. Schweinf.) J.T. Atwood in Ic. Pl. Trop.,* 14 (Orch. Costa Rica, pt. 1): t. 1306 (1989): Maxillaria tigrina.

CHRYSOSPLENIUM (Saxifragac.).
arctomontanum *(Petrovsky) S.S. Kharkevich in Sosud. Rast. Sovet. Dal'nego Vostoka,* 4: 185 (1989): C. alternifolium subsp. arctomontanum.
dezhnevii *(Jurtz.) S.S. Kharkevich in Sosud. Rast. Sovet. Dal'nego Vostoka,* 4: 182 (1989): C. rimosum subsp. dezhnevii.
sibiricum *(Ser. ex DC.) S.S. Kharkevich in Sosud. Rast. Sovet. Dal'nego Vostoka,* 4: 185 (1989): C. alternifolium var. sibiricum.
sibiricum *(Ser. ex DC.) A.P. Khokhryakov, Analiz Fl. Kolỹmskogo Nagor'ya:* 43 (1989): C. alternifolium var. sibiricum.

CHUSQUEA (Gramin.).
sect. Longifoliae *L.G. Clark, Syst. Chusquea (Syst. Bot. Monog., 27):* 105 (1989).
sect. Serpentes *L.G. Clark, Syst. Chusquea (Syst. Bot. Monog., 27):* 91 (1989).
sect. Swallenochloa *(McClure) L.G. Clark, Syst. Chusquea (Syst. Bot. Monog., 27):* 29 (1989): gen. Swallenochloa.
sect. Verticillatae *L.G. Clark, Syst. Chusquea (Syst. Bot. Monog., 27):* 74 (1989).

CHUSQUEA :–
glauca *L.G. Clark, Syst. Chusquea (Syst. Bot. Monog., 27):* 95 (1989)—Mexico.

CICERBITA (Compos.).
filicina *(Duthie ex Stebbins) B.S. Aswal & A.K. Goel in J. Econ. Taxon. Bot.,* 12(1): 12 (1988): Lactuca filicina.

CICHORIUM (Compos.).
intybus
subsp. glabratum *(C. Presl) G. Wagenitz & U. Bedarff in Davis & Hedge Festschrift (ed. Kit Tan):* 12 (1989): C. glabratum.

CINERARIA (Compos.).
albomontana *O.M. Hilliard in Notes Roy. Bot. Gard. Edinburgh,* 45(2): 185 (1988 publ. 1989)—South Africa (Cape Province), Lesotho.
vagans *O.M. Hilliard in Notes Roy. Bot. Gard. Edinburgh,* 45(2): 186 (1988 publ. 1989)—South Africa (Cape Province).

CINNAMOMUM (Laurac.).
alcinii *(Allen) A.J.G.H. Kostermans in Reinwardtia,* 10(5): 439 (1988): Phoebe alcinii.
altissimum *A.J.G.H. Kostermans in Reinwardtia,* 10(5): 439 (1988)—Malaya, Sumatra.
antillanum *(Meissn.) A.J.G.H. Kostermans in Reinwardtia,* 10(5): 441 (1988): Phoebe antillana.
areolatocostae *(Allen) A.J.G.H. Kostermans in Reinwardtia,* 10(5): 441 (1988): Phoebe areolato-costae.
bintulense *A.J.G.H. Kostermans in Reinwardtia,* 10(5): 441 (1988)—Borneo (Sarawak).
breedlovii *(Lundell) A.J.G.H. Kostermans in Reinwardtia,* 10(5): 442 (1988): Phoebe breedlovii.
chavarrianum *(Hammel) A.J.G.H. Kostermans in Reinwardtia,* 10(5): 442 (1988): Phoebe chavarriana.
corsoanum *(Lundell) A.J.G.H. Kostermans in Reinwardtia,* 10(5): 442 (1988): Phoebe corsoana.
erectifolium *(Allen) A.J.G.H. Kostermans in Reinwardtia,* 10(5): 442 (1988): Phoebe erectifolia.
fouilloyi *A.J.G.H. Kostermans in Reinwardtia,* 10(5): 442 (1988)—Borneo (Kalimantan).
fruticosum *(Lundell) A.J.G.H. Kostermans in Reinwardtia,* 10(5): 443 (1988): Phoebe fruticosa.
galeottianum *(Mez) A.J.G.H. Kostermans in Reinwardtia,* 10(5): 443 (1988): Phoebe galeottiana.
grandis *A.J.G.H. Kostermans in Reinwardtia,* 10(5): 443 (1988)—Borneo (Sabah).
hartmannii *(Johnston) A.J.G.H. Kostermans in Reinwardtia,* 10(5): 444 (1988): Phoebe hartmannii.

CINNAMOMUM :–
impressum *(Meissn.) A.J.G.H. Kostermans in Reinwardtia,* 10(5): 444 (1988): Phoebe impressa.
malayanum *A.J.G.H. Kostermans in Reinwardtia,* 10(5): 444 (1988)—Malaya.
mayanum *(Lundell) A.J.G.H. Kostermans in Reinwardtia,* 10(5): 446 (1988): Phoebe mayana.
miaoshanense *S. Lee & F.N. Wei in Guihaia,* 8(4): 302 (1988)—China.
nooteboomii *A.J.G.H. Kostermans in Reinwardtia,* 10(5): 446 (1988)—Borneo (Sarawak), Sumatra.
nunesianum *(de Vattimo) A.J.G.H. Kostermans in Reinwardtia,* 10(5): 447 (1988): Phoebe nunesiana.
pachypes *A.J.G.H. Kostermans in Reinwardtia,* 10(5): 447 (1988)—Malaya.
pachypodum *(Nees) A.J.G.H. Kostermans in Reinwardtia,* 10(5): 447 (1988): Persea pachypoda.
paiei *A.J.G.H. Kostermans in Reinwardtia,* 10(5): 448 (1988)—Borneo (Sarawak).
parvulum *(Lundell) A.J.G.H. Kostermans in Reinwardtia,* 10(5): 448 (1988): Phoebe parvula.
peruvianum *(Meissn.) A.J.G.H. Kostermans in Reinwardtia,* 10(5): 448 (1988): Phoebe peruviana.
pickellii *(Coe-Teixeira) A.J.G.H. Kostermans in Reinwardtia,* 10(5): 448 (1988): Phoebe pickellii.
polyadelphum *(Lour.) A.J.G.H. Kostermans in Reinwardtia,* 10(5): 449 (1988): Laurus polyadelpha.
porphyrospermum *A.J.G.H. Kostermans in Reinwardtia,* 10(5): 450 (1988)—Sumatra.
safrol *A.J.G.H. Kostermans in Reinwardtia,* 10(5): 451 (1988)—Malaya.
subsericeum *A.J.G.H. Kostermans in Reinwardtia,* 10(5): 452 (1988)—Vietnam, Cambodia.
trinerve *(Lundell) A.J.G.H. Kostermans in Reinwardtia,* 10(5): 453 (1988): Phoebe trinervis.
trintaense *A.J.G.H. Kostermans in Reinwardtia,* 10(5): 453 (1988)—Malaya.
vanderwerffii *A.J.G.H. Kostermans in Reinwardtia,* 10(5): 454 (1988), nom. nov.: Phoebe glabra *van der Werff.*
villosulum *S. Lee & F.N. Wei in Guihaia,* 8(4): 301 (1988)—China.
woodii *A.J.G.H. Kostermans in Reinwardtia,* 10(5): 454 (1988)—Borneo (Sabah).

CIPURA (Iridac.).
formosa *P. Ravenna in Onira, Bot. Leafl.,* 1(5): 36 (1988)—Brazil.
insularis *P. Ravenna in Onira, Bot. Leafl.,* 1(5): 42 (1988), nom. nov.: C. cubensis *Wright ex Sauvalle.*
paludosa
subsp. boliviensis *P. Ravenna in Onira, Bot. Leafl.,* 1(5): 40 (1988)—Bolivia.

CIPURA :–
subsp. caatingensis *P. Ravenna in Onira, Bot. Leafl.,* 1(5): 39 (1988)—Brazil.
subsp. decora *P. Ravenna in Onira, Bot. Leafl.,* 1(5): 38 (1988)—Brazil.
subsp. robusta *P. Ravenna in Onira, Bot. Leafl.,* 1(5): 38 (1988)—Brazil.
subsp. warmingii *(Kl.) P. Ravenna in Onira, Bot. Leafl.,* 1(5): 39 (1988): Marica warmingii.
paradisiaca *P. Ravenna in Onira, Bot. Leafl.,* 1(5): 37 (1988)—Brazil.
xanthomelas
subsp. flavescens *P. Ravenna in Onira, Bot. Leafl.,* 1(5): 40 (1988)—Brazil.

CIRSIUM (Compos.).
falconeri
var. horrida *(Hook. f.) B.S. Aswal & A.K. Goel in J. Econ. Taxon. Bot.,* 12(1): 12 (1988): C. involucratus var. horrida.
nipponicum
var. sadoense *T. Yamazaki in J. Jap. Bot.,* 64(2): 49 (1989)—Japan.
wallichii
var. fasciculatum *(Hook. f.) K.S. Manilal & C.R. Suresh in K.S. Manilal, Fl. Silent Valley:* 153 (1988), isonym, see p. 368: Cnicus wallichii var. fasciculata.

CISSUS (Vitac.).
arnottiana *B.V. Shetty & P. Singh in Kew Bull.,* 44(3): 473 (1989)—India.
austroyunnanensis *Y.H. Li & Y. Zhang in Guihaia,* 9(1): 28 (1989), as 'austro-yunnanensis'—China.
cardiophylla *(F. Muell.) B.R. Jackes in Austrobaileya,* 2(5): 491 (1988): Vitis cardiophylla.
elongata
subsp. littoralis *(Talbot) B.V. Shetty & P. Singh in Kew Bull.,* 44(3): 475 (1989): Vitis elongata var. littoralis.
guerkeana *(Büttn.) T. Durand & H. Schinz, Etudes Fl. Congo (in Mem. Cour. Acad. Roy. Belg., 53(4)):* 94 (1896): Vitis guerkeana.
pentaclada *B.R. Jackes in Austrobaileya,* 2(5): 490 (1988)—Australia (Queensland).
vinosa *B.R. Jackes in Austrobaileya,* 2(5): 494 (1988)—Australia (Queensland).

CITRUS (Rutac.).
grandis
var. brunneorugosa *R.K. Karaya in Sborn. Nauch. Tr. Prikl. Bot. Genet. Selek.,* 112: 66 (1987)—Japan.
var. glaberrima *(Hort. ex Tanaka) R.K. Karaya in Sborn. Nauch. Tr. Prikl. Bot. Genet. Selek.,* 112: 67 (1987): C. glaberrima.
var. kotokan *(Hayata) R.K. Karaya in Sborn. Nauch. Tr. Prikl. Bot. Genet. Selek.,* 112: 65 (1987): C. kotokan.

CITRUS :-

var. natsudaidai *(Tanaka) R.K. Karaya in Sborn. Nauch. Tr. Prikl. Bot. Genet. Selek.*, 112: 67 (1987): C. aurantium var. natsudaidai.

var. pyriformis *(Hassk.) R.K. Karaya in Sborn. Nauch. Tr. Prikl. Bot. Genet. Selek.*, 112: 66 (1987): C. pyriformis.

var. sabon *(Seib. ex Hayata) R.K. Karaya in Sborn. Nauch. Tr. Prikl. Bot. Genet. Selek.*, 112: 65 (1987): C. sabon.

var. sulcata *(Takahashi) R.K. Karaya in Sborn. Nauch. Tr. Prikl. Bot. Genet. Selek.*, 112: 67 (1987): C. sulcata.

var. tengu *(Y. Tanaka) R.K. Karaya in Sborn. Nauch. Tr. Prikl. Bot. Genet. Selek.*, 112: 66 (1987): C. tengu.

var. yamabuki *(Hort. ex Y. Tanaka) R.K. Karaya in Sborn. Nauch. Tr. Prikl. Bot. Genet. Selek.*, 112: 65 (1987): C. yamabuki.

CLAVIJA (Theophrastac.).

fusca *B. Ståhl in Nordic J. Bot.*, 9(1): 19 (1989)—Panama.

mezii

subsp. allenii *(Lundell) B. Ståhl in Nordic J. Bot.*, 9(1): 18 (1989): C. allenii.

CLEIDION (Euphorbiac.).

bishnui *T. Chakrabarty & M. Gangopadhyay in J. Econ. Taxon. Bot.*, 12(2): 473 (1988 publ. 1989)—Burma.

javanicum

var. longipedicellatum *T. Chakrabarty & M. Gangopadhyay in J. Econ. Taxon. Bot.*, 12(2): 491 (1988 publ. 1989)—Sikkim.

var. moniliflorum *(Airy Shaw) T. Chakrabarty & M. Gangopadhyay in J. Econ. Taxon. Bot.*, 12(2): 493 (1988 publ. 1989): C. moniliflorum.

CLEISOSTOMA (Orchidac.).

longipaniculatum *P.J. Kores in Allertonia*, 5(1): 202 (1989)—Fiji.

nangongense *Z.H. Tsi in Bull. Bot. Res. North-East. Forest. Inst.*, 9(2): 26 (1989)—China.

pacificum *P.J. Cribb & B. Lewis in Orchid Rev.*, 97(1150): 251 (1989)—Vanuatu.

paludosum *(v. Royen) K. Senghas in Schlechter Orchideen*, 1(22): 1340 (1989): Sarcanthus paludosus.

pauciflorum *(Wight) K. Senghas in Schlechter Orchideen*, 1(22): 1340 (1989): Sarcanthus pauciflorus.

CLEISTOCACTUS (Cactac.).

baumannii

subsp. santacruzensis *(Backeberg) R. Mottram in Cact. Succ. J. (U.S.A.)*, 61(4): 156 (1989): C. santacruzensis.

leonensis *J.E. Madsen in Fl. Ecuador*, 35(45): 18 (1989)—Ecuador.

sepium

var. morleyanus *(Britton & Rose) J.E. Madsen in Fl. Ecuador*, 35(45): 21 (1989): Borzicactus morleyanus.

CLEISTOCACTUS :-

var. ventimigliae *(Riccob.) J.E. Madsen in Fl. Ecuador*, 35(45): 22 (1989): Borzicactus ventimigliae.

winteri *D. Hunt in Bradleya*, 6: 100 (1988), nom. nov.: C. aureispinus *(Ritter) D. Hunt.*

CLEMATIS (Ranunculac.).

sect. Aethusifoliae *(Tamura) V.P. Serov in Bot. Zhurn.*, 73(12): 1739 (1988): C. ser. Aethusifoliae.

sect. Angustifoliae *(Tamura) V.P. Serov in Bot. Zhurn.*, 73(12): 1740 (1988): C. ser. Angustifoliae.

sect. Integrifoliae *V.P. Serov in Bot. Zhurn.*, 73(12): 1740 (1988).

ser. Ispahanicae *V.P. Serov in Bot. Zhurn.*, 73(12): 1739 (1988).

sect. Rectae *(Tamura) V.P. Serov in Bot. Zhurn.*, 73(12): 1739 (1988): C. ser. Rectae.

ser. Songoricae *V.P. Serov in Bot. Zhurn.*, 73(12): 1739 (1988).

acuminata

var. longicaudata *W.T. Wang in Bull. Bot. Res. North-East. Forest. Inst.*, 9(2): 7 (1989)—China.

ladakhiana *C. Grey-Wilson in Kew Bull.*, 44(1): 49 (1989)—India, Xizang.

mashanensis *W.T. Wang in Bull. Bot. Res. North-East. Forest. Inst.*, 9(2): 7 (1989)—China.

orientalis

var. baluchistanica *C. Grey-Wilson in Kew Bull.*, 44(1): 40 (1989)—Pakistan.

var. hindukushensis *C. Grey-Wilson in Kew Bull.*, 44(1): 40 (1989)—Afghanistan.

var. robusta *C. Grey-Wilson in Kew Bull.*, 44(1): 40 (1989)—Afghanistan.

var. tenuifolia *(Royle) C. Grey-Wilson in Kew Bull.*, 44(1): 42 (1989): C. tenuifolia.

pamiralaica *C. Grey-Wilson in Kew Bull.*, 44(1): 54 (1989)—U.S.S.R. (Middle Asia), Xinjiang.

tangutica

subsp. mongolica *C. Grey-Wilson in Kew Bull.*, 44(1): 54 (1989)—Mongolia.

subsp. obtusiuscula *(Rehder & Wilson) C. Grey-Wilson in Kew Bull.*, 44(1): 54 (1989): C. tangutica var. obtusiuscula.

tibetana

var. dentata *C. Grey-Wilson in Kew Bull.*, 44(1): 48 (1989)—Xizang.

var. laciniifolia *C. Grey-Wilson in Kew Bull.*, 44(1): 48 (1989)—Nepal.

subsp. vernayi *(C.E.C. Fischer) C. Grey-Wilson in Kew Bull.*, 44(1): 47 (1989): C. vernayi.

var. vernayi *(C.E.C. Fischer) C. Grey-Wilson in Kew Bull.*, 44(1): 48 (1989): C. vernayi.

villosa *B.M. Yang in Acta Phytotax. Sin.*, 27(3): 230 (1989)—China.

CLEMATOCLETHRA (Actinidiac.).
lasioclada
var. oblonga *C.F. Liang & Y.C. Chen ex C.Y. Chang in Fl. Sichuanica,* 4: 82 (1988)—China.
leiboensis *C.F. Liang & Y.C. Chen ex C.Y. Chang in Fl. Sichuanica,* 4: 67 (1988)—China.
scandens
subsp. actinidioides *(Maxim.) Y.C. Tang & Q.Y. Xiang in Acta Phytotax. Sin.,* 27(2): 91 (1989): C. actinidioides.
subsp. hemsleyi *(Baill.) Y.C. Tang & Q.Y. Xiang in Acta Phytotax. Sin.,* 27(2): 90 (1989): C. hemsleyi.
subsp. tomentella *(Franch.) Y.C. Tang & Q.Y. Xiang in Acta Phytotax. Sin.,* 27(2): 90 (1989): C. tomentella.

CLERODENDRUM (Verbenac.).
buchananii *(Roxb.) W.G. Walpers, Repert. Bot. Syst.,* 4: 108 (1845): Volkameria buchananii.†
chinense *(Osb.) D.J. Mabberley, Plant-book:* 707 (1989 reprint): Cryptanthus chinensis.
grayi *A.A. Munir in J. Adelaide Bot. Gard.,* 11(2): 166 (1989)—Australia (Queensland).
heterophyllum
f. baueri *(Mold.) A.A. Munir in J. Adelaide Bot. Gard.,* 11(2): 117 (1989): C. heterophyllum var. baueri.
longiflorum
var. glabrum *A.A. Munir in J. Adelaide Bot. Gard.,* 11(2): 120 (1989)—Australia (Queensland, N. Terr.), New Guinea (Papua New Guinea).
tatei *(F. Muell.) A.A. Munir in J. Adelaide Bot. Gard.,* 11(2): 143 (1989): Strobilanthes tatei.
tomentosum
var. lanceolatum *(F. Muell.) A.A. Munir in J. Adelaide Bot. Gard.,* 11(2): 157 (1989): C. lanceolatum.

CLIDEMIA (Melastomatac.).
andersonii *J.J. Wurdack in Mem. New York Bot. Gard.,* 51: 104 (1989)—Venezuela.
hammelii *F. Almeda in Proc. Calif. Acad. Sci.,* Ser. 4, 46(5): 140 (1989)—Costa Rica, Panama, Colombia.
marahuacensis *J.J. Wurdack in Mem. New York Bot. Gard.,* 51: 104 (1989)—Venezuela.

CLINOPODIUM (Labiat.).
calamintha *(L.) C.A. Stace in Watsonia,* 17(4): 443 (1989): Melissa calamintha.
grandiflorum *(L.) C.A. Stace in Watsonia,* 17(4): 443 (1989): Melissa grandiflora.
menthifolium *(Host) C.A. Stace in Watsonia,* 17(4): 443 (1989): Calamintha menthifolia.

CNIDIUM (Umbellif.).
kamelinii *V.M. Vinogradova in Novosti Sist. Vÿssh. Rast.,* 25: 122 (1988)—Mongolia.

CNIDOSCOLUS (Euphorbiac.).
urens
subsp. adenophilus *(Pax & K. Hoffm.) G. Breckon in Ann. Missouri Bot. Gard.,* 75(3): 1114 (1988): Jatropha adenophila.

COCCOLOBA (Polygonac.).
ecuadorensis *J. Brandbyge in Nordic J. Bot.,* 9(2): 205 (1989)—Ecuador.
trollii *J. Brandbyge in Nordic J. Bot.,* 9(2): 206 (1989)—Bolivia.

COILOSTIGMA (Ericac.).
zeyherianum
var. tenuifolium *(Klotzsch) E.G.H. Oliver in Bothalia,* 18(2): 185 (1988): C. tenuifolium.

COINCYA (Crucif.).
wrightii *(O. Schulz) C.A. Stace in Watsonia,* 17(4): 443 (1989): Brassicella wrightii.

COIX (Gramin.).
gasteenii *B.K. Simon in Austrobaileya,* 3(1): 1 (1989)—Australia (Queensland).

COLEOCEPHALOCEREUS (Cactac.).
subgen. Buiningia *(F. Buxbaum) P.J. Braun in Bradleya,* 6: 92 (1988): gen. Buiningia.
elongatus *(Buining) P.J. Braun in Bradleya,* 6: 92 (1988): Buiningia brevicylindrica var. elongata.

COLOGANIA (Legumin.).
parviflora *V.M. Badillo in Ernstia,* 52: 9 (1989)—Venezuela.

COLOMBOBALANUS K.C. Nixon & W.L. Crepet in Amer. J. Bot., 76(6): 840 (1989). *FAGACEAE.*
excelsa *(Lozano, Hdz-C. & Henao) K.C. Nixon & W.L. Crepet in Amer. J. Bot.,* 76(6): 841 (1989): Trigonobalanus excelsa.

COMAROSTAPHYLIS (Ericac.).
discolor
subsp. macvaughii *G.M. Diggs in Bull. Torrey Bot. Club,* 115(3): 204 (1988)—Mexico.
subsp. manantlanensis *G.M. Diggs in Bull. Torrey Bot. Club,* 115(3): 205 (1988)—Mexico.

COMBRETUM (Combretac.).
mkuzense *J.D. Carr & E. Retief in Bothalia,* 19(1): 38 (1989)—South Africa (Natal).

COMMICARPUS (Nyctaginac.).
somalensis *(Chiov.) J.P. Lebrun & A.L. Stork in Bull. Mus. Nation. Hist. Nat., B, Adansonia,* Ser. 4, 11(2): 215 (1989): Selinocarpus somalensis.

COMOLIOPSIS J.J. Wurdack in Acta Bot. Venezuel., 14(3): 23 (1984). *MELASTOMATACEAE.*

COMOLIOPSIS :-
neblinae *J.J. Wurdack in Acta Bot. Venezuel.,* 14(3): 23 (1984)—Venezuela.

COMPSONEURA (Myristicac.).
rigidifolia *W.A. Rodrigues in Brittonia,* 41(2): 160 (1989)—Panama.
schultesiana *W.A. Rodrigues in Brittonia,* 41(2): 161 (1989)—Colombia.

CONOPHYTUM (Aizoac.).
sect. Minuscula *(Schwantes) Tischer ex S.A. Hammer in Bradleya,* 6: 104 (1988): C. Reihe Minuscula.
sect. Pellucida *(Schwantes) Tischer ex S.A. Hammer in Bradleya,* 6: 104 (1988): C. Reihe Pellucida.
sect. Saxetana *(Schwantes) S.A. Hammer in Bradleya,* 6: 104 (1988): C. Reihe Saxetana.
sect. Tuberculata *(Schwantes) Tischer ex S.A. Hammer in Bradleya,* 6: 104 (1988): C. Reihe Tuberculata.
sect. Wettsteinia *(Schwantes) Tischer ex S.A. Hammer in Bradleya,* 6: 104 (1988): C. Reihe Wettsteinia.
achabense *S. Hammer in Cact. Succ. J. (U.S.A.),* 60(5): 217 (1988)—South Africa (Cape Province).
armianum *S. Hammer in Aloe,* 25(1): 10 (1988)—South Africa (Cape Province).
buysianum *A.R. Mitchell & S. Hammer in Aloe,* 25(2): 34 (1988)—South Africa (Cape Province).
ernstii *S. Hammer in Cact. Succ. J. (U.S.A.),* 60(6): 262 (1988)—South Africa (Cape Province).
lindenianum *M. Lavis & S. Hammer in Cact. Succ. J. (U.S.A.),* 61(1): 3 (1989)—South Africa (Cape Province).

CONOSTOMIUM (Rubiac.).
gazense *B. Verdcourt in Fl. Zambes.,* 5(1): 103 (1989)—Mozambique.

CONSOLIDA (Ranunculac.).
hispanica *(Costa) W. Greuter & Burdet in Willdenowia,* 19(1): 43 (1989): Delphinium hispanicum.

×**COOKARA** hort. in Orchid Rev., 97(1149): centre page pull-out p. 8 (1989). *ORCHIDACEAE.* (Broughtonia × Cattleya × Diacrium × Laelia).

COPAIFERA (Legumin.).
venezuelana
var. laxa *N. Xena de Enrech & M.T.K. Arroyo in Acta Bot. Venezuel.,* 14(2): 264 (1983)—Venezuela.

COPROSMA (Rubiac.).
waima *A.P. Druce in New Zealand J. Bot.,* 27(1): 119 (1989)—New Zealand.

CORALLODISCUS (Gesneriac.).
flabellatus
var. puberulus *K.Y. Pan in Acta Phytotax. Sin.,* 26(6): 437 (1988)—China, Xizang.

CORDIA (Boraginac.).
crenata
subsp. meridionalis *A.M. Warfa in Nordic J. Bot.,* 8(6): 617 (1989)—Kenya, Uganda, Somali Republic, Tanzania.
gentryi *J.S. Miller in Ann. Missouri Bot. Gard.,* 76(2): 595 (1989)—Colombia.
longipetiolata *A.M. Warfa in Nordic J. Bot.,* 9(3): 251 (1989)—Kenya.

COREOCARPUS (Compos.).
parthenioides
var. involutus *(E.L. Greene) E.B. Smith in Syst. Bot.,* 14(4): 470 (1989): C. involutus.

CORNUS (Cornac.).
schindleri
subsp. poliophylla *(Schneid. & Wanger.) Q.Y. Xiang in Bull. Bot. Res. North-East. Forest. Inst.,* 9(1): 135 (1989): C. poliophylla.

CORONILLA (Legumin.).
somalensis *M. Thulin in Nordic J. Bot.,* 8(5): 485 (1989)—Somali Republic.

CORYANTHES (Orchidac.).
cataniapoensis *G.A. Romero & G. Carnevali in Ann. Missouri Bot. Gard.,* 76(2): 454 (1989)—Venezuela.
verrucolineata *G. Gerlach in Orchidee,* 40(2): 52 (1989)—Peru.

CORYBAS (Orchidac.).
barbarae *D.L. Jones in Austrobaileya,* 2(5): 548 (1988)—Australia (Queensland, New South Wales).
montanus *D.L. Jones in Austrobaileya,* 2(5): 547 (1988)—Australia (Queensland).

CORYDALIS (Papaverac.).
sect. Mucroniferae *Fedde ex M. Lidén in Bull. Brit. Mus. (Nat. Hist.), Bot.,* 18(6): 524 (1989).
brevipedicellata *M. Lidén in Notes Roy. Bot. Gard. Edinburgh,* 45(2): 360 (1988 publ. 1989)—Turkey.
calycina *M. Lidén in Bull. Brit. Mus. (Nat. Hist.), Bot.,* 18(6): 532 (1989)—Nepal.
elegans
subsp. robusta *M. Lidén in Bull. Brit. Mus. (Nat. Hist.), Bot.,* 18(6): 518 (1989)—Nepal.
ellipticarpa *C.Y. Wu & Z.Y. Su in Acta Bot. Yunnanica,* 11(3): 311 (1989)—China.
grandicalyx *B.U. Oh & Y.S. Kim in Korean J. Pl. Taxon.,* 17(1): 21 (1987)—Korea.
grandiflora *C.Y. Wu & Z.Y. Su in Acta Bot. Yunnanica,* 11(3): 314 (1989)—China.

CORYDALIS :-
haussknechtii *M. Lidén in Notes Roy. Bot. Gard. Edinburgh,* 45(2): 358 (1988 publ. 1989)—Turkey, Iraq.
hebephylla *C.Y. Wu & Z.Y. Su in Acta Bot. Yunnanica,* 11(3): 312 (1989)—China.
var. glabrescens *C.Y. Wu & Z.Y. Su in Acta Bot. Yunnanica,* 11(3): 313 (1989)—China.
heracleifolia *C.Y. Wu & Z.Y. Su in Acta Bot. Yunnanica,* 11(3): 315 (1989)—China.
humilis *B.U. Oh & Y.S. Kim in Korean J. Pl. Taxon.,* 17(1): 24 (1987)—Korea.
×hybrida *M.A. Mikhaĭlova in Novosti Sist. Vyssh. Rast.,* 25: 88 (1988)—Cultivation. (C. bracteata × solida).
lowndesii *M. Lidén in Bull. Brit. Mus. (Nat. Hist.), Bot.,* 18(6): 491 (1989)—Nepal, Xizang.
maculata *B.U. Oh & Y.S. Kim in Korean J. Pl. Taxon.,* 17(1): 26 (1987)—Korea.
oppositifolia
subsp. kurdica *(Cullen & Davis) M. Lidén in Notes Roy. Bot. Gard. Edinburgh,* 45(2): 362 (1988 publ. 1989): C. rutifolia subsp. kurdica.
paschei *M. Lidén in Notes Roy. Bot. Gard. Edinburgh,* 45(2): 357 (1988 publ. 1989)—Turkey.
pseudolongipes *M. Lidén in Bull. Brit. Mus. (Nat. Hist.), Bot.,* 18(6): 532 (1989)—Nepal, Sikkim, Bhutan.
shakyae *M. Lidén in Bull. Brit. Mus. (Nat. Hist.), Bot.,* 18(6): 519 (1989)—Nepal.
simplex *M. Lidén in Bull. Brit. Mus. (Nat. Hist.), Bot.,* 18(6): 514 (1989)—Nepal.
solida
subsp. incisa *M. Lidén in Notes Roy. Bot. Gard. Edinburgh,* 45(2): 351 (1988 publ. 1989)—Greece, Albania, Jugoslavia.
subsp. oligantha *(Trinajstić) W. Greuter & Burdet in Willdenowia,* 19(1): 38 (1989): Bulbocapnos solida subsp. oligantha.
spicata *M. Lidén in Bull. Brit. Mus. (Nat. Hist.), Bot.,* 18(6): 528 (1989)—Nepal.
stipulata *M. Lidén in Bull. Brit. Mus. (Nat. Hist.), Bot.,* 18(6): 522 (1989)—Nepal.
tauricola *(Cullen & Davis) M. Lidén in Notes Roy. Bot. Gard. Edinburgh,* 45(2): 357 (1988 publ. 1989): C. solida subsp. tauricola.
terracina *M. Lidén in Bull. Brit. Mus. (Nat. Hist.), Bot.,* 18(6): 522 (1989)—Nepal.
tianzhuensis *M.S. Yan & C.J. Wang in Bull. Bot. Res. North-East. Forest. Inst.,* 9(3): 21 (1989)—China.
uncinata *M. Lidén in Bull. Brit. Mus. (Nat. Hist.), Bot.,* 18(6): 520 (1989)—Nepal.
uncinatella *M. Lidén in Bull. Brit. Mus. (Nat. Hist.), Bot.,* 18(6): 520 (1989)—Nepal.
wendelboi *M. Lidén in Notes Roy. Bot. Gard. Edinburgh,* 45(2): 355 (1988 publ. 1989)—Turkey.

CORYLUS (Corylac.).
heterophylla
var. shenyangensis *L. Zhao & D. Chen in Fl. Liaoningica, chief ed. S.X. Li,* 1: 1361, 251 (1988)—China.

CORYPHANTHA (Cactac.).
tripugionacantha *A.B. Lau in Cact. Suc. Mex.,* 33(1): 21 (1988)—Mexico.

COSTUS (Costac.).
oblongus *S.Q. Tong in Acta Phytotax. Sin.,* 27(4): 291 (1989)—China, Xizang.
viridis *S.Q. Tong in Acta Phytotax. Sin.,* 27(4): 289 (1989)—China.

COTINUS (Anacardiac.).
coggygria
var. chengkouensis *Y.T. Wu in Fl. Sichuanica,* 4: 119 (1988)—China.

COTONEASTER (Rosac.).
assadii *M. Khatamsaz in Iranian J. Bot.,* 4(1): 116 (1988)—Iran.

COTTENDORFIA (Bromeliac.).
arachnoidea *L.B. Smith, J. Steyermark & Robinson in Acta Bot. Venezuel.,* 14(3): 9 (1984)—Venezuela.
geniculata
var. minor *L.B. Smith, J. Steyermark & Robinson in Acta Bot. Venezuel.,* 14(3): 15 (1984)—Venezuela.
marahuacae *L.B. Smith, J. Steyermark & Robinson in Acta Bot. Venezuel.,* 14(3): 9 (1984)—Venezuela.

COUEPIA (Chrysobalanac.).
amaraliae *G.T. Prance in Fl. Neotrop.,* 9S: 72 (1989), as 'amaralae'—Brazil.
carautae *G.T. Prance in Fl. Neotrop.,* 9S: 70 (1989)—Brazil.
caryophylloides
subsp. glabra *G.T. Prance in Fl. Neotrop.,* 9S: 72 (1989)—Brazil.
coarctata *G.T. Prance in Fl. Neotrop.,* 9S: 77 (1989)—Brazil.
monteclarensis *G.T. Prance in Fl. Neotrop.,* 9S: 66 (1989)—Brazil.
scottmorii *G.T. Prance in Fl. Neotrop.,* 9S: 68 (1989)—Panama.

COUSINIA (Compos.).
baluchistanica *K.H. Rechinger in Anz. Österr. Akad. Wiss., Mat.-Naturwiss.,* 1988(4): 53 (1988)—Pakistan.

COUSSAREA (Rubiac.).
evoluta *J.A. Steyermark in Ann. Missouri Bot. Gard.,* 75(1): 334 (1988)—Venezuela.

CRANICHIS (Orchidac.).
pilosissima *K. Senghas in Orchidee,* 40(2): 47 (1989)—Ecuador.
ricartii *J.D. Ackerman in Lindleyana,* 4(1): 43 (1989)—Puerto Rico.

CRASSOCEPHALUM (Compos.).
crepidioides
f. luteum *(Steen.) R.O. Belcher in Kew Bull.*, 44(3): 540 (1989): C. crepidioides var. luteum.

CREPIS (Compos.).
arcuata *G. Kamari & A. Strid in Willdenowia*, 19(1): 79 (1989)—Greece.
dinarica *(Beck) S. Šiljak-Yakovlev & T. Wraber in Acta Bot. Croat.*, 47: 164 (1988): C. incarnata var. dinarica.

CRINUM (Amaryllidac.).
belleymei *F. Hérincq, Horticulteur Français*, 7: 241, t. 19 (1857)—Paraguay.

CRITONIOPSIS (Compos.).
tamana *V.M. Badillo in Ernstia*, 53: 10 (1989)—Venezuela.
turmalensis *V.M. Badillo in Ernstia*, 53: 12 (1989)—Venezuela.

CROCUS (Iridac.).
ligusticus *M.G. Mariotti in Candollea*, 43(2): 674 (1988)—Italy, France.

CROTALARIA (Legumin.).
allophylla *M. Thulin in Nordic J. Bot.*, 8(5): 481 (1989)—Somali Republic.
avonensis *K.R. DeLaney & R.P. Wunderlin in Sida*, 13(3): 315 (1989)—U.S.A. (Florida).
chrysantha *Ahn & S.H. Park in J. Korean Pl. Taxon.*, 7(1–2): 23 (1976)—Korea.
intricata *M. Thulin in Nordic J. Bot.*, 8(5): 482 (1989)—Somali Republic.
warfae *M. Thulin in Nordic J. Bot.*, 8(5): 484 (1989)—Somali Republic.

CROTON (Euphorbiac.).
billbergianus
subsp. pyramidalis *(J.D. Smith) G.L. Webster in Ann. Missouri Bot. Gard.*, 75(3): 1123 (1988): C. pyramidalis.
draco
subsp. panamensis *(Klotzsch) G.L. Webster in Ann. Missouri Bot. Gard.*, 75(3): 1120 (1988): Cyclostigma panamense.
howii *Merr. & Chun ex Y.T. Chang in Acta Phytotax. Sin.*, 27(2): 147 (1989)—Hainan.
pachypodus *G.L. Webster in Ann. Missouri Bot. Gard.*, 75(3): 1119 (1988)—Panama.
santaritensis *M.J. Huft in Ann. Missouri Bot. Gard.*, 75(3): 1123 (1988)—Panama.
socotranus
var. pachyclados *(A.R. Smith) A. Radcliffe-Smith in Kew Bull.*, 44(3): 449 (1989): C. pachyclados.
speciosus
subsp. tacarcunensis *G.L. Webster in Ann. Missouri Bot. Gard.*, 75(3): 1119 (1988)—Panama.
zavaletae *Conzatti ex J. Rzedowski et al. in C. Conzatti, Fl. Taxon. Mexic., ed. 3*, 1: 846 (1988)—Mexico.

CRUCIANELLA (Rubiac.).
gilanica
subsp. carduchorum *F. Ehrendorfer & E. Schönbeck-Temesy in Pl. Syst. Evol.*, 165(1–2): 122 (1989)—Turkey, Iraq, Iran.
subsp. damghanensis *F. Ehrendorfer & E. Schönbeck-Temesy in Pl. Syst. Evol.*, 165(1–2): 116 (1989)—Iran.
subsp. demavendensis *F. Ehrendorfer & E. Schönbeck-Temesy in Pl. Syst. Evol.*, 165(1–2): 109 (1989)—Iran.
subsp. elbursensis *(Ehrend.) F. Ehrendorfer & E. Schönbeck-Temesy in Pl. Syst. Evol.*, 165(1–2): 115 (1989): C. glauca subsp. elbursensis.
subsp. hirsuta *(Ehrend.) F. Ehrendorfer & E. Schönbeck-Temesy in Pl. Syst. Evol.*, 165(1–2): 114 (1989): C. glauca var. hirsuta.
subsp. nezvensis *F. Ehrendorfer & E. Schönbeck-Temesy in Pl. Syst. Evol.*, 165(1–2): 110 (1989)—Iran.
subsp. suleimanica *F. Ehrendorfer & E. Schönbeck-Temesy in Pl. Syst. Evol.*, 165(1–2): 111 (1989)—Iran.
subsp. transcaspica *(Ehrend.) F. Ehrendorfer & E. Schönbeck-Temesy in Pl. Syst. Evol.*, 165(1–2): 125 (1989): C. glauca subsp. transcaspica.
platyphylla *F. Ehrendorfer & E. Schönbeck-Temesy in Pl. Syst. Evol.*, 165(1–2): 104 (1989)—Iran.

CRYPTANTHUS (Bromeliac.).
colnagoi *W. Rauh & E.M.C. Leme in J. Bromeliad Soc.*, 39(6): 258 (1989)—Brazil.
leopoldo-horstii *W. Rauh in Abh. Akad. Wiss. Lit. Mainz, Math.-Naturwiss. Klasse, Trop. Subtrop. Pflanzenwelt*, 65: 68 (1988)—Brazil.

CRYPTOCARYA (Laurac.).
albifrons *A.J.G.H. Kostermans in Reinwardtia*, 10(5): 455 (1988)—New Guinea (Papua New Guinea).
beilschmiediaefolia *A.J.G.H. Kostermans in Reinwardtia*, 10(5): 455 (1988)—Philippine Is.
calandoi *A.J.G.H. Kostermans in Reinwardtia*, 10(5): 456 (1988)—Celebes.
ceramica *A.J.G.H. Kostermans in Reinwardtia*, 10(5): 457 (1988)—Moluccas.
crassinerviopsis *A.J.G.H. Kostermans in Reinwardtia*, 10(5): 457 (1988)—Celebes.
durifolia *A.J.G.H. Kostermans in Reinwardtia*, 10(5): 458 (1988), nom. nov.: C. percoriacea *Kostermans*.
gigaphylla *A.J.G.H. Kostermans in Reinwardtia*, 10(5): 458 (1988), with two types cited—Moluccas.
lancifolia *A.J.G.H. Kostermans in Reinwardtia*, 10(5): 459 (1988)—Celebes.
lyoniifolia *S. Lee & F.N. Wei in Guihaia*, 8(4): 303 (1988)—China.

CRYPTOCARYA :–
parinarifolia *A.J.G.H. Kostermans in Reinwardtia,* 10(5): 459 (1988)—New Guinea (Papua New Guinea).

CRYPTOSTYLIS (Orchidac.).
arachnites
var. taiwaniana *(Masamune) S.S. Ying, Coloured Illustr. Fl. Taiwan,* 3: 622 (1988): C. taiwaniana.

CUATRECASANTHUS H. Robinson in Rev. Acad. Colomb. Cienc. Exact. Fis. Nat., 65: 209 (1989). *COMPOSITAE.*
flexipappus *(Gleason) H. Robinson in Rev. Acad. Colomb. Cienc. Exact. Fis. Nat.,* 65: 210 (1989): Vernonia flexipappa.
jelskii *(Hieron.) H. Robinson in Rev. Acad. Colomb. Cienc. Exact. Fis. Nat.,* 65: 210 (1989): Eremanthus jelskii.
sandemanii *(H. Robinson) H. Robinson in Rev. Acad. Colomb. Cienc. Exact. Fis. Nat.,* 65: 210 (1989): Vernonia sandemanii.

CUATRECASASIELLA H. Robinson in Fl. Neotrop., 2 Suppl.: 14 (1985). *COMPOSITAE.*
argentina *(Cabrera) H. Robinson in Fl. Neotrop.,* 2 Suppl.: 15 (1985): Luciliopsis argentina.
isernii *(Cuatrecasas) H. Robinson in Fl. Neotrop.,* 2 Suppl.: 15 (1985): Luciliopsis isernii.

CUCUMIS (Cucurbitac.).
myriocarpus
subsp. leptodermis *(Schweick.) L.W.D. van Raamsdonk in Pl. Syst. Evol.,* 163(3–4): 144 (1989): C. leptodermis.

CUCURBITA (Cucurbitac.).
argyrosperma
var. callicarpa *L.C. Merrick & D.M. Bates in Baileya,* 23(2): 97 (1989)—Cultivation.
var. palmeri *(L.H. Bailey) L.C. Merrick & D.M. Bates in Baileya,* 23(2): 98 (1989): C. palmeri.
subsp. sororia *(L.H. Bailey) L.C. Merrick & D.M. Bates in Baileya,* 23(2): 98 (1989): C. sororia.
var. stenosperma *(Pangalo) L.C. Merrick & D.M. Bates in Baileya,* 23(2): 96 (1989): C. mixta var. stenosperma.

CUPANIOPSIS (Sapindac.).
grandiflora *F. Adema in Bull. Mus. Nation. Hist. Nat., B, Adansonia,* Ser. 4, 10(3): 263 (1988 publ. 1989)—New Caledonia.
mackeeana *F. Adema in Bull. Mus. Nation. Hist. Nat., B, Adansonia,* Ser. 4, 10(3): 269 (1988 publ. 1989)—New Caledonia.
phalacrocarpa *F. Adema in Bull. Mus. Nation. Hist. Nat., B, Adansonia,* Ser. 4, 10(3): 266 (1988 publ. 1989)—New Caledonia.
squamosa *F. Adema in Bull. Mus. Nation. Hist. Nat., B, Adansonia,* Ser. 4, 10(3): 264 (1988 publ. 1989)—New Caledonia.

CUPHEA (Lythrac.).
spraguei *A. Lourteig in Mutisia,* 70: 19 (1988)—Panama.

CUPRESSUS (Cupressac.).
austrotibetica *J. Silba in Phytologia,* 65(5): 334 (1988), as 'austro-tibetica'—Xizang.

CURCUMA (Zingiberac.).
coriacea *J.K. Mangaly & M. Sabu in Notes Roy. Bot. Gard. Edinburgh,* 45(3): 429 (1988 publ. 1989)—India.
peethapushpa *N. Sasidharan & V.V. Sivarajan in Notes Roy. Bot. Gard. Edinburgh,* 45(3): 425 (1988 publ. 1989)—India.
raktakanta *J.K. Mangaly & M. Sabu in J. Econ. Taxon. Bot.,* 12(2): 475 (1988 publ. 1989)—India.

CYBIANTHUS (Myrsinac.).
agostinianus *J.J. Pipoly in Ernstia,* 50: 33 (1988)—Venezuela.
sipapoensis *J.J. Pipoly & Agostini in Ernstia,* 50: 36 (1988)—Venezuela.

CYCLAMEN (Primulac.).
somalense *M. Thulin & A.M. Warfa in Pl. Syst. Evol.,* 166(3–4): 249 (1989)—Somali Republic.

CYCLEA (Menispermac.).
longgangensis *J.Y. Luo in Guihaia,* 9(3): 197 (1989)—China.

CYCLOCARYA (Juglandac.).
paliurus
var. micropaliurus *(Tsoong) P.S. Hsu, X.Z. Feng & L.G. Xu in Guihaia,* 8(4): 322 (1988): Pterocarya micropaliurus.

CYCLOPOGON (Orchidac.).
pringlei *(S. Watson) M.A. Soto Arenas in Orquídea (México),* 11: 270 (1988): Spiranthes pringlei.

CYMBIDIUM (Orchidac.).
gongshanense *H. Li & G.H. Feng in Acta Bot. Yunnanica,* 11(1): 39 (1989)—China.

CYMBOPOGON (Gramin.).
dependens *B.K. Simon in Austrobaileya,* 3(1): 80 (1989)—Australia (N. Terr., W. Australia).
nardus
var. luridus *(Hook. f.) N. Rama Rao in J. Econ. Taxon. Bot.,* 12(2): 378 (1988 publ. 1989): Andropogon nardus var. luridus.

CYNANCHUM (Asclepiadac.).
amplexicaule
f. multicaurium *(Chung) W.T. Lee in J. Korean Pl. Taxon.,* 12(2): 85 (1982): C. multicaurium.
barbadense *(Schlechter) R.A. Howard in Phytologia,* 65(4): 285 (1988): Metastelma barbadense.

CYNANCHUM :-
brachystelmoides *P.I. Forster in Austrobaileya,* 2(5): 451 (1988)—Australia (Queensland, N. Terr.), New Guinea (Papua New Guinea).
buntingii *G. Morillo in Ernstia,* 51: 2 (1989)—Venezuela.
carautanum *(Font. & Schw.) G. Morillo in Ernstia,* 51: 3 (1989): Telminostelma carautanum.
christineae *P.I. Forster in Austrobaileya,* 3(1): 112 (1989)—Australia (N. Terr.).
franciscoi *G. Morillo in Ernstia,* 51: 3 (1989)—Venezuela, Colombia.
liebiana *(F. Muell.) P.I. Forster in Austrobaileya,* 3(1): 110 (1989): Tylophora liebiana.
martinicense *(Schlechter) R.A. Howard in Phytologia,* 65(4): 285 (1988): Metastelma martinicense.
neei *G. Morillo in Ernstia,* 51: 5 (1989)—Venezuela.
prevostii *G. Morillo in Ernstia,* 51: 5 (1989)—French Guiana.
readii *(Schlechter) R.A. Howard in Phytologia,* 65(4): 285 (1988): Metastelma readii.
stannardii *G. Morillo in Ernstia,* 51: 6 (1989)—Venezuela.

CYNOGLOSSUM (Boraginac.).
germanicum
subsp. pellucidum *(Lapeyrouse) K. Sutorý in Čas. Morav. Muz., Vědy Přír.,* 73(1–2): 150 (1988): C. pellucidum.
subsp. rotundum *K. Sutorý in Čas. Morav. Muz., Vědy Přír.,* 73(1–2): 150 (1988)—Switzerland, Germany, Belgium, France.

CYNOMETRA (Legumin.).
lukei *H.J. Beentje in Utafiti,* 1(3): 125 (1988)—Kenya.

CYPERUS (Cyperac.).
altochrysocephalus *K.A. Lye in Candollea,* 43(2): 508 (1988)—Zambia.
austrochrysanthus *K.A. Lye in Candollea,* 43(2): 505 (1988)—Angola.
macropachycephalus *P. Goetghebeur in Blumea,* 34(1): 161 (1989)—New Guinea (Papua New Guinea).
subcastaneus
subsp. graomogolii *D.A. Simpson in Kew Bull.,* 44(2): 279 (1989)—Brazil.

CYPHOSTEMMA (Vitac.).
megabotrys *(Collett & Hemsley) B.V. Shetty in Kew Bull.,* 44(3): 477 (1989): Vitis megabotrys.

CYPRIPEDIUM (Orchidac.).
sect. Acaulia *(Brieger) E.F. Hennessy in E.F. Hennessy & T.A. Hedge, Slipper Orchids:* 31 (1989), without basionym ref.: Basionym not stated.

CYPRIPEDIUM :-
sect. Flabellinervia *(Pfitzer) E.F. Hennessy in E.F. Hennessy & T.A. Hedge, Slipper Orchids:* 31 (1989), without basionym ref.: Basionym not stated.
parviflorum
f. albolabium *L.K. Magrath & J.L. Norman in Sida,* 13(3): 372 (1989)—U.S.A. (Oklahoma).

CYRTANDRA (Gesneriac.).
acriserrata *H. St. John in Phytologia,* 65(3): 199 (1988)—Hawaiian Is.
acuminifolia *H. St. John & W.N. Takeuchi in Phytologia,* 65(3): 207 (1988)—Hawaiian Is.
aequalis *H. St. John in Phytologia,* 65(3): 195 (1988)—Hawaiian Is.
basirotundata *H. St. John & W.N. Takeuchi in Phytologia,* 65(3): 207 (1988)—Hawaiian Is.
biformalis *H. St. John & W.N. Takeuchi in Phytologia,* 65(3): 211 (1988)—Hawaiian Is.
calpidicarpa
var. distincta *H. St. John & W.N. Takeuchi in Phytologia,* 65(3): 213 (1988)—Hawaiian Is.
catenulata *H. St. John & W.N. Takeuchi in Phytologia,* 65(3): 211 (1988)—Hawaiian Is.
caudiflora *H. St. John in Phytologia,* 65(3): 201 (1988)—Hawaiian Is.
christensenii *H. St. John in Phytologia,* 65(3): 195 (1988)—Hawaiian Is.
cuspidata *H. St. John & W.N. Takeuchi in Phytologia,* 65(3): 207 (1988)—Hawaiian Is.
dasygyna *H. St. John & W.N. Takeuchi in Phytologia,* 65(3): 211 (1988)—Hawaiian Is.
deorsa *H. St. John & W.N. Takeuchi in Phytologia,* 65(3): 213 (1988)—Hawaiian Is.
depressa *H. St. John & W.N. Takeuchi in Phytologia,* 65(3): 209 (1988)—Hawaiian Is.
discors *H. St. John & W.N. Takeuchi in Phytologia,* 65(3): 212 (1988)—Hawaiian Is.
disgrega *H. St. John in Phytologia,* 65(3): 201 (1988)—Hawaiian Is.
elliptiapicis *H. St. John in Phytologia,* 65(3): 195 (1988)—Hawaiian Is.
ferrea *H. St. John & W.N. Takeuchi in Phytologia,* 65(3): 209 (1988)—Hawaiian Is.
fusata *H. St. John & W.N. Takeuchi in Phytologia,* 65(3): 209 (1988)—Hawaiian Is.
gagnei *H. St. John in Phytologia,* 65(3): 196 (1988)—Hawaiian Is.
glabriflora *H. St. John & W.N. Takeuchi in Phytologia,* 65(3): 209 (1988)—Hawaiian Is.
habenosa *H. St. John in Phytologia,* 65(3): 200 (1988)—Hawaiian Is.
haikuensis *H. St. John & W.N. Takeuchi in Phytologia,* 65(3): 212 (1988)—Hawaiian Is.
hanaensis *H. St. John in Phytologia,* 65(3): 202 (1988)—Hawaiian Is.
harrisoniae *H. St. John in Phytologia,* 65(3): 196 (1988), as 'harrisonae'—Hawaiian Is.
hemisphaerica *H. St. John in Phytologia,* 65(3): 200 (1988), nom. nov.: C. ustulata *St. John in Phytologia, 63: 483 (1987).*
honokawaiensis *H. St. John in Phytologia,* 65(3): 196 (1988)—Hawaiian Is.

CYRTANDRA :-
honomanuensis *H. St. John in Phytologia,* 65(3): 197 (1988)—Hawaiian Is.
humiliserrulata *H. St. John in Phytologia,* 65(3): 197 (1988)—Hawaiian Is.
hylematos *H. St. John in Phytologia,* 65(3): 202 (1988)—Hawaiian Is.
imparilis *H. St. John & W.N. Takeuchi in Phytologia,* 65(3): 210 (1988)—Hawaiian Is.
indiscissa *H. St. John in Phytologia,* 65(3): 197 (1988)—Hawaiian Is.
infera *H. St. John in Phytologia,* 65(3): 202 (1988)—Hawaiian Is.
infraligulata *H. St. John in Phytologia,* 65(3): 205 (1988)—Hawaiian Is.
kalamanthe *H. St. John in Phytologia,* 65(3): 203 (1988)—Hawaiian Is.
kamooaliiensis *H. St. John & W.N. Takeuchi in Phytologia,* 65(3): 207 (1988)—Hawaiian Is.
kauensis *H. St. John in Phytologia,* 65(3): 203 (1988)—Hawaiian Is.
kawelaensis *H. St. John in Phytologia,* 65(3): 200 (1988)—Hawaiian Is.
kormanthike *H. St. John in Phytologia,* 65(3): 203 (1988)—Hawaiian Is.
kremnes *H. St. John & W.N. Takeuchi in Phytologia,* 65(3): 208 (1988)—Hawaiian Is.
laevicalycis *H. St. John in Phytologia,* 65(3): 199 (1988)—Hawaiian Is.
manukaensis *H. St. John in Phytologia,* 65(3): 205 (1988)—Hawaiian Is.
noie *H. St. John in Phytologia,* 65(3): 197 (1988)—Hawaiian Is.
opposita *H. St. John & W.N. Takeuchi in Phytologia,* 65(3): 210 (1988)—Hawaiian Is.
ovalifolia *H. St. John & W.N. Takeuchi in Phytologia,* 65(3): 212 (1988)—Hawaiian Is.
ovalis *H. St. John & W.N. Takeuchi in Phytologia,* 65(3): 210 (1988)—Hawaiian Is.
oxybapha *W.L. Wagner & D.R. Herbst in Bishop Mus. Occas. Pap.,* 29: 100 (1989)—Hawaiian Is.
paludosa
var. montana *H. St. John in Phytologia,* 65(3): 199 (1988)—Hawaiian Is.
pentenourou *H. St. John in Phytologia,* 65(3): 204 (1988)—Hawaiian Is.
piaensis *H. St. John & W.N. Takeuchi in Phytologia,* 65(3): 208 (1988)—Hawaiian Is.
pilosicalycis *H. St. John & W.N. Takeuchi in Phytologia,* 65(3): 210 (1988)—Hawaiian Is.
prolixa *H. St. John in Phytologia,* 65(3): 204 (1988)—Hawaiian Is.
prorsiflora *H. St. John & W.N. Takeuchi in Phytologia,* 65(3): 212 (1988)—Hawaiian Is.
pukeleensis *H. St. John & W.N. Takeuchi in Phytologia,* 65(3): 208 (1988)—Hawaiian Is.
quinqueflora *H. St. John & W.N. Takeuchi in Phytologia,* 65(3): 212 (1988)—Hawaiian Is.
rotundata *H. St. John & W.N. Takeuchi in Phytologia,* 65(3): 208 (1988)—Hawaiian Is.
scapiflora *H. St. John & W.N. Takeuchi in Phytologia,* 65(3): 210 (1988)—Hawaiian Is.
semiorbicularis *H. St. John in Phytologia,* 65(3): 200 (1988)—Hawaiian Is.

CYRTANDRA :-
septentrionalis *H. St. John in Phytologia,* 65(3): 199 (1988)—Hawaiian Is.
subaequalis *H. St. John & W.N. Takeuchi in Phytologia,* 65(3): 209 (1988)—Hawaiian Is.
subrubra *H. St. John & W.N. Takeuchi in Phytologia,* 65(3): 209 (1988)—Hawaiian Is.
tantalusensis *H. St. John & W.N. Takeuchi in Phytologia,* 65(3): 211 (1988)—Hawaiian Is.
triados *H. St. John & W.N. Takeuchi in Phytologia,* 65(3): 213 (1988)—Hawaiian Is.
triens *H. St. John & W.N. Takeuchi in Phytologia,* 65(3): 214 (1988)—Hawaiian Is.
trifoliata *H. St. John in Phytologia,* 65(3): 205 (1988)—Hawaiian Is.
trinalis *H. St. John & W.N. Takeuchi in Phytologia,* 65(3): 214 (1988)—Hawaiian Is.
trionanthe *H. St. John & W.N. Takeuchi in Phytologia,* 65(3): 213 (1988)—Hawaiian Is.
tripla *H. St. John in Phytologia,* 65(3): 206 (1988)—Hawaiian Is.
trisse *H. St. John in Phytologia,* 65(3): 206 (1988)—Hawaiian Is.
vallecularis *H. St. John in Phytologia,* 65(3): 204 (1988)—Hawaiian Is.
vespertina *H. St. John in Phytologia,* 65(3): 198 (1988), nom. nov.: C. occidentalis *St. John.*
villifera *H. St. John & W.N. Takeuchi in Phytologia,* 65(3): 213 (1988)—Hawaiian Is.
villosula *H. St. John in Phytologia,* 65(3): 198 (1988)—Hawaiian Is.
waikamoiensis *H. St. John in Phytologia,* 65(3): 198 (1988)—Hawaiian Is.
wailauensis *H. St. John in Phytologia,* 65(3): 198 (1988)—Hawaiian Is.
wailupeensis *H. St. John & W.N. Takeuchi in Phytologia,* 65(3): 213 (1988)—Hawaiian Is.

CYRTOCYMURA H. Robinson in Proc. Biol. Soc. Wash., 100(4): 849 (1987). *COMPOSITAE.*
cincta *(Griseb.) H. Robinson in Proc. Biol. Soc. Wash.,* 100(4): 851 (1987): Vernonia cincta.
harleyi *(H. Robinson) H. Robinson in Proc. Biol. Soc. Wash.,* 100(4): 852 (1987): Vernonia harleyi.
lanuginosa *(Gardn.) H. Robinson in Proc. Biol. Soc. Wash.,* 100(4): 852 (1987): Vernonia lanuginosa.
mattos-silvae *(H. Robinson) H. Robinson in Proc. Biol. Soc. Wash.,* 100(4): 852 (1987): Vernonia mattos-silvae.
saepia *(Ekman) H. Robinson in Proc. Biol. Soc. Wash.,* 100(4): 853 (1987): Vernonia saepia.
scorpioides *(Lamarck) H. Robinson in Proc. Biol. Soc. Wash.,* 100(4): 852 (1987): Conyza scorpioides.

CYRTORCHIS (Orchidac.).
praetermissa
subsp. zuluensis *(Harrison) H.P. Linder in Kew Bull.,* 44(2): 319 (1989): C. praetermissa var. zuluensis.

CYTISOPSIS (Legumin.).
pseudocytisus
subsp. reeseana *(Guyot) Lassen ex W. Greuter & T. Raus in Willdenowia,* 19(1): 31 (1989): C. dorycniifolia f. reeseana.

CYTISUS (Legumin.).
scoparius
subsp. bourgaei *(Boiss.) S. Rivas-Martínez, F. Fernández González & D. Sánchez-Mata in Opusc. Bot. Pharm. Complutensis,* 2: 107 (1986): Sarothamnus bourgaei.

D

DABOECIA (Ericac.).
cantabrica
subsp. azorica *(Tutin & E.F. Warburg) D. McClintock in Bot. J. Linn. Soc.,* 101(3): 280 (1989): D. azorica.

DACRYODES (Burserac.).
cuspidata *(Cuatrec.) D.C. Daly in Brittonia,* 41(1): 25 (1989): Trattinnickia cuspidata.

DACTYLIS (Gramin.).
glomerata
subsp. hylodes *(Parker) J. Holub in Folia Geobot. Phytotax.,* 23(4): 413 (1988): D. smithii subsp. hylodes.
spicata *(Pamp.) A. Guglielmo & C. Scalia in Boll. Accad. Gioenia Sci. Nat. Catania,* 18(325): 234 (1985): D. glomerata var. spicata.

DACTYLOCARDAMUM I.A. Al-Shehbaz in J. Arnold Arbor., 70(4): 519 (1989). *CRUCIFERAE.*
imbricatifolium *I.A. Al-Shehbaz in J. Arnold Arbor.,* 70(4): 519 (1989)—Peru.

DACTYLORHIZA (Orchidac.).
sect. Romanae *E. Klein & D. Strack in Phytochemistry,* 28(8): 2136 (1989).
bohemica *R. Businský in Preslia,* 61(4): 296 (1989)—Czechoslovakia.
fuchsii
var. alpina *(Landwehr) R.M. Bateman & I. Denholm in Watsonia,* 17(3): 341 (1989): D. fuchsii f. alpina.
var. hebridensis *(Wilmott) R.M. Bateman & I. Denholm in Watsonia,* 17(3): 340 (1989): Orchis hebridensis.
var. okellyi *(Druce) R.M. Bateman & I. Denholm in Watsonia,* 17(3): 342 (1989): Orchis maculata var. okellyi.
×jestrebiensis *R. Businský in Preslia,* 61(4): 306 (1989)—Czechoslovakia. (D. bohemica × majalis subsp. majalis).

DACTYLORHIZA :–
×koutsourana *B. & E. Willing in Ber. Arbeitskr. Heim. Orch.,* 6(1): 36 (1989)—Greece. (D. baumanniana × smolikana).
×prochazkana *R. Businský in Preslia,* 61(4): 305 (1989)—Czechoslovakia. (D. bohemica × maculata subsp. maculata).
×salictina *B. & E. Willing in Ber. Arbeitskr. Heim. Orch.,* 6(1): 35 (1989)—Greece. (D. pindica × smolikana).
smolikana *B. & E. Willing in Ber. Arbeitskr. Heim. Orch.,* 6(1): 33 (1989)—Greece.

DAEMONOROPS (Palmae).
polita *E.S. Fernando in Gard. Bull. Singapore,* 41(2): 56 (1988 publ. 1989)—Philippine Is.

DAENIKERANTHUS M.G. Baumann-Bodenheim, Syst. Fl. Neu-Caledonien, 5: 76 (1989), nom. nov., without replaced synonym ref.: Dracophyllum author not stated (Epacridac.).
alticolus *(Däniker) M.G. Baumann-Bodenheim, Syst. Fl. Neu-Caledonien,* 5: 76 (1989), generic name not validly publ.: Dracophyllum alticolum.

DAISWA (Trilliac.).
fargesii
var. brevipetalata *T.C. Huang & K.C. Yang in Taiwania,* 33: 123 (1988)—Taiwan, China.

DALBERGIA (Legumin.).
cultrata
f. fusca *(Pierre ex Prain) K. Thothathri, Taxon. Revis. Tribe Dalbergieae in Indian Subcont.:* 94 (1987): D. fusca.
horrida
var. concanensis *K. Thothathri, Taxon. Revis. Tribe Dalbergieae in Indian Subcont.:* 110 (1987)—India.
palo-escrito *J. Rzedowski & L.I. Guridi-Gómez in Acta Bot. Mex.,* 4: 2 (1988)—Mexico.

DALECHAMPIA (Euphorbiac.).
armbrusteri *G.L. Webster in Brittonia,* 41(1): 3 (1989)—Brazil.
attenuistylus *W.S. Armbruster in Brittonia,* 41(1): 44 (1989)—Venezuela.
canescens
subsp. friedrichsthalii *(Muell. Arg.) G.L. Webster & M.J. Huft in Ann. Missouri Bot. Gard.,* 75(3): 1109 (1988): D. friedrichsthalii.
fernandesii *G.L. Webster in Brittonia,* 41(1): 1 (1989)—Brazil.
megacarpa *W.S. Armbruster in Brittonia,* 41(1): 47 (1989)—Venezuela.
papillistigma *W.S. Armbruster in Brittonia,* 41(1): 49 (1989)—Venezuela.
trifoliata
var. trifida *A. Radcliffe-Smith in Kew Bull.,* 44(3): 450 (1989)—Somali Republic, Ethiopia, Kenya.
viridissima *G.L. Webster in Brittonia,* 41(1): 6 (1989)—Brazil.

DAPHNE (Thymelaeac.).
chingshuishaniana *S.S. Ying, Coloured Illustr. Fl. Taiwan,* 3: 531 (1988)—Taiwan.
jinzhaiensis *D.C. Zhang & J.Z. Shao in Bull. Bot. Res. North-East. Forest. Inst.,* 9(3): 37 (1989)—China.

DAPHNIPHYLLUM (Daphniphyllac.).
glaucescens
var. taiwanianum *S.S. Ying, Coloured Illustr. Fl. Taiwan,* 3: 124 (1988), nom. nov.: D. reticulatum *Keng.*

DATURA (Solanac.).
nanakii *S.C. Pandeya & A.B. Bhatt in Beitr. Biol. Pfl.,* 64(2): 266 (1989), as 'nanakai'—India.

DAVIDSEA T.R. Soderstrom & R.P. Ellis in Smithsonian Contrib. Bot., 72: 59 (1988). *GRAMINEAE.*
attenuata *(Thwaites) T.R. Soderstrom & R.P. Ellis in Smithsonian Contrib. Bot.,* 72: 59 (1988): Bambusa attenuata.

DEHERAINIA (Theophrastac.).
smaragdina
subsp. occidentalis *B. Ståhl in Nordic J. Bot.,* 9(1): 20 (1989)—Mexico.

DEIREGYNE (Orchidac.).
nelsonii *(Greenman) P. Burns-Balogh in Orquídea (México),* 11: 140 (1988): Spiranthes nelsonii.
subsp. obtusa *(Schltr.) P. Burns-Balogh in Orquídea (México),* 11: 143 (1988): Spiranthes obtusa.
subsp. oestlundii *P. Burns-Balogh in Orquídea (México),* 11: 143 (1988)—Mexico.
subsp. pulchra *(Schltr.) P. Burns-Balogh in Orquídea (México),* 11: 142, 212 (1988): Spiranthes pulchra.
riodelayensis *P. Burns-Balogh in Orquídea (México),* 11: 151 (1988)—Mexico.

×**DEISELARA** hort. in Orchid Rev., 97(1144): centre page pull-out p. 8 (1989). *ORCHIDACEAE.* (Laelia × Schomburgkia × Sophronitis).

DELPHINIUM (Ranunculac.).
albocoeruleum
var. latilobum *Y.Z. Zhao in Acta Sci. Nat. Univ. Intramongolicae,* 19(4): 676 (1988)—China.
fissum
subsp. albiflorum *(DC.) W. Greuter & Burdet in Willdenowia,* 19(1): 44 (1989): D. albiflorum.
subsp. pallasii *(Nevski) W. Greuter & Burdet in Willdenowia,* 19(1): 44 (1989): D. pallasii.

DELPHINIUM :–
grandiflorum
var. pilosum *Y.Z. Zhao in Acta Sci. Nat. Univ. Intramongolicae,* 19(4): 677 (1988)—China.
f. roseolum *Y.Z. Zhao in Acta Sci. Nat. Univ. Intramongolicae,* 19(4): 677 (1988)—China.
kaschgaricum *C.Y. Yang & B. Wang in Bull. Bot. Res. North-East. Forest. Inst.,* 9(1): 21 (1989)—Xinjiang.
kunlunshanicum *C.Y. Yang & B. Wang in Bull. Bot. Res. North-East. Forest. Inst.,* 9(1): 23 (1989)—Xinjiang.
lingbaoense *S.Y. Wang & Q.S. Yang in Acta Bot. Bor.-Occid. Sin.,* 9(1): 42 (1989)—China.
tarbagataicum *C.Y. Yang & B. Wang in Bull. Bot. Res. North-East. Forest. Inst.,* 9(1): 24 (1989)—Xinjiang.
yuchuanii *Y.Z. Zhao in Acta Sci. Nat. Univ. Intramongolicae,* 20(2): 248 (1989)—China.

DENDRANTHEMA (Compos.).
miyatojimense *(Kitamura) D.J.N. Hind in Kew Mag.,* 6(1): 48 (1989): Chrysanthemum miyatojimense.

DENDROBIUM (Orchidac.).
carnicarinum *P.J. Kores in Allertonia,* 5(1): 92 (1989)—Fiji.
diodon
subsp. kodayarensis *R. Gopalan & A.N. Henry in J. Econ. Taxon. Bot.,* 12(2): 487 (1988 publ. 1989)—India.
greenianum *P.J. Cribb & B. Lewis in Orchid Rev.,* 97(1150): 251 (1989)—Vanuatu.
khasianum *N.C. Deori in J. Orchid Soc. India,* 2(1–2): 73 (1988)—India.
minutiflorum *S.C. Chen & Z.H. Tsi in Bull. Bot. Res. North-East. Forest. Inst.,* 9(2): 27 (1989)—China.
nobile
var. alboluteum *D.D. Huyen & L.V. Aver'yanov in Bot. Zhurn.,* 74(7): 1039 (1989)—Vietnam.
tosaense
var. chingshuishanianum *S.S. Ying in J. Jap. Bot.,* 64(5): 151 (1989)—Taiwan.
trilobulatum *P.J. Kores in Allertonia,* 5(1): 91 (1989)—Fiji.
unicarinatum *P.J. Kores in Allertonia,* 5(1): 97 (1989)—Fiji.

DENDROCALAMUS (Gramin.).
sect. Patellares *C.J. Hsueh & D.Z. Li in J. Bamboo Res.,* 8(1): 36 (1989).
cinctus *R.B. Majumder ex T.R. Soderstrom & R.P. Ellis in Smithsonian Contrib. Bot.,* 72: 45 (1988)—Sri Lanka.
liboensis *C.J. Hsueh & D.Z. Li in J. Bamboo Res.,* 8(1): 37 (1989)—China.

DENDROCALAMUS :-
minor
var. amoenus *(Q.H. Dai & C.F. Huang) C.J. Hsueh & D.Z. Li in J. Bamboo Res.*, 8(1): 39 (1989): Sinocalamus minor var. amoenus.
pachystachyus *C.J. Hsueh & D.Z. Li in J. Bamboo Res.*, 8(1): 25 (1989)—China.
peculiaris *C.J. Hsueh & D.Z. Li in J. Bamboo Res.*, 8(1): 32 (1989)—China.
semiscandens *C.J. Hsueh & D.Z. Li in J. Bamboo Res.*, 8(1): 28 (1989)—China.
tomentosus *C.J. Hsueh & D.Z. Li in J. Bamboo Res.*, 8(1): 34 (1989)—China.

DENDROCHILUM (Orchidac.).
javieriense *L.K. Magrath, G. Bulmer & I. Shafer in Lindleyana*, 4(3): 135 (1989)—Philippine Is.

DENDROPANAX (Araliac.).
bracteatus *M.J. & J.F.M. Cannon in Bull. Brit. Mus. (Nat. Hist.), Bot.*, 19: 8 (1989)—Panama.
capillaris *M.J. & J.F.M. Cannon in Bull. Brit. Mus. (Nat. Hist.), Bot.*, 19: 20 (1989)—Panama.
globosus *M.J. & J.F.M. Cannon in Bull. Brit. Mus. (Nat. Hist.), Bot.*, 19: 14 (1989)—Costa Rica, Panama.
hondurensis *M.J. & J.F.M. Cannon in Bull. Brit. Mus. (Nat. Hist.), Bot.*, 19: 14 (1989)—Honduras, Mexico.
latilobus *M.J. & J.F.M. Cannon in Bull. Brit. Mus. (Nat. Hist.), Bot.*, 19: 8 (1989)—Panama, Costa Rica.
maritimus *M.J. & J.F.M. Cannon in Bull. Brit. Mus. (Nat. Hist.), Bot.*, 19: 17 (1989)—Panama.
pallidus *M.J. & J.F.M. Cannon in Bull. Brit. Mus. (Nat. Hist.), Bot.*, 19: 20 (1989)—Guatemala, Mexico.
panamensis *M.J. & J.F.M. Cannon in Bull. Brit. Mus. (Nat. Hist.), Bot.*, 19: 17 (1989)—Panama.
punctatus *M.J. & J.F.M. Cannon in Bull. Brit. Mus. (Nat. Hist.), Bot.*, 19: 11 (1989)—Panama.
ravenii *M.J. & J.F.M. Cannon in Bull. Brit. Mus. (Nat. Hist.), Bot.*, 19: 11 (1989)—Costa Rica.

DENDROSIDA (Malvac.).
cuatrecasasii *J. Fuertes in Rev. Acad. Colomb. Cienc. Exact. Fis. Nat.*, 65: 376 (1989)—Colombia.

DESCHAMPSIA (Gramin.).
hispanica
subsp. gredensis *(Vivant) S. Rivas-Martínez, F. Fernández González & D. Sánchez-Mata in Opusc. Bot. Pharm. Complutensis*, 2: 108 (1986): D. caespitosa subsp. gredensis.

DESFONTAINIA (Loganiac.).
plowmanii *R.E. Schultes in Rev. Acad. Colomb. Cienc. Exact. Fis. Nat.*, 65: 316 (1989)—Colombia.
spinosa
var. acutangula *(Dun.) R.E. Schultes in Rev. Acad. Colomb. Cienc. Exact. Fis. Nat.*, 65: 318 (1989): D. acutangula.

DESMODIUM (Legumin.).
cordifoliolatum *P.C. Li in Acta Bot. Yunnanica*, 11(3): 301 (1989)—China.
pendenticarpum *C.Z. Gao & Q.R. Lai in Bull. Bot. Res. North-East. Forest. Inst.*, 9(4): 41 (1989), with two types cited—China.

DESMONCUS (Palmae).
cirrhifera *A.H. Gentry & Zardini in Ann. Missouri Bot. Gard.*, 75(4): 1436 (1988 publ. 1989)—Colombia.

DEUTEROCOHNIA (Bromeliac.).
brevispicata *W. Rauh & L. Hromadnik in Abh. Akad. Wiss. Lit. Mainz, Math.-Naturwiss. Klasse, Trop. Subtrop. Pflanzenwelt*, 65: 5 (1988)—Bolivia.

DIANELLA (Phormiac.).
adenanthera *(G. Forster) R.J.F. Henderson in Austrobaileya*, 2(5): 421 (1988): Anthericum adenanthera.

DIANTHOVEUS B.E. Hammel & G.J. Wilder in Ann. Missouri Bot. Gard., 76(1): 113 (1989). *CYCLANTHACEAE.*
cremnophilus *B.E. Hammel & G.J. Wilder in Ann. Missouri Bot. Gard.*, 76(1): 113 (1989)—Ecuador, Colombia.

DIANTHUS (Caryophyllac.).
langeanus
subsp. gredensis *(Pau ex Caballero) S. Rivas-Martínez, F. Fernández González & D. Sánchez-Mata in Opusc. Bot. Pharm. Complutensis*, 2: 108 (1986): D. gredensis.
repens
var. scabripilosus *Y.Z. Zhao in Acta Sci. Nat. Univ. Intramongolicae*, 20(1): 110 (1989)—China.

DIAPHANANTHE (Orchidac.).
caffra *(H. Bol.) H.P. Linder in Kew Bull.*, 44(2): 318 (1989): Angraecum caffrum.
millarii *(H. Bol.) H.P. Linder in Kew Bull.*, 44(2): 318 (1989): Mystacidium millarii.

DIASCIA (Scrophulariac.).
austromontana *K.E. Steiner in S. Afr. J. Bot.*, 55(2): 250 (1989)—South Africa (Natal), Lesotho.

DICERANDRA (Labiat.).
christmanii *R.B. Huck & W.S. Judd in Syst. Bot.*, 14(2): 198 (1989)—U.S.A. (Florida).

DICHANTHELIUM (Gramin.).
ensifolium
var. breve *(A. Hitchcock & Chase) B.F. Hansen & R.P. Wunderlin in Ann. Missouri Bot. Gard.,* 75(4): 1646 (1988 publ. 1989): Panicum breve.
var. unciphyllum *(Trinius) B.F. Hansen & R.P. Wunderlin in Ann. Missouri Bot. Gard.,* 75(4): 1647 (1988 publ. 1989): Panicum unciphyllum.
portoricense *(Desvaux ex Hamilton) B.F. Hansen & R.P. Wunderlin in Ann. Missouri Bot. Gard.,* 75(4): 1649 (1988 publ. 1989): Panicum portoricense.

DICHANTHIUM (Gramin.).
sericeum
subsp. humilius *(J. Black) B.K. Simon in Austrobaileya,* 3(1): 83 (1989): D. humilius.
subsp. polystachyum *(Benth.) B.K. Simon in Austrobaileya,* 3(1): 84 (1989): Andropogon sericeus var. polystachyus.

DICHAPETALUM (Dichapetalac.).
bernalii *G.T. Prance in Brittonia,* 40(4): 441 (1988)—Colombia.
mexicanum *G.T. Prance in Brittonia,* 40(4): 442 (1988)—Mexico.

DICTYOPHLEBA (Apocynac.).
setosa *B. de Hoogh in Bull. Jard. Bot. Nation. Belg.,* 59(1–2): 220 (1989)—Cameroun.

DIDISSANDRA (Gesneriac.).
porphyrantha *A. Weber & R. Kiew in Gard. Bull. Singapore,* 41(1): 1 (1988)—Malaya.

DIDYMOCARPUS (Gesneriac.).
breviflorus *(Ridl.) A. Weber & R. Kiew in Gard. Bull. Singapore,* 41(1): 7 (1988): Didissandra breviflora.
geitleri *A. Weber in Pl. Syst. Evol.,* 165(1–2): 95 (1989)—Malaya.
nitidus *R. Kiew & A. Weber in Gard. Bull. Singapore,* 41(1): 4 (1988)—Malaya.

DIEFFENBACHIA (Arac.).
duidae *(Steyerm.) G.S. Bunting in Phytologia,* 65(5): 390 (1988): Spathicarpa duidae.

DIELITZIA P.S. Short in Muelleria, 7(1): 103 (1989). *COMPOSITAE.*
tysonii *P.S. Short in Muelleria,* 7(1): 105 (1989)—Australia (W. Australia).

DIGITALIS (Scrophulariac.).
mariana
subsp. heywoodii *(Pinto da Silva & Silva) P.A. Hinz in Candollea,* 44(1): 168 (1989): D. purpurea subsp. heywoodii.
purpurea
subsp. carpetana *(Rivas Mateos) S. Rivas-Martínez, F. Fernández González & D. Sánchez-Mata in Opusc. Bot. Pharm. Complutensis,* 2: 108 (1986): D. purpurea var. carpetana.

DIGITALIS :-
subsp. nevadensis *(Kunze) S. Rivas-Martínez, F. Fernández González & D. Sánchez-Mata in Opusc. Bot. Pharm. Complutensis,* 2: 108 (1986): D. nevadensis.

DIGITARIA (Gramin.).
brunoana *F.M. Raimondo in Giorn. Bot. Ital.,* 122(5–6): 252 (1988 publ. 1989)—Somali Republic.
neghellensis *J.P. Lebrun in Bull. Mus. Nation. Hist. Nat., B, Adansonia,* Ser. 4, 10(3): 271 (1988 publ. 1989)—Ethiopia.

DILOMILIS (Orchidac.).
filicina *D.D. Dod in Moscosoa,* 5: 228 (1989)—Hispaniola (Haiti).

DIMORPHORCHIS (Orchidac.).
rossii *J.A. Fowlie in Orchid Dig.,* 53(1): 14 (1989)—Borneo (Sabah).

DINOCHLOA (Gramin.).
darvelana *S. Dransfield in Kew Bull.,* 44(3): 435 (1989)—Borneo (Sabah, Kalimantan).

DIONYSIA (Primulac.).
subsect. Afghanicae *C. Grey-Wilson, Genus Dionysia:* 168 (1989).
caespitosa
subsp. bolivarii *(Pau) C. Grey-Wilson, Genus Dionysia:* 99 (1989): D. bolivarii.
sarvestanica *Z. Jamzad & C. Grey-Wilson in Kew Bull.,* 44(1): 124 (1989)—Iran.

DIOSCOREA (Dioscoreac.).
abysmophila *B. Maguire & J.A. Steyermark in Mem. New York Bot. Gard.,* 51: 105 (1989)—Venezuela.
berenicea *R. McVaugh, Fl. Novo-Galiciana,* 15: 362 (1989)—Mexico.
chamela *R. McVaugh, Fl. Novo-Galiciana,* 15: 364 (1989)—Mexico.
macvaughii *B.G. Schubert in R. McVaugh, Fl. Novo-Galiciana,* 15: 369 (1989), as 'mcvaughii'—Mexico.
neblinensis *B. Maguire & J.A. Steyermark in Mem. New York Bot. Gard.,* 51: 108 (1989)—Venezuela.
oreodoxa *B.G. Schubert in R. McVaugh, Fl. Novo-Galiciana,* 15: 378 (1989)—Mexico.
sessiliflora *R. McVaugh, Fl. Novo-Galiciana,* 15: 384 (1989)—Mexico.

DIOSMA (Rutac.).
aristata *I. Williams in J. S. Afr. Bot.,* 41(3): 180 (1975), in Supp. 16 as 'aristala' in error—South Africa (Cape Province).†

DIOSPYROS (Ebenac.).
abyssinica
subsp. attenuata *F. White in Bull. Jard. Bot. Nation. Belg.,* 58(3–4): 381 (1988)—Zimbabwe, Malawi.
subsp. reticulata *F. White in Bull. Jard. Bot. Nation. Belg.,* 58(3–4): 383 (1988)—Tanzania.

DIOSPYROS :–
capricornuta *F. White in Bull. Jard. Bot. Nation. Belg.*, 58(3–4): 405 (1988)—Tanzania.
greenwayi *F. White in Bull. Jard. Bot. Nation. Belg.*, 58(3–4): 392 (1988)—Kenya, Tanzania.
kabuyeana *F. White in Bull. Jard. Bot. Nation. Belg.*, 58(3–4): 401 (1988)—Tanzania, Kenya.
kanurii *F. White in Bull. Jard. Bot. Nation. Belg.*, 58(3–4): 364 (1988)—Kenya.
magogoana *F. White in Bull. Jard. Bot. Nation. Belg.*, 58(3–4): 398 (1988)—Tanzania.
occulta *F. White in Bull. Jard. Bot. Nation. Belg.*, 58(3–4): 399 (1988)—Tanzania.
shimbaensis *F. White in Bull. Jard. Bot. Nation. Belg.*, 58(3–4): 385 (1988)—Kenya.
wajirensis *F. White in Bull. Jard. Bot. Nation. Belg.*, 58(3–4): 359 (1988)—Kenya, Somali Republic.

DIPLOSPORA (Rubiac.).
parkinsonii *M. Gangopadhyay & T. Chakrabarty in J. Econ. Taxon. Bot.*, 12(2): 497 (1988 publ. 1989), nom. nov.: Pubistylis andamanensis *Thoth.*

DIPLUSODON (Lythrac.).
adpressipilus *A. Lourteig in Bradea*, 5(19): 236 (1989)—Brazil.
appendiculosus *A. Lourteig in Bradea*, 5(19): 219 (1989)—Brazil.
argenteus *A. Lourteig in Bradea*, 5(19): 221 (1989)—Brazil.
cordifolius *A. Lourteig in Bradea*, 5(19): 222 (1989)—Brazil.
ericoides *A. Lourteig in Bradea*, 5(19): 213 (1989)—Brazil.
fastigiatus *A. Lourteig in Bradea*, 5(19): 236 (1989)—Brazil.
hatschbachii *A. Lourteig in Bradea*, 5(19): 226 (1989)—Brazil.
heringeri *A. Lourteig in Bradea*, 5(19): 239 (1989)—Brazil.
irwinii *A. Lourteig in Bradea*, 5(19): 231 (1989)—Brazil.
leucocalycinus *A. Lourteig in Bradea*, 5(19): 232 (1989)—Brazil.
minasensis *A. Lourteig in Bradea*, 5(19): 229 (1989)—Brazil.
paraisoensis *A. Lourteig in Bradea*, 5(19): 240 (1989)—Brazil.
psammophilus *A. Lourteig in Bradea*, 5(19): 212 (1989)—Brazil.
saxatilis *A. Lourteig in Bradea*, 5(19): 218 (1989)—Brazil.
sigillatus *A. Lourteig in Bradea*, 5(19): 237 (1989)—Brazil.
urceolatus *A. Lourteig in Bradea*, 5(19): 233 (1989)—Brazil.
vidalii *A. Lourteig in Bradea*, 5(19): 216 (1989)—Brazil.
xerampelinus *A. Lourteig in Bradea*, 5(19): 230 (1989)—Brazil.

DIPSACUS (Dipsacac.).
lushanensis *C.Y. Cheng & T.M. Ai in J. Wuhan Bot. Res.*, 7(1): 27 (1989)—China.

DISOCACTUS (Cactac.).
macranthus
var. glaucocladus *E. Ewald in Epiphytes*, 13(52): 107 (1989)—Mexico.

DISPERIS (Orchidac.).
purpurata
subsp. pallescens *P. Bruyns in S. Afr. J. Bot.*, 55(5): 494 (1989)—South Africa (Cape Province).

DISPOROPSIS (Convallariac.).
taiwanensis *S.S. Ying in J. Jap. Bot.*, 64(5): 153 (1989)—Taiwan.

DISPORUM (Convallariac.).
×hishiyamanum *K. Suzuki in J. Phytogeogr. Taxon.*, 36(1): 13 (1988)—Japan. (D. sessile × smilacinum).
ovale
var. tamnanum *Y.N. Lee & N.S. Lee in Korean J. Pl. Taxon.*, 14(3): 168 (1984)—Korea.
taiwanense *S.S. Ying in J. Jap. Bot.*, 64(5): 151 (1989)—Taiwan.

DITASSA (Asclepiadac.).
auyantepuiensis *(Stey.) G. Morillo in Ernstia*, 51: 7 (1989): Cynanchum auyantepuiense.
carnevalii *(Morillo) G. Morillo in Ernstia*, 51: 7 (1989): Cynanchum carnevalii.
ciliata *(Moldenke) G. Morillo in Ernstia*, 51: 7 (1989): Blepharodon ciliatum.
colellae *(Morillo) G. Morillo in Ernstia*, 51: 7 (1989): Cynanchum colellae.
cordata *(Turcz.) J. Fontella Pereira in Eugeniana*, 16: 24 (1989): Metastelma cordatum.
var. abortiva *(Fourn.) J. Fontella Pereira in Eugeniana*, 16: 26 (1989): D. abortiva.
var. virgata *(Fourn.) J. Fontella Pereira in Eugeniana*, 16: 25 (1989): D. virgata.
foldatsii *G. Morillo in Ernstia*, 51: 7 (1989)—Venezuela.
julianii *(Morillo) G. Morillo in Ernstia*, 51: 9 (1989): Cynanchum julianii.
liesneri *G. Morillo in Ernstia*, 51: 9 (1989)—Venezuela.
lourteigiae *J. Fontella Pereira in Eugeniana*, 16: 23 (1989), as 'lourteigii'—Brazil.
multinervia *(Morillo) G. Morillo in Ernstia*, 51: 9 (1989): Cynanchum multinervium.
obovata *(Holm) G. Morillo in Ernstia*, 51: 9 (1989): Cynanchum obovatum.
oliva-estevae *G. Morillo in Ernstia*, 51: 9 (1989)—Venezuela.
peruviana *G. Morillo in Ernstia*, 51: 11 (1989)—Peru.
sipapoana *G. Morillo in Ernstia*, 51: 11 (1989)—Venezuela.
venamensis *G. Morillo in Ernstia*, 51: 12 (1989)—Venezuela.

DITHYROSTEGIA (Compos.).
gracilis *P.S. Short in Muelleria,* 7(1): 106 (1989)—Australia (W. Australia).

DODECAHEMA J.L. Reveal & C.B. Hardham in Phytologia, 66(2): 86 (1989). *POLYGONACEAE.*
leptoceras *(A. Gray) J.L. Reveal & C.B. Hardham in Phytologia,* 66(2): 87 (1989): Centrostegia leptoceras.

DOGA (Legumin.).
pancheri *(Baill.) M.G. Baumann-Bodenheim, Syst. Fl. Neu-Caledonien,* 5: 77 (1989): Storckiella pancheri.

DOLICHORRHIZA (Compos.).
persica *(Boiss.) B. Nordenstam in Fl. Iran.,* 164: 52 (1989): Senecio persicus.

DORSTENIA (Morac.).
cuspidata
var. brinkmaniana *M.E.E. Hijman in Fl. Trop. E. Afr., Morac.:* 40 (1989)—Tanzani, Kenya, Zambia.
hildebrandtii
var. schlechteri *(Engl.) M.E.E. Hijman in Fl. Trop. E. Afr., Morac.:* 34 (1989): D. schlechteri.
holstii
var. longestipulata *M.E.E. Hijman in Fl. Trop. E. Afr., Morac.:* 30 (1989)—Tanzania.
tayloriana
var. laikipiensis *(Rendle) M.E.E. Hijman in Fl. Trop. E. Afr., Morac.:* 26 (1989): D. laikipiensis.
thikaensis *M.E.E. Hijman in Fl. Trop. E. Afr.:* 27 (1989)—Kenya.

DRABA (Crucif.).
argentifolia *I.A. Al-Shehbaz in J. Arnold Arbor.,* 70(3): 427 (1989), nom. nov.: D. argentea *O.E. Schulz.*
boyacana *I.A. Al-Shehbaz in J. Arnold Arbor.,* 70(3): 427 (1989)—Colombia.
cocuyana *I.A. Al-Shehbaz in J. Arnold Arbor.,* 70(3): 429 (1989)—Colombia.
cocuyensis *E. Santana C. & J.O. Rangel Ch. in Rev. Acad. Colomb. Cienc. Exact. Fis. Nat.,* 65: 351 (1989)—Colombia.
cuatrecasana *J.O. Rangel Ch. & E. Santana C. in Rev. Acad. Colomb. Cienc. Exact. Fis. Nat.,* 65: 348 (1989)—Colombia.
ecuadoriana *I.A. Al-Shehbaz in J. Arnold Arbor.,* 70(3): 431 (1989), nom. nov.: D. affinis *J.D. Hooker.*
pseudocheiranthoides *I.A. Al-Shehbaz in J. Arnold Arbor.,* 70(3): 431 (1989)—Colombia.
ritacuvana *I.A. Al-Shehbaz in J. Arnold Arbor.,* 70(3): 435 (1989)—Colombia.
rositae *E. Santana C. & J.O. Rangel Ch. in Rev. Acad. Colomb. Cienc. Exact. Fis. Nat.,* 65: 348 (1989)—Colombia.

DRABA :–
sericea *E. Santana C. & J.O. Rangel Ch. in Rev. Acad. Colomb. Cienc. Exact. Fis. Nat.,* 65: 353 (1989)—Colombia.

DRACOCEPHALUM (Labiat.).
nuratavicum *T.A. Adylov in Opred. Rast. Sred. Azii,* 9: 361 (1987)—U.S.S.R. (Middle Asia).

DRACULA (Orchidac.).
subgen. Sodiroa *(Luer) C.A. Luer in Amer. Orchid Soc. Bull.,* 58(10): 1000 (1989): D. sect. Sodiroa.
subgen. Xenosia *C.A. Luer in Amer. Orchid Soc. Bull.,* 58(10): 1001 (1989).
aphrodes *C.A. Luer & Escobar in Amer. Orchid Soc. Bull.,* 58(10): 1002 (1989)—Colombia.
carcinopsis *C.A. Luer & Escobar in Amer. Orchid Soc. Bull.,* 58(10): 1003 (1989)—Colombia.
dalessandroi *C.A. Luer in Amer. Orchid Soc. Bull.,* 58(10): 1004 (1989)—Ecuador.
exasperata *C.A. Luer & Escobar in Amer. Orchid Soc. Bull.,* 58(10): 1005 (1989)—Colombia.
gastrophora *C.A. Luer & Hirtz in Amer. Orchid Soc. Bull.,* 58(10): 1005 (1989)—Ecuador.
incognita *C.A. Luer & Escobar in Amer. Orchid Soc. Bull.,* 58(10): 1006 (1989)—Colombia.
insolita *C.A. Luer & Escobar in Amer. Orchid Soc. Bull.,* 58(10): 1007 (1989)—Colombia.
ligiae *C.A. Luer & Escobar in Amer. Orchid Soc. Bull.,* 58(10): 1008 (1989)—Colombia.
nosferatu *C.A. Luer & Escobar in Amer. Orchid Soc. Bull.,* 58(10): 1009 (1989)—Colombia.
verticulosa *C.A. Luer & Escobar in Amer. Orchid Soc. Bull.,* 58(10): 1010 (1989)—Colombia.
xenos *C.A. Luer & Escobar in Amer. Orchid Soc. Bull.,* 58(10): 1011 (1989)—Colombia.

DRESSLERELLA (Orchidac.).
subgen. Pilositas *C.A. Luer, Ic. Pleurothallid. V (Monogr. Syst. Bot. Missouri Bot. Gard. 26):* 3 (1988).

DRESSLEROTHAMNUS (Compos.).
gentryi *H. Robinson in Syst. Bot.,* 14(3): 384 (1989)—Colombia.
peperomioides *H. Robinson in Syst. Bot.,* 14(3): 386 (1989)—Panama.

DROSERA (Droserac.)
×hybrida *J.M. Macfarlane in Trans. & Proc. Bot. Soc. Pennsylvania,* 1: 90, 94, 95, 99, pl 12 (1899)—U.S.A. (New Jersey). (D. filiformis × intermedia).

DRYMOCALLIS (Rosac.).
calycina *(Boiss. & Bal.) J. Soják in Čas. Nár. Muz. (Prague),* 154(3–4): 118 (1985 publ. 1989): Potentilla calycina.
geoides *(M.-Bieb.) J. Soják in Čas. Nár. Muz. (Prague),* 154(3–4): 118 (1985 publ. 1989): Potentilla geoides.

DRYMOCALLIS :–
inquinans *(Turcz.) J. Soják in Čas. Nár. Muz. (Prague),* 154(3–4): 118 (1985 publ. 1989): Potentilla inquinans.
jailae *(Juz.) J. Soják in Čas. Nár. Muz. (Prague),* 154(3–4): 118 (1985 publ. 1989): Potentilla jailae.
kulabensis *(Wolf) J. Soják in Čas. Nár. Muz. (Prague),* 154(3–5): 118 (1985 publ. 1989): Potentilla kulabensis.
poteriifolia *(Boiss.) J. Soják in Čas. Nár. Muz. (Prague),* 154(3–4): 118 (1985 publ. 1989): Potentilla poteriifolia.
regis-borisii *(Stoj.) J. Soják in Čas. Nár. Muz. (Prague),* 154(3–4): 118 (1985 publ. 1989): Potentilla regis-borisii.
rupestris *(L.) J. Soják in Čas. Nár. Muz. (Prague),* 154(3–4): 118 (1985 publ. 1989): Potentilla rupestris.
subsp. corsica *(Sol. ex Lehm.) J. Soják in Čas. Nár. Muz. (Prague),* 154(3–4): 118 (1985 publ. 1989): Potentilla corsica.
tianschanica *(Wolf) J. Soják in Čas. Nár. Muz. (Prague),* 154(3–4): 118 (1985 publ. 1989): Potentilla tianschanica.

DRYPETES (Euphorbiac.).
balakrishnanii *T. Chakrabarty & M. Gangopadhyay in J. Econ. Taxon. Bot.,* 12(2): 393 (1988 publ. 1989)—Burma, Bangladesh.

DUPERREA (Rubiac.).
pavettaefolia
f. mollissisima *H. Chu in Acta Sci. Nat. Univ. Sunyatseni,* 28(2): 66 (1989)—China.

DURANDEA (Linac.).
neo-caledonica *(Vieill.) M.G. Baumann-Bodenheim, Syst. Fl. Neu-Caledonien,* 4: 103 (1988 publ. Jan. 1989), without basionym ref.; *et Syst. Fl. Neu-Caledonien,* 5: 76, 98 (May 1989), with basionym ref.: Penicillanthemum neo-caledonicum.

DUVALIA (Asclepiadac.).
immaculata *(Luckhoff) Bayer ex L.C. Leach in S. Afr. J. Bot.,* 55(2): 268 (1989): D. maculata var. immaculata.
vestita *U. Meve in Kakt. And. Sukk.,* 39(9): 197 (1988)—South Africa (Cape Province).

DYCKIA (Bromeliac.).
brachystachya *W. Rauh & E. Gross in Abh. Akad. Wiss. Lit. Mainz, Math.-Naturwiss. Klasse, Trop. Subtrop. Pflanzenwelt,* 65: 8 (1988)—Brazil.
granmogulensis *W. Rauh in Abh. Akad. Wiss. Lit. Mainz, Math.-Naturwiss. Klasse, Trop. Subtrop. Pflanzenwelt,* 65: 11 (1988)—Brazil.
lindevaldae *W. Rauh in Abh. Akad. Wiss. Lit. Mainz, Math.-Naturwiss. Klasse, Trop. Subtrop. Pflanzenwelt,* 65: 12 (1988)—Brazil.

DYCKIA :–
oligantha
var. cristallina *W. Rauh in Abh. Akad. Wiss. Lit. Mainz, Math.-Naturwiss. Klasse, Trop. Subtrop. Pflanzenwelt,* 65: 17 (1988)—Brazil.
polycladus *L.B. Smith in J. Bromeliad Soc.,* 39(5): 206 (1989)—Brazil.

DYSOXYLUM (Meliac.).
rufescens
subsp. dzumacense *(Guillaumin) D.J. Mabberley in Fl. N. Caléd. & Dépend.,* 15: 39 (1988): D. dzumacense.

DYSTOVOMITA (Guttif.).
paniculata *(J.D. Smith) B. Hammel in Ann. Missouri Bot. Gard.,* 76(3): 927 (1989): Rheedia paniculata.

E

ECHEANDIA (Anthericac.).
bolivarensis *R.W. Cruden in Ann. Missouri Bot. Gard.,* 76(1): 350 (1989)—Venezuela.

ECHIDNOPSIS (Asclepiadac.).
sect. Glabra *P.V. Bruyns in Bradleya,* 6: 8 (1988).
sect. Profundicorona *P.V. Bruyns in Bradleya,* 6: 31 (1988).
sect. Vadosicorona *P.V. Bruyns in Bradleya,* 6: 14 (1988), as 'Vadosigorona'.
malum *(Lavranos) P.V. Bruyns in Bradleya,* 6: 43 (1988): Pseudopectinaria malum.
scutellata
subsp. australis *P.V. Bruyns in Bradleya,* 6: 19 (1988)—Kenya.
subsp. dhofarensis *P.V. Bruyns in Bradleya,* 6: 18 (1988)—Oman.
subsp. planiflora *(Bally) P.V. Bruyns in Bradleya,* 6: 19 (1988): E. planiflora.
sharpei
subsp. ciliata *(Bally) P.V. Bruyns in Bradleya,* 6: 38 (1988): E. ciliata.
subsp. repens *(R.A. Dyer & Verdoorn) P.V. Bruyns in Bradleya,* 6: 37 (1988): E. repens.

ECHINOCEREUS (Cactac.).
cinerascens
var. septentrionalis *N.P. Taylor in Bradleya,* 6: 68 (1988)—Mexico.
var. tulensis *(H. Bravo-H.) N.P. Taylor in Bradleya,* 6: 69 (1988): E. tulensis.
coccineus
var. arizonicus *(Rose ex Orcutt) D.J. Ferguson in Cact. Succ. J. (U.S.A.),* 61(5): 221 (1989): E. arizonicus.

ECHINOCEREUS :–
var. paucispinus *(Engelm.) D.J. Ferguson in Cact. Succ. J. (U.S.A.),* 61(5): 222 (1989): Cereus paucispinus.
longisetus
var. delaetii *(M. Gürke) N.P. Taylor in Bradleya,* 6: 79 (1988): Cephalocereus delaetii.
mombergerianus *G.R.W. Frank in Kakt. And. Sukk.,* 40(11): 274 (1989)—Mexico.
parkeri *N.P. Taylor in Bradleya,* 6: 73 (1988)—Mexico.
var. gonzalezii *N.P. Taylor in Bradleya,* 6: 74 (1988)—Mexico.
polyacanthus
var. huitcholensis *(F.A.C. Weber) N.P. Taylor in Bradleya,* 6: 82 (1988): Cereus huitcholensis.
rayonesensis *N.P. Taylor in Bradleya,* 6: 75 (1988)—Mexico.
scheeri
var. obscuriensis *A.B. Lau in Kakt. And. Sukk.,* 40(2): 36 (1989)—Mexico.
stramineus
var. occidentalis *N.P. Taylor in Bradleya,* 6: 70 (1988)—Mexico.

ECHINOPS (Compos.).
bampsianus *S. Lisowski in Bull. Jard. Bot. Nation. Belg.,* 59(1–2): 255 (1989)—Zaïre.

ECHINOPSIS (Cactac.).
aurea
var. sierragrandensis *(Rausch) J. Lambert in Succulenta,* 67(9): 182 (1988), without basionym page: Lobivia aurea var. sierragrandensis.
fallax
var. albiflora *(Rausch) J. Lambert in Succulenta,* 67(9): 182 (1988): Lobivia aurea var. albiflora.
var. callochrysea *(Ritter) J. Lambert in Succulenta,* 67(9): 182 (1988): Hymenorebutia aurea var. callochrysea.
var. catamarcensis *(Ritter) J. Lambert in Succulenta,* 67(9): 182 (1988): Hymenorebutia aurea var. catamarcensis.
var. depressicostata *(Ritter) J. Lambert in Succulenta,* 67(9): 182 (1988): Hymenorebutia aurea var. depressicostata.
var. shaferi *(Britton & Rose) J. Lambert in Succulenta,* 67(9): 182 (1988): Lobivia shaferi.
var. tortuosa *(Rausch) J. Lambert in Succulenta,* 67(9): 182 (1988), without basionym page: Lobivia aurea var. tortuosa.

EDMONDIA (Compos.).
fasciculata *(Andr.) O.M. Hilliard in Fl. S. Afr.,* 33(7:2): 316 (1983): Xeranthemum fasciculatum.

EGGELINGIA (Orchidac.).
gabonensis *P.J. Cribb & F.M. Van der Laan in Kew Bull.,* 44(3): 479 (1989)—Gabon, Cameroun.

EHRHARTA (Gramin.).
ramosa
subsp. aphylla *(Schrad.) G.E. Gibbs Russell in Bothalia,* 19(2): 191 (1989): E. aphylla.
rehmannii
subsp. filiformis *(Stapf) G.E. Gibbs Russell in Bothalia,* 19(2): 193 (1989): E. rehmannii var. filiformis.
subsp. subspicata *(Stapf) G.E. Gibbs Russell in Bothalia,* 19(2): 194 (1989): E. subspicata.

EIRMOCEPHALA H. Robinson in Proc. Biol. Soc. Wash., 100(4): 853 (1987). *COMPOSITAE.*
brachiata *(Bentham ex Oersted) H. Robinson in Proc. Biol. Soc. Wash.,* 100(4): 854 (1987): Vernonia brachiata.
cainarachiensis *(Hieron.) H. Robinson in Proc. Biol. Soc. Wash.,* 100(4): 854 (1987): Vernonia cainarachiensis.
megaphylla *(Hieron.) H. Robinson in Proc. Biol. Soc. Wash.,* 100(4): 854 (1987): Vernonia megaphylla.

ELAEAGNUS (Elaeagnac.).
macrophylla
var. kotoensis *(Hayata) S.S. Ying, Colour. Illustr. Pl. Taiwan, Pt. 3; Colour. Illustr. Ligneous Pl. Taiwan,* 1: 455 (1987): E. kotoensis.

ELAEOCARPUS (Elaeocarpac.).
floribundus
var. tahanensis *(Hend.) F.S.P. Ng in Tree Fl. Malaya,* 4: 88 (1989): E. tahanensis.
nitidus
var. salicifolius *(King) F.S.P. Ng in Tree Fl. Malaya,* 4: 91 (1989): E. salicifolius.
var. velutinus *F.S.P. Ng in Gard. Bull. Singapore,* 41(2): 44 (1988 publ. 1989)—Malaya.
var. wrayi *(King) F.S.P. Ng in Tree Fl. Malaya,* 4: 91 (1989): E. wrayi.
sallehiana *F.S.P. Ng in Gard. Bull. Singapore,* 41(2): 43 (1988 publ. 1989)—Malaya.
symingtonii *F.S.P. Ng in Gard. Bull. Singapore,* 41(2): 43 (1988 publ. 1989)—Malaya.

ELAPHANTHERA N. Hallé in Fl. N. Caléd. & Dépend., 15: 134 (1988). *SANTALACEAE.*
baumannii *(Stauffer) N. Hallé in Fl. N. Caléd. & Dépend.,* 15: 136 (1988): Exocarpos baumannii.

ELATOSTEMA (Urticac.).
ser. Auriculata *W.T. Wang in Bull. Bot. Res. North-East. Forest. Inst.,* 9(2): 72 (1989).
acutitepalum *W.T. Wang in Bull. Bot. Res. North-East. Forest. Inst.,* 9(2): 74 (1989)—China.
biglomeratum *W.T. Wang in Bull. Bot. Res. North-East. Forest. Inst.,* 9(2): 67 (1989)—China.

ELATOSTEMA :-
dulongense *W.T. Wang in Bull. Bot. Res. North-East. Forest. Inst.*, 9(2): 72 (1989)—China.
ferrugineum *W.T. Wang in Bull. Bot. Res. North-East. Forest. Inst.*, 9(2): 73 (1989)—China.
mabienense
var. sexbracteatum *W.T. Wang in Bull. Bot. Res. North-East. Forest. Inst.*, 9(2): 69 (1989)—China.
salvinioides
var. robustum *W.T. Wang in Bull. Bot. Res. North-East. Forest. Inst.*, 9(2): 69 (1989)—China.
sinense
var. trilobatum *W.T. Wang in Bull. Bot. Res. North-East. Forest. Inst.*, 9(2): 69 (1989)—China.
tenuicaudatum
var. lasiocladum *W.T. Wang in Bull. Bot. Res. North-East. Forest. Inst.*, 9(2): 67 (1989)—China.
yui *W.T. Wang in Bull. Bot. Res. North-East. Forest. Inst.*, 9(2): 70 (1989)—China.

ELISANTHE (Caryophyllac.).
quadriloba *(Turcz. ex Kar. & Kir.) S.S. Ikonnikov in Novosti Sist. Vÿssh. Rast.*, 25: 180 (1988): Silene quadriloba.

ELYMUS (Gramin.).
glaucescens *O. Seberg in Pl. Syst. Evol.*, 166(1–2): 99 (1989), nom. nov.: Triticum pubiflorum *Steudel.*
harsukhii *H.S. Dubey & S.N. Dixit in J. Econ. Taxon. Bot.*, 12(1): 227 (1988)—India, Pakistan.
longisetus *(Hitchc.) J.F. Veldkamp in Blumea*, 34(1): 74 (1989): Brachypodium longisetum.
vaillantianus *(Wulf. & Schreb.) K.B. Jensen in Genome*, 32(4): 645 (1989): Triticum vaillantianum.

ELYTRIGIA (Gramin.).
sartorii *(Boiss. & Heldr.) J. Holub in Folia Geobot. Phytotax.*, 23(4): 413 (1988): Agropyron junceum var. sartorii.

EMBLEMANTHA B.C. Stone in Proc. Acad. Nat. Sci. Philadelphia, 140(2): 275 (1988). *MYRSINACEAE.*
urnulata *B.C. Stone in Proc. Acad. Nat. Sci. Philadelphia*, 140(2): 276 (1988)—Sumatra.

ENCELIA (Compos.).
canescens *Lamarck, Encycl.*, 2: 356 (1786)—Peru.

ENCEPHALARTOS (Zamiac.).
middelburgensis *P. Vorster, P.J. Robbertse & S. van der Westhuizen in S. Afr. J. Bot.*, 55(1): 122 (1989)—South Africa (Transvaal).
sclavoi *P. De Luca, D.W. Stevenson & A. Moretti in Delpinoa*, 29–30: 3 (1987–1988 publ. 1989)—Tanzania.

ENCEPHALARTOS :-
voiensis *A. Moretti, D.W. Stevenson & J.P. Sclavo in Ann. Missouri Bot. Gard.*, 76(3): 935 (1989)—Kenya.

ENCHOLIRIUM (Bromeliac.).
suzannae *W. Rauh in Abh. Akad. Wiss. Lit. Mainz, Math.-Naturwiss. Klasse, Trop. Subtrop. Pflanzenwelt*, 65: 19 (1988)—Brazil.

ENCYCLIA (Orchidac.).
favoris *(Reichenbach f.) M.A. Soto Arenas in Orquídea (México)*, 11: 270 (1988): Epidendrum favoris.

ENDIANDRA (Laurac.).
chartacea *A.J.G.H. Kostermans in Reinwardtia*, 10(5): 460 (1988)—Celebes.
deomalica *(Bennet & Chandra) A.J.G.H. Kostermans in Reinwardtia*, 10(5): 461 (1988): Beilschmiedia deomalica.
oviformis *A.J.G.H. Kostermans in Reinwardtia*, 10(5): 461 (1988)—New Guinea (Papua New Guinea).

ENGLEROCHARIS (Crucif.).
pauciflora *I.A. Al-Shehbaz in Nordic J. Bot.*, 8(6): 624 (1989)—Peru.

ENNEAPOGON (Gramin.).
caerulescens
var. occidentalis *E.K.Z. Kakudidi in Austral. Syst. Bot.*, 1(4): 333 (1988 publ. 1989)—Australia (W. Australia).
decipiens *E.K.Z. Kakudidi in Austral. Syst. Bot.*, 1(4): 336 (1988 publ. 1989)—Australia (N. Terr., Queensland, W. Australia).
eremophilus *E.K.Z. Kakudidi in Austral. Syst. Bot.*, 1(4): 338 (1988 publ. 1989)—Australia (N. Terr., Queensland).
truncatus *E.K.Z. Kakudidi in Austral. Syst. Bot.*, 1(4): 350 (1988 publ. 1989)—Australia (Queensland, New South Wales).
virens *(Lindl.) E.K.Z. Kakudidi in Austral. Syst. Bot.*, 1(4): 351 (1988 publ. 1989): Pappophorum virens.

EPACRIS (Epacridac.).
curtisiae *S.J. Jarman in Pap. Proc. Roy. Soc. Tasmania*, 122(2): 115 (1988)—Tasmania.

EPIDANTHUS (Orchidac.).
talamancana *J.T. Atwood in Ic. Pl. Trop.*, 14 (Orch. Costa Rica, pt. 1): t. 1315 (1989)—Costa Rica.

EPIDENDRUM (Orchidac.).
candelabrum *E. Hágsater in Orquídea (México)*, 11: 23 (1988)—Costa Rica, Panama.
citrosmum *E. Hágsater in Orquídea (México)*, 11: 25 (1988)—Mexico.
kerichilum *E. Hágsater in Orquídea (México)*, 11: 26 (1988)—Panama.
mantis-religiosae *E. Hágsater in Orquídea (México)*, 11: 27 (1988)—Panama.

EPIDENDRUM :-
melistagum *E. Hágsater in Orquídea (México),* 11: 28 (1988)—Mexico, Guatemala, Salvador, Honduras.
odontochilum *E. Hágsater in Orquídea (México),* 11: 30 (1988)—Costa Rica.
rosilloi *E. Hágsater in Orquídea (México),* 11: 9 (1988)—Mexico.
succulentum *E. Hágsater in Orquídea (México),* 11: 31 (1988)—Mexico.

EPIPACTIS (Orchidac.).
danubialis *K. Robatsch & J. Rydlo in Linz. Biol. Beitr.,* 21(1–2): 296 (1989)—Romania.
lusitanica *D. Tyteca in Orchidophile,* 84: 218 (1988)—Portugal.
nauosaensis *K. Robatsch in Linz. Biol. Beitr.,* 21(1–2): 298 (1989)—Greece.

ERAGROSTIS (Gramin.).
cilianensis *(All.) Vign. ex M.E. Janchen in Mitt. Naturw. Ver. Univ. Wien,* 5: 110 (1907): Poa cilianensis.
deqenensis *B.S. Sun & S. Wang in Acta Bot. Yunnanica,* 11(3): 306 (1989)—China.
lincangensis *B.S. Sun & S. Wang in Acta Bot. Yunnanica,* 11(3): 305 (1989)—China.
mairei
var. eglandis *B.S. Sun & S. Wang in Acta Bot. Yunnanica,* 11(3): 307 (1989)—China.
minor
var. minima *B.S. Sun & S. Wang in Acta Bot. Yunnanica,* 11(3): 307 (1989)—China.
quinquenervis *B.S. Sun & S. Wang in Acta Bot. Yunnanica,* 11(3): 303 (1989)—China.
rubida *B.S. Sun & S. Wang in Acta Bot. Yunnanica,* 11(3): 304 (1989)—China.

EREMOSTACHYS (Labiat.).
baburii *T. Adylov in Dokl. Akad. Nauk Uzb. SSR,* 1984(9): 47 (1984)—U.S.S.R. (Middle Asia).

ERIA (Orchidac.).
bilobulata
subsp. meghasaniensis *S. Misra in J. Orchid Soc. India,* 2(1–2): 49 (1988)—India.
bogoriensis
var. kurokawae *T. Hashimoto in Ann. Tsukuba Bot. Gard.,* 7: 165 (1988)—New Guinea (Papua New Guinea).
glandulifera *N.C. Deori & S.J. Phukan in J. Orchid Soc. India,* 2(1–2): 55 (1988)—India.

ERICA (Ericac.).
mackaiana
subsp. andevalensis *(Cabezudo & Rivera) D. McClintock & E.C. Nelson in Bot. J. Linn. Soc.,* 101(3): 282 (1989): E. andevalensis.
rivularis *L.E. Davidson in S. Afr. J. Bot.,* 55(5): 473 (1989)—South Africa (Transvaal).
scoparia
subsp. maderincola *D. McClintock in Heather Soc. Year Book,* 3(7): 35 (1989)—Madeira.

ERICA :-
vagans
f. viridula *D. McClintock in Heather Soc. Year Book,* 3(7): 54 (1989)—Cultivation.

ERICAMERIA (Compos.).
cooperi
var. bajacalifornica *(Urbatsch & Wussow) L.E. Urbatsch in Phytologia,* 67(2): 109 (1989): E. cooperi subsp. bajacalifornica.
juarezensis *(Moran) L.E. Urbatsch in Phytologia,* 67(2): 109 (1989): Haplopappus juarezensis.
palmeri
var. pachylepis *(H.M. Hall) G.L. Nesom in Phytologia,* 67(1): 104 (1989): Haplopappus palmeri subsp. pachylepis.
parishii
var. peninsularis *(R. Moran) G.L. Nesom in Phytologia,* 67(1): 104 (1989): Haplopappus arborescens subsp. peninsularis.

ERIGERON (Compos.).
sect. Arenarioides *(Rydb.) G.L. Nesom in Phytologia,* 67(1): 70 (1989): E. (sp.-group) Arenarioides.
sect. Cincinnactis *G.L. Nesom in Phytologia,* 67(1): 73 (1989).
sect. Imbarba *G.L. Nesom in Phytologia,* 67(1): 78 (1989).
sect. Karvinskia *G.L. Nesom in Phytologia,* 67(1): 79 (1989).
sect. Linearifolii *(G. Don) G.L. Nesom in Phytologia,* 67(1): 79 (1989): Aster sect. Linearifolii.
sect. Osteocaulis *G.L. Nesom in Phytologia,* 67(1): 82 (1989).
sect. Polyactis *(Less.) G.L. Nesom in Phytologia,* 66(5): 416 (1989): gen. Polyactis.
sect. Scopulincola *G.L. Nesom in Phytologia,* 67(1): 84 (1989).
sect. Spathifolium *G.L. Nesom in Phytologia,* 67(1): 84 (1989).
sect. Spinosi *(Alexander) G.L. Nesom & Sundberg in Phytologia,* 67(1): 85 (1989): Aster (sp.-group) Spinosi.
acomanus *R. Spellenberg & P. Knight in Madroño,* 36(2): 115 (1989)—U.S.A. (New Mexico).
annuactis *G.L. Nesom in Phytologia,* 66(5): 422 (1989)—Mexico.
basaseachensis *G.L. Nesom in Phytologia,* 66(5): 426 (1989)—Mexico.
caulinifolius *G.L. Nesom in Phytologia,* 66(5): 428 (1989)—Mexico.
cavernensis *S.L. Welsh & N.D. Atwood in Great Basin Nat.,* 48(4): 495 (1988)—U.S.A. (Nevada).
circulis *G.L. Nesom in Phytologia,* 66(5): 430 (1989)—Mexico.
coroniglandifer *G.L. Nesom in Phytologia,* 66(5): 432 (1989)—Mexico.

ERIGERON :–
dactyloides *(Greenm.) G.L. Nesom in Phytologia,* 66(5): 435 (1989): E. oreophilus var. dactyloides.
eruptens *G.L. Nesom in Phytologia,* 66(5): 438 (1989)—Mexico.
fundus *G.L. Nesom in Phytologia,* 66(6): 471 (1989)—Mexico.
hintoniorum *G.L. Nesom in Phytologia,* 66(6): 489 (1989)—Mexico.
mohinorensis *G.L. Nesom in Phytologia,* 66(6): 492 (1989)—Mexico.
nacoriensis *G.L. Nesom in Phytologia,* 66(5): 442 (1989)—Mexico.
onofrensis *G.L. Nesom in Phytologia,* 66(6): 491 (1989)—Mexico.
piscaticus *G.L. Nesom in Phytologia,* 67(4): 304 (1989)—U.S.A. (Arizona).
podophyllus *G.L. Nesom in Phytologia,* 66(5): 447 (1989)—Mexico.
rhizomactis *G.L. Nesom in Phytologia,* 66(5): 449 (1989)—Mexico.
scoparioides *G.L. Nesom in Phytologia,* 66(6): 488 (1989)—Mexico.
veracruzensis *G.L. Nesom in Phytologia,* 67(1): 74 (1989), nom. nov.: Aster scaberrimus *Less.*
vichrenensis *B. Pawłowski in Acta Bot. Croat.,* 28: 287 (1969)—Bulgaria.

ERINACEA (Legumin.).
schoenenbergeri *(Ray.) C. Raynaud in Nat. Monspel., Bot.,* 53: 49 (1988 publ. 1989): E. anthyllis subsp. schoenenbergeri.

ERIOGONUM (Polygonac.).
sect. Pterogonum *(H. Gross) J.L. Reveal in Phytologia,* 66(3): 262 (1989): gen. Pterogonum.
capistratum *J.L. Reveal in Phytologia,* 66(3): 254 (1989)—U.S.A. (Idaho).
var. muhlickii *J.L. Reveal in Phytologia,* 66(3): 256 (1989)—U.S.A. (Montana).
var. welshii *J.L. Reveal in Phytologia,* 66(3): 256 (1989)—U.S.A. (Idaho).
deflexum
var. rectum *J.L. Reveal in Phytologia,* 66(3): 263 (1989)—U.S.A. (California).
gracile
var. incultum *J.L. Reveal in Phytologia,* 66(3): 265 (1989)—U.S.A. (California).
henricksonii *J.L. Reveal in Phytologia,* 66(3): 261 (1989)—Mexico.
lonchophyllum
var. fendlerianum *(Benth.) J.L. Reveal in Phytologia,* 66(3): 252 (1989): E. microthecum var. fendlerianum.
var. intermontanum *(Rev.) J.L. Reveal in Phytologia,* 66(3): 252 (1989): E. intermontanum.
luteolum
var. saltuarium *J.L. Reveal in Phytologia,* 66(3): 264 (1989)—U.S.A. (California).
meledonum *J.L. Reveal in Phytologia,* 66(3): 254 (1989)—U.S.A. (Idaho).

ERIOGONUM :–
nudum
var. paralinum *J.L. Reveal in Phytologia,* 66(3): 258 (1989)—U.S.A. (Oregon, California).
var. regirivum *J.L. Reveal & J.C. Stebbins in Phytologia,* 66(3): 247 (1989)—U.S.A. (California).
ochrocephalum
var. sceptrum *J.L. Reveal in Phytologia,* 66(3): 252 (1989)—U.S.A. (Idaho).
ovalifolium
var. pansum *J.L. Reveal in Phytologia,* 66(3): 259 (1989)—U.S.A. (Idaho, Montana).
prattenianum
var. avium *J.L. Reveal & J.R. Shevock in Phytologia,* 66(3): 249 (1989)—U.S.A. (California).
shockleyi
var. packardiae *J.L. Reveal in Phytologia,* 66(3): 257 (1989), as 'packardae'—U.S.A. (Idaho).
thompsoniae
var. matthewsiae *(Rev.) J.L. Reveal in Phytologia,* 66(3): 252 (1989), as 'E. thomsonae var. matthewsae': E. corymbosum var. matthewsiae.
trichopes
var. hooveri *J.L. Reveal in Phytologia,* 66(3): 262 (1989)—U.S.A. (California).
umbellatum
var. argus *J.L. Reveal in Phytologia,* 66(3): 261 (1989)—U.S.A. (California, Oregon).
var. goodmanii *J.L. Reveal in Phytologia,* 66(3): 259 (1989)—U.S.A. (Oregon, California).
var. humistratum *J.L. Reveal in Phytologia,* 66(3): 260 (1989)—U.S.A. (California).
verrucosum *J.L. Reveal in Phytologia,* 66(3): 253 (1989)—U.S.A. (Idaho).

ERIOPHORUM (Cyperac.).
angustifolium
subsp. minus *(Dalla Torre) J. Holub in Folia Geobot. Phytotax.,* 23(4): 413 (1988): E. minus.

ERIOSPERMUM (Eriospermac.).
erinum *P.L. Perry in S. Afr. J. Bot.,* 55(1): 83 (1989)—South Africa (Cape Province).
eriophorum *Marloth ex P.L. Perry in S. Afr. J. Bot.,* 55(1): 85 (1989)—South Africa (Cape Province).
lammarginatum *Marloth ex P.L. Perry in S. Afr. J. Bot.,* 55(1): 85 (1989)—South Africa (Cape Province).
laxiracemosum *P.L. Perry in S. Afr. J. Bot.,* 55(1): 89 (1989)—South Africa (Cape Province).
marginatum *Marloth ex P.L. Perry in S. Afr. J. Bot.,* 55(1): 89 (1989)—South Africa (Cape Province).
ratelpoortianum *P.L. Perry in S. Afr. J. Bot.,* 55(1): 93 (1989)—South Africa (Cape Province).

ERIOSPERMUM :–
subincanum *P.L. Perry in S. Afr. J. Bot.*, 55(1): 93 (1989)—South Africa (Cape Province).
subtile *P.L. Perry in S. Afr. J. Bot.*, 55(1): 97 (1989)—South Africa (Cape Province).
undulatum *P.L. Perry in S. Afr. J. Bot.*, 55(1): 99 (1989)—South Africa (Cape Province).
vermiforme *Marloth ex P.L. Perry in S. Afr. J. Bot.*, 55(1): 99 (1989)—South Africa (Cape Province).

ERODIUM (Geraniac.).
subsect. Ciconia *A. El-Oqlah in Feddes Repert.*, 100(3–4): 109 (1989), without latin descr.
sect. Oxyrrhynchea *A. El-Oqlah in Feddes Repert.*, 100(3–4): 108 (1989), without latin descr.
sect. Pelargoniflorea *(Brumh.) A. El-Oqlah in Feddes Repert.*, 100(3–4): 109 (1989): E. subsect. Pelargoniflorea.
oxyrrhynchum
var. bryonifolium *(Boiss.) A. El-Oqlah in Feddes Repert.*, 100(3–4): 111 (1989), without basionym ref.: Basionym not stated.
paularense *F. Fernández-González & J. Izco in Candollea*, 44(1): 241 (1989)—Spain.
sibthorpianum
var. somanum *(Pesmen) A. El-Oqlah in Feddes Repert.*, 100(3–4): 115 (1989), without exact basionym page: E. somanum.

ERYMOPHYLLUM Paul G. Wilson in Nuytsia, 7(1): 105 (1989). *COMPOSITAE*.
compactum *Paul G. Wilson in Nuytsia*, 7(1): 110 (1989)—Australia (W. Australia).
glossanthus *Paul G. Wilson in Nuytsia*, 7(1): 106 (1989)—Australia (W. Australia).
hemisphaericum *Paul G. Wilson in Nuytsia*, 7(1): 113 (1989)—Australia (W. Australia).
ramosum *(A. Gray) Paul G. Wilson in Nuytsia*, 7(1): 108 (1989): Pteropogon ramosus.
subsp. involucratum *(F. Muell.) Paul G. Wilson in Nuytsia*, 7(1): 109 (1989): Helichrysum involucratum.
tenellum *(Turcz.) Paul G. Wilson in Nuytsia*, 7(1): 114 (1989): Helipterum tenellum.

ERYNGIUM (Umbellif.).
bourgatii
subsp. hispanicum *(Willk. & Lange) S. Rivas-Martínez, F. Fernández González & D. Sánchez-Mata in Opusc. Bot. Pharm. Complutensis*, 2: 109 (1986): E. bourgatii var. hispanicum.
davisii *Kit Tan & B. Yildiz in Notes Roy. Bot. Gard. Edinburgh*, 45(3): 441 (1988 publ. 1989)—Turkey.
hedgeanum *Kit Tan & B. Yildiz in Notes Roy. Bot. Gard. Edinburgh*, 45(3): 443 (1988 publ. 1989)—Turkey.

ERYSIMUM (Crucif.).
michaelis *T.A. Adylov in Dokl. Akad. Nauk Uzb. SSR*, 1980(6): 58 (1980), nom. nov.: E. popovii *Botsch. & Vved.*

ERYTHRODES (Orchidac.).
laticalcari *D.D. Dod in Moscosoa*, 5: 231 (1989)—Hispaniola (Dominican Republic).
parvula *P.J. Kores in Allertonia*, 5(1): 30 (1989)—Fiji, Tonga, Samoa.

ERYTHROXYLUM (Erythroxylac.).
macrophyllum
var. ecuadorense *T. Plowman in Fl. Ecuador*, 36(93): 18 (1989)—Ecuador.

ESPOSTOA (Cactac.).
frutescens *J.E. Madsen in Fl. Ecuador*, 35(45): 36 (1989)—Ecuador.

EUCALYPTUS (Myrtac.).
ser. Rigentes *M.I.H. Brooker & S.D. Hopper in Nuytsia*, 7(1): 7 (1989).
cadens *J.D. Briggs & M.D. Crisp in Muelleria*, 7(1): 7 (1989)—Australia (Victoria).
decolor *A.R. Bean & M.I.H. Brooker in Austrobaileya*, 3(1): 41 (1989)—Australia (Queensland).
dissimulata *M.I.H. Brooker in Nuytsia*, 6(3): 332 (1988)—Australia (W. Australia).
famelica *M.I.H. Brooker & S.D. Hopper in Nuytsia*, 7(1): 12 (1989)—Australia (W. Australia).
hypochlamydea *M.I.H. Brooker in Nuytsia*, 6(3): 328 (1988)—Australia (W. Australia).
kumarlensis *M.I.H. Brooker in Nuytsia*, 6(3): 333 (1988)—Australia (W. Australia).
latens *M.I.H. Brooker in Nuytsia*, 6(3): 332 (1988)—Australia (W. Australia).
litorea *M.I.H. Brooker & S.D. Hopper in Nuytsia*, 7(1): 10 (1989)—Australia (W. Australia).
perangusta *M.I.H. Brooker in Nuytsia*, 6(3): 330 (1988)—Australia (W. Australia).
rigens *M.I.H. Brooker & S.D. Hopper in Nuytsia*, 7(1): 8 (1989)—Australia (W. Australia).
salicola *M.I.H. Brooker in Nuytsia*, 6(3): 329 (1988)—Australia (W. Australia).
xanthope *A.R. Bean & M.I.H. Brooker in Austrobaileya*, 3(1): 39 (1989)—Australia (Queensland).

EUCHARIS (Amaryllidac.).
subgen. Heterocharis *A.W. Meerow in Ann. Missouri Bot. Gard.*, 76(1): 203 (1989).
bouchei
var. darienensis *A.W. Meerow in Ann. Missouri Bot. Gard.*, 76(1): 187 (1989)—Panama, Guatemala.
var. dressleri *A.W. Meerow in Ann. Missouri Bot. Gard.*, 76(1): 187 (1989)—Panama.
caucana *A.W. Meerow in Ann. Missouri Bot. Gard.*, 76(1): 191 (1989)—Colombia.
plicata
subsp. brevidentata *A.W. Meerow in Ann. Missouri Bot. Gard.*, 76(1): 200 (1989)—Bolivia, Peru.

EUGENIA (Myrtac.).
subgen. Phyllocalyx *(Berg) J.R. Mattos in Loefgrenia,* 94: 1 (1989): gen. Phyllocalyx.
subgen. Stenocalyx *(Berg) J.R. Mattos in Loefgrenia,* 94: 3 (1989): gen. Stenocalyx.
alagoensis *(Berg) J.R. Mattos in Loefgrenia,* 94: 3 (1989): Stenocalyx alagoensis.
angelyana *J.R. Mattos in Loefgrenia,* 94: 2 (1989)—Brazil.
angustisepala *J.R. Mattos in Loefgrenia,* 94: 3 (1989), nom. nov.: Stenocalyx ligustrina *Miquel.*
bahiana *J.R. Mattos in Loefgrenia,* 94: 4 (1989), nom. nov., Stenocalyx bahiensis *Berg.*
calystegia *(Berg) J.R. Mattos in Loefgrenia,* 94: 1 (1989): Phyllocalyx calystegius.
cavalcanteana *J.R. Mattos in Loefgrenia,* 94: 1 (1989), nom. nov.: Phyllocalyx riedelianus *Berg.*
concolor *J.R. Mattos in Loefgrenia,* 94: 5 (1989)—Brazil.
cuprea *(Berg) J.R. Mattos in Loefgrenia,* 94: 1 (1989): Phyllocalyx cupreus.
dentata *(Berg) J.R. Mattos in Loefgrenia,* 94: 3 (1989): Stenocalyx dentatus.
glandulosopunctata *P.E. Sánchez & L.J. Poveda in Brenesia,* 24: 413 (1985 publ. 1986), as 'glanduloso-punctata'—Costa Rica.
grandiflora *(Berg) J.R. Mattos in Loefgrenia,* 94: 3 (1989): Stenocalyx grandiflorus.
hartmanniae *J.R. Mattos in Loefgrenia,* 94: 6 (1989), as 'hartmannii'—Brazil.
hermesiana *J.R. Mattos in Loefgrenia,* 94: 1 (1989)—Brazil.
impressa *(Berg) J.R. Mattos in Loefgrenia,* 94: 4 (1989): Stenocalyx impressus.
leonanii *J.R. Mattos in Loefgrenia,* 94: 5 (1989)—Brazil.
macrosepala *(Berg) J.R. Mattos in Loefgrenia,* 94: 1 (1989): Phyllocalyx macrosepala.
matagalpensis *P.E. Sánchez-Vindas in Brenesia,* 25–26: 307 (1986 publ. 1988)—Nicaragua.
minensis *(Berg) J.R. Mattos in Loefgrenia,* 94: 3 (1989): Stenocalyx ligustrinus var. minensis.
oblongifolia *(Berg) J.R. Mattos in Loefgrenia,* 94: 3 (1989): Stenocalyx oblongifolius.
paranapiacabensis *J.R. Mattos in Loefgrenia,* 94: 4 (1989)—Brazil.
pinhaesensis *J.R. Mattos in Loefgrenia,* 94: 5 (1989), nom. nov.: E. oblongata *Legr. & Mattos.*
piresii *J.R. Mattos in Loefgrenia,* 94: 3 (1989), nom. nov.: Stenocalyx gardnerianus *Berg.*
prostrata *(Berg) J.R. Mattos in Loefgrenia,* 94: 4 (1989): Stenocalyx prostratus.
sarapiquensis *P.E. Sánchez-Vindas in Brenesia,* 25–26: 317 (1986 publ. 1988)—Costa Rica.
stipularis *(Berg) J.R. Mattos in Loefgrenia,* 94: 4 (1989): Stenocalyx stipularis.

EULALIA (Gramin.).
annua *B.K. Simon in Austrobaileya,* 3(1): 84 (1989)—Australia (Queensland, N. Terr.).

EUONYMUS (Celastrac.).
fertilis
var. euryanthus *(Hand.-Mazz.) C.Y. Chang in Fl. Sichuanica,* 4: 261 (1988): E. dielsianus var. euryanthus.
fortunei
f. albifructus *G. Murata in Acta Phytotax. Geobot.,* 39(4–6): 149 (1988)—Japan.

EUPATORIASTRUM (Compos.).
corvi *(McVaugh) B.L. Turner in Phytologia,* 67(4): 337 (1989): Eupatorium corvi.

EUPATORIUM (Compos.).
breedlovei *(King & H. Robins.) B.L. Turner in Phytologia,* 67(2): 112 (1989): Bartlettina breedlovei.
calderonii *B.L. Turner in Phytologia,* 67(2): 111 (1989)—Mexico.
cronquistii *(King & H. Robins.) B.L. Turner in Phytologia,* 67(2): 112 (1989): Bartlettina cronquistii.
cymuliferum *(B.L. Robins.) B.L. Turner in Phytologia,* 67(4): 337 (1989): Brickellia cymulifera.
dibollii *(King & H. Robins.) B.L. Turner in Phytologia,* 67(4): 338 (1989): Kyrsteniopsis dibollii.
enigmaticum *B.L. Turner in Phytologia,* 67(4): 337 (1989), nom. nov.: Fleischmannia guatemalensis *King & H. Robins.*
hintonii *(King & H. Robins.) B.L. Turner in Phytologia,* 67(2): 112 (1989): Bartlettina hintonii.
macdougallii *(King & H. Robins.) B.L. Turner in Phytologia,* 67(2): 112 (1989): Bartlettina macdougallii.
matudae *(King & H. Robins.) B.L. Turner in Phytologia,* 67(4): 338 (1989): Fleischmannia matudae.
ravenii *(King & H. Robins.) B.L. Turner in Phytologia,* 67(4): 338 (1989): Koanophyllon ravenii.
reyrobinsonii *(B. Turner) B.L. Turner in Phytologia,* 67(4): 338 (1989): Koanophyllon reyrobinsonii.
richardsonii *(B. Turner) B.L. Turner in Phytologia,* 67(4): 338 (1989): Koanophyllon richardsonii.
zacuapanum *B.L. Turner in Phytologia,* 67(4): 337 (1989), nom. nov.: Fleischmannia purpusii *King & H. Robins.*

EUPHORBIA (Euphorbiac.).
aaron-rossii *A.H. Holmgren & N.H. Holmgren in Brittonia,* 40(4): 357 (1988)—U.S.A. (Arizona).
adenoptera
subsp. gundlachii *(Urban) R.C.H.M. Oudejans in Phytologia,* 67(1): 44 (1989): E. gundlachii.

EUPHORBIA :-

subsp. pergamena *(Small) R.C.H.M. Oudejans in Phytologia,* 67(1): 44 (1989): E. pergamena.

alainii *R.C.H.M. Oudejans in Phytologia,* 67(1): 44 (1989), nom. nov.: Chamaesyce montana *Alain.*

amandi *R.C.H.M. Oudejans in Phytologia,* 67(1): 44 (1989), nom. nov.: E. minuta *R.A. Philippi.*

amygdaloides

subsp. robbiae *(Turrill) C.A. Stace in Watsonia,* 17(4): 443 (1989): E. robbiae.

antonii *R.C.H.M. Oudejans in Phytologia,* 67(1): 44 (1989), nom. nov.: E. lanceolata *D.A. Larrañaga.*

barnesii *(Millspaugh) R.C.H.M. Oudejans in Phytologia,* 67(1): 44 (1989): Adenopetalum barnesii.

bayeri *L.C. Leach in S. Afr. J. Bot.,* 54(6): 539 (1988)—South Africa (Cape Province).

blatteri *R.C.H.M. Oudejans in Phytologia,* 67(1): 44 (1989), nom. nov.: E. helioscopioides *E. Blatter.*

bruntii *(Proctor) R.C.H.M. Oudejans in Phytologia,* 67(1): 44 (1989): Chamaesyce bruntii.

celastroides

var. tomentella *(Boissier) R.C.H.M. Oudejans in Phytologia,* 67(1): 44 (1989): E. multiformis var. tomentella.

chapmanii *R.C.H.M. Oudejans in Phytologia,* 67(1): 44 (1989), nom. nov.: E. nudicaulis *A.W. Chapman.*

chiogenes *(Small) R.C.H.M. Oudejans in Phytologia,* 67(1): 44 (1989): Chamaesyce chiogenes.

conferta *(Small) R.C.H.M. Oudejans in Phytologia,* 67(1): 44 (1989): Chamaesyce conferta.

cordatella *R.C.H.M. Oudejans in Phytologia,* 67(1): 45 (1989), nom. nov.: E. cordata *D.A. Larrañaga.*

cowellii *(Millspaugh ex Britton) R.C.H.M. Oudejans in Phytologia,* 67(1): 45 (1989): Chamaesyce cowellii.

cumulicola *(Small) R.C.H.M. Oudejans in Phytologia,* 67(1): 45 (1989): Chamaesyce cumulicola.

damasoi *R.C.H.M. Oudejans in Phytologia,* 67(1): 45 (1989), nom. nov.: E. ovata *D.A. Larrañaga.*

deltoidea

var. serpyllum *(Small) R.C.H.M. Oudejans in Phytologia,* 67(1): 45 (1989): Chamaesyce serpyllum.

denisii *R.C.H.M. Oudejans in Phytologia,* 67(1): 45 (1989), nom. nov.: E. obcordata *M. Denis.*

donii *R.C.H.M. Oudejans in Phytologia,* 67(1): 45 (1989), nom. nov.: E. longifolia *D. Don.*

dubovikii *R.C.H.M. Oudejans in Phytologia,* 67(1): 45 (1989), nom. nov.: E. pinetorum *Dubovik.*

EUPHORBIA :-

duckei *(Croizat) R.C.H.M. Oudejans in Phytologia,* 67(1): 45 (1989): Chamaesyce duckei.

gentilis

subsp. tanquana *L.C. Leach in S. Afr. J. Bot.,* 54(6): 538 (1988)—South Africa (Cape Province).

georgei *R.C.H.M. Oudejans in Phytologia,* 67(1): 45 (1989), nom. nov.: E. acuta *G. Engelmann.*

graminifolia

subsp. tommasiniana *(Bertoloni) R.C.H.M. Oudejans in Phytologia,* 67(1): 45 (1989): E. tommasiniana.

subsp. zhiguliensis *(Prokhanov) R.C.H.M. Oudejans in Phytologia,* 67(1): 45 (1989): Galarhoeus zhiguliensis.

hamiltonii *R.C.H.M. Oudejans in Phytologia,* 67(1): 46 (1989), nom. nov.: E. angustifolia *Hamilton ex D. Don.*

heraldiana *(Millspaugh) R.C.H.M. Oudejans in Phytologia,* 67(1): 46 (1989): Chamaesyce heraldiana.

herbstii *(Wagner) R.C.H.M. Oudejans in Phytologia,* 67(1): 46 (1989): Chamaesyce herbstii.

hiernii *(Croizat) R.C.H.M. Oudejans in Phytologia,* 67(1): 46 (1989): Elaeophorbia hiernii.

humifusa

f. glabra *(Thell.) S.Z. Liou in Fl. Liaoningica, chief ed. S.X. Li,* 1: 1016 (1988): E. humifusa var. glabra.

f. pilosa *(Thell.) S.Z. Liou in Fl. Liaoningica, chief ed. S.X. Li,* 1: 1016 (1988): E. humifusa var. pilosa.

hurusawae *R.C.H.M. Oudejans in Phytologia,* 67(1): 46 (1989), nom. nov.: E. barbellata *I. Hurusawa.*

var. imaii *(Hurusawa) R.C.H.M. Oudejans in Phytologia,* 67(1): 46 (1989): E. imaii.

hypericifolia

var. coimbatorensis *M. Chandrabose in M. Chandrabose & N.C. Nair, Fl. Coimbatore:* 266 (1988)—India.

inaguaensis *R.C.H.M. Oudejans in Phytologia,* 67(1): 46 (1989), nom. nov.: E. abbreviata *D.S. Correll.*

josei *R.C.H.M. Oudejans in Phytologia,* 67(1): 46 (1989), nom. nov.: E. litoralis *M. Sessé & J.M. Moçiño.*

klotzschii *R.C.H.M. Oudejans in Phytologia,* 67(1): 46 (1989), nom. nov.: E. ovalifolia *(Engelmann ex Klotzsch) Boissier.*

kondoi *W. Rauh & Razafindratsira in Cact. Succ. J. (U.S.A.),* 61(3): 113 (1989)—Madagascar.

larranagae *R.C.H.M. Oudejans in Phytologia,* 67(1): 46 (1989), nom. nov.: E. rupestris *D.A. Larrañaga.*

leoncroizatii *R.C.H.M. Oudejans in Phytologia,* 67(1): 46 (1989), nom. nov.: E. croizatii *(Hurusawa) M. Kitagawa.*

EUPHORBIA :-

limaensis *R.C.H.M. Oudejans in Phytologia,* 67(1): 47 (1989), nom. nov.: E. inornata *(Dressler) A. Radcliffe-Smith.*

lutulenta *(Croizat) R.C.H.M. Oudejans in Phytologia,* 67(1): 47 (1989): Chamaesyce lutulenta.

marianoi *R.C.H.M. Oudejans in Phytologia,* 67(1): 47 (1989), nom. nov.: E. pulchella *M. Sessé & J.M. Moçiño.*

mazatlamensis *R.C.H.M. Oudejans in Phytologia,* 67(1): 47 (1989), nom. nov.: E. triflora *M. Sessé & J.M. Moçiño.*

mocinoi *R.C.H.M. Oudejans in Phytologia,* 67(1): 47 (1989), nom. nov.: E. peltata *M. Sessé & J.M. Moçiño.*

nicaeensis

subsp. maleevii *(Tamamshyan) R.C.H.M. Oudejans in Phytologia,* 67(1): 47 (1989): E. maleevii.

nicholasii *R.C.H.M. Oudejans in Phytologia,* 67(1): 47 (1989), nom. nov.: Elaeophorbia acuta *N.E. Brown.*

ocellata

subsp. arenicola *(Parish) R.C.H.M. Oudejans in Phytologia,* 67(1): 47 (1989): E. arenicola.

subsp. rattanii *(Watson) R.C.H.M. Oudejans in Phytologia,* 67(1): 47 (1989): E. rattanii.

orbifolia *(Alain) R.C.H.M. Oudejans in Phytologia,* 67(1): 47 (1989): Chamaesyce orbifolia.

paredonensis *(Millspaugh) R.C.H.M. Oudejans in Phytologia,* 67(1): 47 (1989): Chamaesyce paredonensis.

parodii *R.C.H.M. Oudejans in Phytologia,* 67(1): 47 (1989), nom. nov.: E. angustifolia *D. Parodi.*

porteriana *(Small) R.C.H.M. Oudejans in Phytologia,* 67(1): 48 (1989): Chamaesyce porteriana.

var. keyensis *(Small) R.C.H.M. Oudejans in Phytologia,* 67(1): 48 (1989): Chamaesyce keyensis.

var. scoparia *(Small) R.C.H.M. Oudejans in Phytologia,* 67(1): 48 (1989): Chamaesyce scoparia.

pterococca

var. pilosa *J.J. Lazare & A. Charpin in Saussurea,* 19: 17 (1988 publ. 1989)—Spain, Portugal.

quaitensis *S. Carter in Candollea,* 43(2): 564 (1988)—South Yemen.

radlkoferi *R.C.H.M. Oudejans in Phytologia,* 67(1): 48 (1989), nom. nov.: E. gracilis *Radlkofer.*

raphilippii *R.C.H.M. Oudejans in Phytologia,* 67(1): 48 (1989), nom. nov.: E. lanceolata *R.A. Philippi.*

rochaensis *(Croizat) R.C.H.M. Oudejans in Phytologia,* 67(1): 48 (1989): Chamaesyce rochaensis.

EUPHORBIA :-

rockii

var. grandiflora *(Hillebrand) R.C.H.M. Oudejans in Phytologia,* 67(1): 48 (1989), typographic error; *see Phytologia,* 67(3): 236 (1989): E. clusiifolia var. grandiflora.

var. grandifolia *(Hillebrand) R.C.H.M. Oudejans in Phytologia,* 67(3): 236 (1989): E. clusiifolia var. grandifolia.

roschianica *(Ikonnikov) R.C.H.M. Oudejans in Phytologia,* 67(1): 48 (1989): Tithymalus roschianus.

rothrockii *(Millspaugh) R.C.H.M. Oudejans in Phytologia,* 67(1): 48 (1989): Chamaesyce rothrockii.

rubriseminalis *S. Carter in Candollea,* 43(2): 566 (1988)—South Yemen.

serpyllifolia

subsp. hirtula *(Engelmann ex Watson) R.C.H.M. Oudejans in Phytologia,* 67(1): 48 (1989): E. hirtula.

sessei *R.C.H.M. Oudejans in Phytologia,* 67(1): 48 (1989), nom. nov.: E. imbricata *M. Sessé & J.M. Moçiño.*

smallii *R.C.H.M. Oudejans in Phytologia,* 67(1): 48 (1989), nom. nov., without replaced synonym date; *et in Phytologia,* 67(3): 236 (1989),with replaced synonym date: Chamaesyce pinetorum *J.K. Small.*

standleyi *(Millspaugh) R.C.H.M. Oudejans in Phytologia,* 67(1): 49 (1989): Chamaesyce standleyi.

sultan-hassei *A. Strid, B. Bentzer, R. von Bothmer, L. Engstrand & M. Gustafsson in Willdenowia,* 19(1): 63 (1989)—Crete.

tamaulipasana *(Millspaugh) R.C.H.M. Oudejans in Phytologia,* 67(1): 49 (1989): Chamaesyce tamaulipasana.

triphylla *(Klotzsch & Garcke ex Klotzsch) R.C.H.M. Oudejans in Phytologia,* 67(1): 49 (1989): Eumecanthus triphyllus.

tulearensis *(Rauh) W. Rauh in Cact. Succ. J. (U.S.A.),* 60(4): 184 (1988): E. capsaintemariensis var. tulearensis.

urbanii *(Millspaugh) R.C.H.M. Oudejans in Phytologia,* 67(1): 49 (1989): Chamaesyce urbanii.

vaalputsiana *L.C. Leach in S. Afr. J. Bot.,* 54(6): 534 (1988)—South Africa (Cape Province).

waldsteinii

subsp. jaxartica *(Prokhanov) R.C.H.M. Oudejans in Phytologia,* 67(1): 49 (1989): Tithymalus jaxarticus.

wangii *R.C.H.M. Oudejans in Phytologia,* 67(1): 49 (1989), nom. nov.: E. villifera. *Wang.*

wootonii *R.C.H.M. Oudejans in Phytologia,* 67(1): 49 (1989), nom. nov.: Zygophyllidium delicatulum *E.O. Wooton & P.C. Standley.*

yinshanica *S.Q. Zhou & G.H. Liu in Acta Phytotax. Sin.,* 27(1): 77 (1989)—China.

EXOCARPOS (Santalac.).
montanus *(Stauff.) M.G. Baumann-Bodenheim, Syst. Fl. Neu-Caledonien,* 5: 76 (1989): E. phyllanthoides var. montanus.

F

FACHEIROA (Cactac.).
chaetacantha *(Ritter) P.J. Braun & E. Esteves Pereira in Kakt. And. Sukk.,* 40(8): 202 (1989): Zehntnerella chaetacantha.
var. montealtoi *(Ritter) P.J. Braun & E. Esteves Pereira in Kakt. And. Sukk.,* 40(8): 202 (1989): Zehntnerella chaetacantha var. montealtoi.
squamosa *(Guerke) P.J. Braun & E. Esteves Pereira in Kakt. And. Sukk.,* 40(8): 199 (1989): Cereus squamosus.
var. polygona *(Ritter) P.J. Braun & E. Esteves Pereira in Kakt. And. Sukk.,* 40(8): 202 (1989): Zehntnerella polygona.

FAGARA (Rutac.).
aculeata *(Ruíz & Pavón) J. Cuatrecasas in Fl. Neotrop.,* 2 Suppl.: 98 (1985): Brunellia aculeata.

FAGOPYRUM (Polygonac.).
snowdenii *(Hutch. & Dandy) S.P. Hong in Grana,* 27(4): 295 (1988): Harpagocarpus snowdenii.

FALCATIFOLIUM (Podocarpac.).
sleumeri *D.J. De Laubenfels & J. Silba in Phytologia,* 65(5): 329 (1988)—New Guinea (W. Irian).

FALLOPIA (Polygonac.).
sect. Reynoutria *(Houtt.) L.P. Ronse Decraene in Bot. J. Linn. Soc.,* 98(4): 346 (1988): gen. Reynoutria.
×bohemica *(Chrtek & Chrtková) J.P. Bailey in Watsonia,* 17(4): 443 (1989): Reynoutria × bohemica.
japonica *(Houtt.) L.P. Ronse Decraene in Bot. J. Linn. Soc.,* 98(4): 369 (1988): Reynoutria japonica.
var. compacta *(Hook. f.) J.P. Bailey in Watsonia,* 17(4): 443 (1989): Polygonum compactum.
sachalinensis *(F. Schmidt Petrop.) L.P. Ronse Decraene in Bot. J. Linn. Soc.,* 98(4): 369 (1988): Polygonum sachalinense.

FARAMEA (Rubiac.).
boomii *J.A. Steyermark in Ann. Missouri Bot. Gard.,* 75(1): 335 (1988)—Venezuela.

FARAMEA :–
morilloi *J.A. Steyermark in Ann. Missouri Bot. Gard.,* 75(1): 337 (1988)—Venezuela.
paludicola *J.A. Steyermark & Boom in Ann. Missouri Bot. Gard.,* 75(1): 338 (1988)—Venezuela.
yavitensis *J.A. Steyermark in Ann. Missouri Bot. Gard.,* 75(1): 338 (1988)—Venezuela.
yutajensis *J.A. Steyermark in Ann. Missouri Bot. Gard.,* 75(1): 339 (1988)—Venezuela.

FARGESIA (Gramin.).
concinna *T.P. Yi in Acta Bot. Yunnanica,* 10(4): 437 (1988)—China.
mali *T.P. Yi in Acta Bot. Yunnanica,* 11(1): 37 (1989)—China.
stenoclada *T.P. Yi in J. Bamboo Res.,* 8(2): 30 (1989)—China.
ungulata *T.H. Wen in J. Bamboo Res.,* 8(1): 22 (1989)—China.
wuliangshanensis *T.P. Yi in Acta Bot. Yunnanica,* 10(4): 438 (1988)—China.

FELDSTONIA P.S. Short in Muelleria, 7(1): 108 (1989). *COMPOSITAE.*
nitens *P.S. Short in Muelleria,* 7(1): 109 (1989)—Australia (W. Australia).

FERDINANDUSA (Rubiac.).
boomii *J.A. Steyermark in Ann. Missouri Bot. Gard.,* 75(1): 339 (1988)—Venezuela.

FESTUCA (Gramin.).
sect. Fauria *E.B. Alekseev in Novosti Sist. Vÿssh. Rast.,* 25: 13 (1988).
sect. Obtusae *(E. Alexeev) E.B. Alekseev in Novosti Sist. Vÿssh. Rast.,* 25: 13 (1988): F. subgen. Obtusae.
eriobasis *H. Scholz in Willdenowia,* 18(2): 431 (1989)—Nepal.
forrestii
subsp. nepalensis *H. Scholz in Willdenowia,* 18(2): 432 (1989)—Nepal.

×**FIALAARA** hort. in Orchid Rev., 97(1143): centre page pull-out p. 8 (1989). *ORCHIDACEAE.* (Broughtonia × Cattleya × Laelia × Laeliopsis).

FICUS (Morac.).
formosana
f. lageniformis *(Lévl. & Van.) C.Y. Wu in Acta Bot. Yunnanica,* 11(1): 33 (1989): F. lageniformis.
lanyuensis *S.S. Ying, Colour. Illustr. Pl. Taiwan, Pt. 3; Colour. Illustr. Ligneous Pl. Taiwan,* 1: 435 (1987)—Taiwan.
microcarpa
f. crassifolia *(Shieh) S.S. Ying, Coloured Illustr. Fl. Taiwan,* 3: 383 (1988): F. retusa var. crassifolia.
pubinervis
var. cuneatonervosa *(Yamamoto) S.S. Ying, Colour. Illustr. Pl. Taiwan, Pt. 3; Colour. Illustr. Ligneous Pl. Taiwan,* 1: 386 (1987): Ficus cuneato-nervosa.

FICUS :–
ruficaulis
var. antaoensis *(Hayata) Hatusima & J.C. Liao in Quart. J. Chinese Forest.*, 22(1): 135 (1989): F. antaoensis.

FILAGO (Compos.).
petro-ianii *J. Rita & M. Dittrich in Davis & Hedge Festschrift (ed. Kit Tan)*: 1 (1989)—Balearic Is.

FILIPENDULA (Rosac.).
purpurea
f. alba *(Nakai) C.Y. Li in Fl. Liaoningica, chief ed. S.X. Li*, 1: 774 (1988): F. multijuga var. alba.

FIRMIANA (Sterculiac.).
subglabra *(Abraham & Dutt) A.J.G.H. Kostermans in Blumea*, 34(1): 118 (1989): F. colorata var. subglabra.

FISCHERIA (Asclepiadac.).
billbergiana *(Beurling) G. Morillo in Ernstia*, 50: 13 (1988): Gonolobus billbergianus.

FITZWILLIA P.S. Short in Muelleria, 7(1): 111 (1989). *COMPOSITAE*.
axilliflora *(W.V. Fitzg. ex Ewart & J. White) P.S. Short in Muelleria*, 7(1): 111 (1989): Angianthus axilliflorus.

FORDIA (Legumin.).
microphylla *Dunn ex Z. Wei in Acta Phytotax. Sin.*, 27(1): 75 (1989)—China.

FORDIOPHYTON (Melastomatac.).
maculatum *C.Y. Wu ex Z. Wei & Y.B. Chang in Bull. Bot. Res. North-East. Forest. Inst.*, 9(2): 35 (1989)—China.

FORMANODENDRON K.C. Nixon & W.L. Crepet in Amer. J. Bot., 76(6): 840 (1989). *FAGACEAE*.
doichangensis *(Camus) K.C. Nixon & W.L. Crepet in Amer. J. Bot.*, 76(6): 840 (1989): Quercus doichangensis.

FORSYTHIA (Oleac.).
saxatilis
var. lanceolata *S. Lee in Korean J. Pl. Taxon.*, 14(2): 91 (1984)—Korea.
var. pilosa *S. Lee in Korean J. Pl. Taxon.*, 14(2): 92 (1984)—Korea.

FRAXINUS (Oleac.).
americana
f. barrii *W.H. Wagner in Mich. Bot.*, 27(4): 130 (1988)—U.S.A. (Michigan).
angustifolia
var. ramosa *B. Kevey in Bot. Közlem.*, 74–75(1–2): 97 (1987–88 publ. 1988), without type—Hungary.
nanchuanensis *S.S. Sun & J.L. Wu in Bull. Bot. Res. North-East. Forest. Inst.*, 9(3): 33 (1989)—China.

FRITILLARIA (Liliac.).
meleagroides
var. plena *X.Z. Duan & X.J. Zheng in Acta Phytotax. Sin.*, 27(4): 309 (1989)—Xinjiang.
var. rhodantha *X.Z. Duan & X.J. Zheng in Acta Phytotax. Sin.*, 27(4): 309 (1989)—Xinjiang.
pallidiflora
var. plena *X.Z. Duan & X.J. Zheng in Acta Phytotax. Sin.*, 27(4): 306 (1989)—Xinjiang.
parvialbiflora
var. viridicaulinis *X.Z. Duan & X.J. Zheng in Acta Phytotax. Sin.*, 27(4): 308 (1989)—Xinjiang.
tortifolia
var. albiflora *X.Z. Duan & X.J. Zheng in Acta Phytotax. Sin.*, 27(4): 307 (1989)—Xinjiang.
var. barlikensis *X.Z. Duan & X.J. Zheng in Acta Phytotax. Sin.*, 27(4): 308 (1989)—Xinjiang.
var. parviflora *X.Z. Duan & X.J. Zheng in Acta Phytotax. Sin.*, 27(4): 307 (1989)—Xinjiang.
var. plena *X.Z. Duan & X.J. Zheng in Acta Phytotax. Sin.*, 27(4): 307 (1989)—Xinjiang.
var. wusunica *X.Z. Duan & X.J. Zheng in Acta Phytotax. Sin.*, 27(4): 308 (1989)—Xinjiang.
walujewii
var. plena *X.Z. Duan & X.J. Zheng in Acta Phytotax. Sin.*, 27(4): 306 (1989)—Xinjiang.
var. shawanensis *X.Z. Duan & X.J. Zheng in Acta Phytotax. Sin.*, 27(4): 306 (1989)—Xinjiang.
yuminensis
var. albiflora *X.Z. Duan & X.J. Zheng in Acta Phytotax. Sin.*, 27(4): 308 (1989)—China.
var. roseoflora *X.Z. Duan & X.J. Zheng in Acta Phytotax. Sin.*, 27(4): 309 (1989)—Xinjiang.

FUCHSIA (Onagrac.).
brevilobis *P.E. Berry in Ann. Missouri Bot. Gard.*, 76(2): 552 (1989)—Brazil.
hatschbachii *P.E. Berry in Ann. Missouri Bot. Gard.*, 76(2): 559 (1989)—Brazil.
regia
subsp. reitzii *P.E. Berry in Ann. Missouri Bot. Gard.*, 76(2): 577 (1989)—Brazil.
subsp. serrae *P.E. Berry in Ann. Missouri Bot. Gard.*, 76(2): 573 (1989)—Brazil.

FURCATELLA M.G. Baumann-Bodenheim, Syst. Fl. Neu-Caledonien, 4: 103 (1988 publ. Jan. 1989), nom. nov., without replaced synonym ref.; et Syst. Fl. Neu-Caledonien, 5: 98 (May 1989), without replaced synonym ref.: Uragoga § Forcipella Baill. (Rubiac.).

FURCATELLA :-
rupicola *(Baill.) M.G. Baumann-Bodenheim, Syst. Fl. Neu-Caledonien,* 4: 103 (1988 publ. Jan. 1989), without basionym ref.; *et Syst. Fl. Neu-Caledonien,* 5: 98 (May 1989), with basionym ref., generic name not validly publ.: Uragoga rupicola.

G

GAGEA (Liliac.).
fragifera *(Vill.) E. Bayer & G. López González in Taxon,* 38(4): 643 (1989): Ornithogalum fragiferum.
soleirolii
subsp. nevadensis *(Boiss.) E. Bayer & G. López González in Taxon,* 38(4): 644 (1989): G. nevadensis.

GAHNIA (Cyperac.).
vitiensis
subsp. kauaiensis *(Benl) T. Koyama in Bishop Mus. Occas. Pap.,* 29: 126 (1989): G. kauaiensis.

GALEOBDOLON (Labiat.).
caucasicum *A.P. Khokhryakov in Byull. Glavn. Bot. Sada (Moscow),* 152: 62 (1989)—U.S.S.R. (Caucasus), Turkey, Iran.

GALEOMMA (Compos.).
stenolepis *(S. Moore) O.M. Hilliard in Fl. S. Afr.,* 33(7:2): 13 (1983): Gnaphalium stenolepis.

GALEOTTIA (Orchidac.).
acuminata *(C. Schweinf.) Dressler & E.A. Christenson in Lindleyana,* 3(4): 221 (1988 publ. 1989): Zygopetalum acuminatum.
antioquiana *(Kränzl.) Dressler & E.A. Christenson in Lindleyana,* 3(4): 221 (1988 publ. 1989): Batemannia antioquiana.
burkei *(Reichb. f.) Dressler & E.A. Christenson in Lindleyana,* 3(4): 221 (1988 publ. 1989): Zygopetalum burkei.
ciliata *(Morel) Dressler & E.A. Christenson in Lindleyana,* 3(4): 221 (1988 publ. 1989): Zygopetalum ciliatum.
colombiana *(Garay) Dressler & E.A. Christenson in Lindleyana,* 3(4): 222 (1988 publ. 1989): Mendoncella colombiana.
marginata *(Garay) Dressler & E.A. Christenson in Lindleyana,* 3(4): 222 (1988 publ. 1989): Zygopetalum marginatum.
prainiana *(Rolfe) Dressler & E.A. Christenson in Lindleyana,* 3(4): 222 (1988 publ. 1989): Zygopetalum prainianum.

GALIUM (Rubiac.).
irinae *M.G. Pakhomova in Opred. Rast. Sred. Azii,* 9: 367 (1987)—U.S.S.R. (Middle Asia).

GALIUM :-
kasachstanicum *M.G. Pakhomova in Opred. Rast. Sred. Azii,* 9: 368 (1987)—U.S.S.R. (Middle Asia).
paradoxum
subsp. duthiei *F. Ehrendorfer & E. Schönbeck-Temesy in Davis & Hedge Festschrift (ed. Kit Tan):* 113 (1989)—Nepal, Sikkim, India, China.
subsp. franchetianum *F. Ehrendorfer & E. Schönbeck-Temesy in Davis & Hedge Festschrift (ed. Kit Tan):* 113 (1989)—Japan.

GAMBELIA (Scrophulariac.).
glabrata *(T.S. Brandegee) D.A. Sutton, Revis. tribe Antirrhineae:* 511 (1988): Galvezia glabrata.
juncea *(Bentham) D.A. Sutton, Revis. tribe Antirrhineae:* 513 (1988): Maurandya juncea.
rupicola *(T.S. Brandegee) D.A. Sutton, Revis. tribe Antirrhineae:* 513 (1988): Galvezia rupicola.

GAMBEYA (Sapotac.).
gorungosana *(Engl.) L. Liben in Bull. Jard. Bot. Nation. Belg.,* 58(3–4): 556 (1988): Chrysophyllum gorungosanum.

GAMOCHAETA (Compos.).
ambatensis *L. Ariza Espinar in Kurtziana,* 20: 169 (1989)—Argentina.

GARCINIA (Guttif.).
intermedia *(Pittier) B. Hammel in Ann. Missouri Bot. Gard.,* 76(3): 927 (1989): Rheedia intermedia.
madruno *(Kunth) B. Hammel in Ann. Missouri Bot. Gard.,* 76(3): 928 (1989): Calophyllum madruno.
magnifolia *(Pittier) B. Hammel in Ann. Missouri Bot. Gard.,* 76(3): 928 (1989): Rheedia magnifolia.

GARDENIA (Rubiac.).
actinocarpa *C.F. Puttock in Austrobaileya,* 2(5): 435 (1988)—Australia (Queensland).
psidioides *C.F. Puttock in Austrobaileya,* 2(5): 440 (1988)—Australia (Queensland).
rupicola *C.F. Puttock in Austrobaileya,* 2(5): 441 (1988)—Australia (Queensland).
scabrella *C.F. Puttock in Austrobaileya,* 2(5): 443 (1988)—Australia (Queensland).
tessellaris *C.F. Puttock in Austrobaileya,* 2(5): 444 (1988)—Australia (Queensland).

GASTRIDIUM (Gramin.).
ventricosum
subsp. laxum *(Boiss. & Reut.) J. Holub in Folia Geobot. Phytotax.,* 23(4): 413 (1988): G. laxum.

GASTROCHILUS (Orchidac.).
acinacifolius *Z.H. Tsi in Bull. Bot. Res. North-East. Forest. Inst.,* 9(2): 25 (1989)—Hainan.

GASTROCHILUS :–
corymbosus *A.P. Das & S. Chanda in J. Econ. Taxon. Bot.*, 12(2): 401 (1988 publ. 1989)—India, Sikkim.
hainanensis *Z.H. Tsi in Bull. Bot. Res. North-East. Forest. Inst.*, 9(2): 21 (1989)—Hainan.
sinensis *Z.H. Tsi in Bull. Bot. Res. North-East. Forest. Inst.*, 9(2): 23 (1989)—China.

GAULTHERIA (Ericac.).
alnifolia
var. grata *(A.C. Smith) J.L. Luteyn in Brittonia,* 41(1): 15 (1989): G. grata.
buxifolia
var. elassantha *(A.C. Smith) J.L. Luteyn in Brittonia,* 41(1): 13 (1989): G. elassantha.
var. secunda *(E.J. Remy) J.L. Luteyn in Brittonia,* 41(1): 13 (1989): G. secunda.
eriophylla
var. mucronata *(E.J. Remy) J.L. Luteyn in Brittonia,* 41(1): 15 (1989): G. mucronata.
lanigera
var. rufolanata *(Sleumer) J.L. Luteyn in Brittonia,* 41(1): 14 (1989): G. rufolanata.
sclerophylla
var. hirsuta *J.L. Luteyn in Brittonia,* 41(1): 11 (1989)—Ecuador, Colombia.
serrata *(Vell. Conc.) Kinoshita-Gouvêa ex J.L. Luteyn in Brittonia,* 41(1): 16 (1989): Andromeda serrata.
sinensis
var. layaensis *S.J. Rae & D.G. Long in Notes Roy. Bot. Gard. Edinburgh,* 45(2): 334 (1988 publ. 1989)—Bhutan.
strigosa
var. revoluta *(A.C. Smith) J.L. Luteyn in Brittonia,* 41(1): 14 (1989): G. revoluta.
trichophylla
var. eciliata *S.J. Rae & D.G. Long in Notes Roy. Bot. Gard. Edinburgh,* 45(2): 334 (1988 publ. 1989)—Bhutan, China.

GEISSORHIZA (Iridac.).
callista *P. Goldblatt in Flow. Pl. Afr.,* 50(2): t. 1996 (1989)—South Africa (Cape Province).

GENIOSPORUM (Labiat.).
menthoides
var. prostratum *(L.) J.R. Press & V.V. Sivarajan in Bull. Brit. Mus. (Nat. Hist.), Bot.,* 19: 116 (1989): Ocimum prostratum.

GENLISEA (Lentibulariac.).
lobata *E. Fromm-Trinta in Bradea,* 5(14): 152 (1989)—Brazil.

GENOPLESIUM (Orchidac.).
acuminatum *(R. Rogers) D.L. Jones & M.A. Clements in Lindleyana,* 4(3): 142 (1989): Prasophyllum acuminatum.
apostasioides *(Fitzg.) D.L. Jones & M.A. Clements in Lindleyana,* 4(3): 142 (1989): Corunastylis apostasioides.
archeri *(J.D. Hook.) D.L. Jones & M.A. Clements in Lindleyana,* 4(3): 142 (1989): Prasophyllum archeri.

GENOPLESIUM :–
brachystachyum *(Lindley) D.L. Jones & M.A. Clements in Lindleyana,* 4(3): 142 (1989): Prasophyllum brachystachyum.
calopterum *(H.G. Reichb.) D.L. Jones & M.A. Clements in Lindleyana,* 4(3): 142 (1989): Prasophyllum calopterum.
ciliatum *(Ewart & B. Rees) D.L. Jones & M.A. Clements in Lindleyana,* 4(3): 142 (1989): Prasophyllum ciliatum.
despectans *(J.D. Hook.) D.L. Jones & M.A. Clements in Lindleyana,* 4(3): 142 (1989): Prasophyllum despectans.
eriochilum *(Fitzg.) D.L. Jones & M.A. Clements in Lindleyana,* 4(3): 143 (1989): Prasophyllum eriochilum.
filiforme *(Fitzg.) D.L. Jones & M.A. Clements in Lindleyana,* 4(3): 143 (1989): Prasophyllum filiforme.
fimbriatum *(R. Br.) D.L. Jones & M.A. Clements in Lindleyana,* 4(3): 143 (1989): Prasophyllum fimbriatum.
morrisii *(Nicholls) D.L. Jones & M.A. Clements in Lindleyana,* 4(3): 143 (1989): Prasophyllum morrisii.
nigricans *(R. Br.) D.L. Jones & M.A. Clements in Lindleyana,* 4(3): 143 (1989): Prasophyllum nigricans.
nudiscapum *(J.D. Hook.) D.L. Jones & M.A. Clements in Lindleyana,* 4(3): 143 (1989): Prasophyllum nudiscapum.
nudum *(J.D. Hook.) D.L. Jones & M.A. Clements in Lindleyana,* 4(3): 144 (1989): Prasophyllum nudum.
parvicallum *(Rupp) D.L. Jones & M.A. Clements in Lindleyana,* 4(3): 144 (1989): Prasophyllum parvicallum.
plumosum *(Rupp) D.L. Jones & M.A. Clements in Lindleyana,* 4(3): 144 (1989): Prasophyllum plumosum.
pumilum *(J.D. Hook.) D.L. Jones & M.A. Clements in Lindleyana,* 4(3): 144 (1989): Prasophyllum pumilum.
rufum *(R. Br.) D.L. Jones & M.A. Clements in Lindleyana,* 4(3): 144 (1989): Prasophyllum rufum.
sagittiferum *(Rupp) D.L. Jones & M.A. Clements in Lindleyana,* 4(3): 145 (1989): Prasophyllum sagittiferum.
woollsii *(F. Muell.) D.L. Jones & M.A. Clements in Lindleyana,* 4(3): 145 (1989): Prasophyllum woollsii.

GENTIANA (Gentianac.).
fukienensis *Y. Ling in Contrib. Inst. Bot. Nation. Acad. Peiping,* 6(4): 217 (1949)—China.
gannanensis *Y. Wang & Z.C. Lou in Acta Phytotax. Sin.,* 26(6): 461 (1988)—China.
kitadakensis *N. Yonezawa in Acta Phytotax. Geobot.,* 40(1–4): 7 (1989)—Japan.

GENTIANODES (Gentianac.).
acuminata *(Clarke) S. Omer, S.I. Ali & M. Qaiser in Pakistan J. Bot.,* 20(1): 5 (1988): Gentiana pygmaea var. acuminata.

GENTIANODES :–
algida
var. nubigena *(Edgew.) S. Omer, S.I. Ali & M. Qaiser in Pakistan J. Bot.*, 20(1): 16 (1988): Gentiana nubigena.
alii *S. Omer & M. Qaiser in Pakistan J. Bot.*, 20(1): 6 (1988)—Pakistan.
aprica *(Decne.) S. Omer, S.I. Ali & M. Qaiser in Pakistan J. Bot.*, 20(1): 15 (1988): Gentiana aprica.
argentea *(Royle ex D. Don) S. Omer, S.I. Ali & M. Qaiser in Pakistan J. Bot.*, 20(1): 6 (1988): Ericala argentea.
burkillii *(H. Smith) S. Omer, S.I. Ali & M. Qaiser in Pakistan J. Bot.*, 20(1): 9 (1988): Gentiana burkillii.
cachemirica *(Decne.) S. Omer, S.I. Ali & M. Qaiser in Pakistan J. Bot.*, 20(1): 16 (1988): Gentiana cachemirica.
decemfida *(Buch.-Ham. ex D. Don) S. Omer, S.I. Ali & M. Qaiser in Pakistan J. Bot.*, 20(1): 14 (1988): Gentiana decemfida.
huxleyi *(Kusn.) S. Omer, S.I. Ali & M. Qaiser in Pakistan J. Bot.*, 20(1): 5 (1988): Gentiana huxleyi.
intermedia *(Clarke) S. Omer, S.I. Ali & M. Qaiser in Pakistan J. Bot.*, 20(1): 14 (1988): Gentiana carinata var. intermedia.
kaufmaniana *(Reg. & Schmalh.) S. Omer, S.I. Ali & M. Qaiser in Pakistan J. Bot.*, 20(1): 17 (1988): Gentiana kaufmaniana.
subsp. affghanica *(Kusn.) S. Omer, S.I. Ali & M. Qaiser in Pakistan J. Bot.*, 20(1): 17 (1988): Gentiana kaufmaniana var. affghanica.
kurroo *(Royle) S. Omer, S.I. Ali & M. Qaiser in Pakistan J. Bot.*, 20(1): 16 (1988): Gentiana kurroo.
lowndesii *(Blatter) S. Omer, S.I. Ali & M. Qaiser in Pakistan J. Bot.*, 20(1): 17 (1988): Gentiana lowndesii.
marginata *(G. Don) S. Omer, S.I. Ali & M. Qaiser in Pakistan J. Bot.*, 20(1): 9 (1988): Ericala marginata.
var. harrissii *S. Omer, S.I. Ali & M. Qaiser in Pakistan J. Bot.*, 20(1): 10 (1988)—Pakistan, India.
nasirii *S. Omer, S.I. Ali & M. Qaiser in Pakistan J. Bot.*, 20(1): 10 (1988)—Pakistan.
olivieri *(Griseb.) S. Omer, S.I. Ali & M. Qaiser in Pakistan J. Bot.*, 20(1): 18 (1988): Gentiana olivieri.
pedicellata *(D. Don) S. Omer, S.I. Ali & M. Qaiser in Pakistan J. Bot.*, 20(1): 14 (1988): Ericala pedicellata.
tianschanica *(Rupr. ex Kusn.) S. Omer, S.I. Ali & M. Qaiser in Pakistan J. Bot.*, 20(1): 18 (1988): Gentiana tianschanica.

GENYORCHIS (Orchidac.).
apetala *(Lindl.) K. Senghas in Orchidee*, 40(1): centre page pull-out, Orchideenkartei 552 (1989): Bulbophyllum apetalum.

GEONOMA (Palmae).
solitaria *(Engel) Jahn ex A.W. Hill in Index Kewensis*, Supp. 7: 102 (1921-1925 publ 1929): Roebelia solitaria.

×**GEORGEBLACKARA** hort. in Orchid Rev., 97(1146): centre page pull-out p. 8 (1989). *ORCHIDACEAE*. (Comparettia × Leochilus × Oncidium × Rodriguezia).

GERANIUM (Geraniac.).
chelikii *Kit Tan & B. Yildiz in Notes Roy. Bot. Gard. Edinburgh*, 45(3): 440 (1988 publ. 1989)—Turkey.
nepalense
f. albiflora *(Chung) W.T. Lee in J. Korean Pl. Taxon.*, 12(2): 85 (1982): G. thunbergii f. albiflora.
viscosum *A.J. Cavanilles, Diss.*, 4: 246 (1787)—South Africa (Cape Province).

GILIA (Polemoniac.).
tenuis *F.J. Smith & E.C. Neese in Great Basin Nat.*, 49(3): 461 (1989)—U.S.A. (Utah).

GLADIOLUS (Iridac.).
aquamontanus *P. Goldblatt in S. Afr. J. Bot.*, 55(2): 261 (1989)—South Africa (Cape Province).
uitenhagensis *P. Goldblatt & J.H.J. Vlok in S. Afr. J. Bot.*, 55(2): 263 (1989)—South Africa (Cape Province).

GLANDULARIA (Verbenac.).
cheitmaniana *(Moldenke) S.M. Botta & L. Poggio in Parodiana*, 5(2): 334 (1989): Verbena cheitmaniana.

GLEKIA O.M. Hilliard in Notes Roy. Bot. Gard. Edinburgh, 45(3): 482 (1988 publ. 1989). *SCROPHULARIACEAE*.
krebsiana *(Benth.) O.M. Hilliard in Notes Roy. Bot. Gard. Edinburgh*, 45(3): 482 (1988 publ. 1989): Phyllopodium krebsianum.

GLIA (Umbellif.).
capensis *(Houtt.) B.L. Burtt in Notes Roy. Bot. Gard. Edinburgh*, 45(2): 199 (1988 publ. 1989): Oenanthe capensis.

GLOMERA (Orchidac.).
emarginata *P.J. Kores in Allertonia*, 5(1): 113 (1989)—Fiji.

GLOSSARION (Compos.).
bilabiatum *(Maguire) J.F. Pruski in Brittonia*, 41(1): 39 (1989): Guaicaia bilabiata.

GLYCINE (Legumin.).
albicans *M.D. Tindale & L.A. Craven in Austral. Syst. Bot.*, 1(4): 401 (1988 publ. 1989)—Australia (W. Australia).
hirticaulis *M.D. Tindale & L.A. Craven in Austral. Syst. Bot.*, 1(4): 406 (1988 publ. 1989)—Australia (N. Terr.).

GLYCINE :–
lactovirens *M.D. Tindale & L.A. Craven in Austral. Syst. Bot.*, 1(4): 404 (1988 publ. 1989)—Australia (W. Australia).

GLYCYRRHIZA (Legumin.).
glabra
var. caduca *X.Y. Li in Bull. Bot. Res. North-East. Forest. Inst.*, 9(1): 31 (1989)—Xinjiang.
var. glandulosa *X.Y. Li in Bull. Bot. Res. North-East. Forest. Inst.*, 9(1): 32 (1989)—Xinjiang.
var. laxifoliolata *X.Y. Li in Bull. Bot. Res. North-East. Forest. Inst.*, 9(1): 32 (1989)—Xinjiang.
macrophylla *X.Y. Li in Bull. Bot. Res. North-East. Forest. Inst.*, 9(1): 30 (1989)—Xinjiang.
shiheziensis *X.Y. Li in Bull. Bot. Res. North-East. Forest. Inst.*, 9(1): 29 (1989)—Xinjiang.

GNIDIA (Thymelaeac.).
singularis *O.M. Hilliard in Notes Roy. Bot. Gard. Edinburgh*, 45(2): 196 (1988 publ. 1989)—Lesotho.

×**GOHARTIA** hort. in Orchid Rev., 97(1145): centre page pull-out p. 9 (1989). *ORCHIDACEAE.* (Gomesa × Lockhartia).

×**GOMETTIA** hort. in Orchid Rev., 97(1145): centre page pull-out p. 9 (1989). *ORCHIDACEAE.* (Comparettia × Gomesa).

GOMPHOSTEMMA (Labiat.).
javanicum
var. brevidens *G. Murata in Acta Phytotax. Geobot.*, 40(1–4): 101 (1989)—Thailand.

GONGORA (Orchidac.).
hirtzii *C.H. Dodson & N.H. Williams in Lindleyana*, 4(1): 30 (1989)—Ecuador.

GONOLOBUS (Asclepiadac.).
germanianus *G. Morillo in Ernstia*, 50: 14 (1988)—Venezuela.
hildegardiae *G. Morillo in Ernstia*, 50: 16 (1988)—Venezuela.
marmoreus *(Woods.) G. Morillo in Ernstia*, 50: 17 (1988): Exolobus marmoreus.
sagasteguii *G. Morillo in Ernstia*, 50: 17 (1988)—Peru.

GOODYERA (Orchidac.).
vitiensis *(L.O. Williams) P.J. Kores in Allertonia*, 5(1): 25 (1989): G. waitziana var. vitiensis.

GOUANIA (Rhamnac.).
wurdackii *J.A. Steyermark in Ann. Missouri Bot. Gard.*, 75(2): 1065 (1988)—Venezuela.

GRAFFENRIEDA (Melastomatac.).
kralii *J.J. Wurdack in Mem. New York Bot. Gard.*, 51: 102 (1989)—Venezuela.

GRAZIELIA (Compos.).
brevipetiolata *R.M. King & H. Robinson in Phytologia*, 65(6): 487 (1989), nom. nov.: Eupatorium brevipetiolatum *Schultz-Bip. ex Hieron.*

GREVILLEA (Proteac.).
acanthifolia
subsp. paludosa *R.O. Makinson & D.E. Albrecht in Muelleria*, 7(1): 89 (1989)—Australia (New South Wales).
donaldiana *K.F. Kenneally in West. Austral. Nat.*, 17(5): 115 (1988)—Australia (W. Australia).

GREWIA (Tiliac.).
fabreguesii *E. Boudouresque in Bull. Mus. Nation. Hist. Nat., B, Adansonia*, Ser. 4, 10(4): 397 (1988 publ. 1989)—Niger, Zimbabwe, Senegal, Namibia, Chad, Kenya, Mauritania, Cameroun.

GRINDELIA (Compos.).
cabrerae
var. alatocarpa *L. Ariza Espinar in Kurtziana*, 20: 170 (1989)—Argentina.

GUADUA (Gramin.).
angustifolia
var. bicolor *X. Londoño in Rev. Acad. Colomb. Cienc. Exact. Fis. Nat.*, 65: 379 (1989)—Colombia.

GUAPIRA (Nyctaginac.).
neblinensis *B. Maguire & J.A. Steyermark in Mem. New York Bot. Gard.*, 51: 109 (1989)—Venezuela.

GUAREA (Meliac.).
pendula *R. da Silva Ramalho, A.L. Pinheiro & T.D. Pennington in Rev. Arvore*, 11(2): 185 (1987)—Brazil.

GUETTARDA (Rubiac.).
andamanica *A.K. Goel & B.N. Mehrotra in Indian J. Forest.*, 11(3): 241 (1988)—Andaman Is.
turrialbana *N. Zamora V. & L.J. Poveda A. in Ann. Missouri Bot. Gard.*, 75(3): 1157 (1988)—Costa Rica.

GUZMANIA (Bromeliac.).
butcheri *W. Rauh in Abh. Akad. Wiss. Lit. Mainz, Math.-Naturwiss. Klasse, Trop. Subtrop. Pflanzenwelt*, 65: 42 (1988)—Panama.
madisonii *H.E. Luther in Selbyana*, 11: 56 (1989)—Ecuador.
morreniana
var. tenuifolia *H.E. Luther in Selbyana*, 11: 57 (1989)—Peru.
musaica
var. discolor *H.E. Luther in Selbyana*, 11: 56 (1989)—Panama.
var. rosea *H.E. Luther in Selbyana*, 11: 56 (1989)—Panama.

GUZMANIA :-
sanguinea
var. comosa *H.E. Luther in Selbyana,* 11: 54 (1989)—Colombia.
zahnii
var. longiscapa *W. Rauh in Abh. Akad. Wiss. Lit. Mainz, Math.-Naturwiss. Klasse, Trop. Subtrop. Pflanzenwelt,* 65: 45 (1988)—Panama.

GYMNANTHES (Euphorbiac.).
dressleri *G.L. Webster in Ann. Missouri Bot. Gard.,* 75(3): 1130 (1988)—Panama.
farinosa *(Griseb.) G.L. Webster in Ann. Missouri Bot. Gard.,* 75(3): 1131 (1988): Excoecaria farinosa.

GYMNEMA (Asclepiadac.).
acuminatum
var. glabrum *M.A. Rahman & C.C. Wilcock in Blumea,* 34(1): 100 (1989)—Bangladesh, Burma.

GYMNOCALYCIUM (Cactac.).
bruchii
var. niveum *W. Rausch in Succulenta,* 68(9): 180 (1989)—Argentina.
pflanzii
subspec. argentinense *H. Till & W. Till in Kakt. And. Sukk.,* 39(12): 277 (1988)—Argentina.
platense
var. ventanicola *(Speg.) R. Kiesling in Succulenta,* 67(11): 237 (1988): Echinocactus gibbosus var. ventanicola.

GYMNOSPERMIUM (Berberidac.).
kiangnanensis *(Chiu) H. Loconte in Canad. J. Bot.,* 67(8): 2315 (1989): Leontice kiangnanensis.

GYMNOSTOMA (Casuarinac.).
australianum *L.A.S. Johnson in Fl. Australia,* 3: 202, 103 (1989)—Australia (Queensland).

GYNOCHTHODES (Rubiac.).
australiensis *J.T. Johansson in Austral. Syst. Bot.,* 1(4): 369 (1988 publ. 1989), as 'Gynochtodes'—Australia (N. Terr., Queensland).

GYNOSTEMMA (Cucurbitac.).
guangxiense *X.X. Chen & D.H. Qin in Acta Bot. Yunnanica,* 10(4): 495 (1988)—China.
pentagynum *Z.P. Wang in Acta Bot. Yunnanica,* 11(2): 165 (1989)—China.

GYNOXYS (Compos.).
tetroici *V.A. Funk & H. Robinson in Rev. Acad. Colomb. Cienc. Exact. Fis. Nat.,* 65: 243 (1989)—Peru.

GYNURA (Compos.).
cusimbua
var. petiolata *(Hook. f.) R. Mathur in R.R. Rao et al., Fl. Ind. Enumerat. - Asterac.:* 42 (1988): G. angulosa var. petiolata.

GYNURA :-
drymophila
var. glabrifolia *P.I. Forster & A. Thongpukdee in Austrobaileya,* 2(5): 564 (1988)—Australia (Queensland).
hmopaengensis *H. Koyama in Acta Phytotax. Geobot.,* 39(4–6): 152 (1988)—Thailand.
malaccensis *R.O. Belcher in Kew Bull.,* 44(3): 538 (1989)—Malaya.

H

HABENARIA (Orchidac.).
falcicornis
subsp. caffra *(Schltr.) J.C. Manning in S. Afr. J. Bot.,* 55(2): 192 (1989): H. caffra.
mossii *(G. Williamson) J.C. Manning in S. Afr. J. Bot.,* 55(2): 193 (1989): H. lithophila subsp. mossii.
pseudociliosa *Schelpe ex J.C. Manning in S. Afr. J. Bot.,* 55(2): 193 (1989)—South Africa (Natal, Cape Province, Transvaal).

HABRACANTHUS (Acanthac.).
puniceus *D.C. Wasshausen in Acta Bot. Venezuel.,* 15(2): 135 (1987)—Venezuela.
tachirensis *D.C. Wasshausen in Acta Bot. Venezuel.,* 15(2): 133 (1987)—Venezuela.
zuliensis *D.C. Wasshausen in Acta Bot. Venezuel.,* 15(2): 136 (1987)—Venezuela.

HABRANTHUS (Amaryllidac.).
albispiritus *P. Ravenna in Onira, Bot. Leafl.,* 1(8): 54 (1988)—Argentina.
bahiensis *P. Ravenna in Onira, Bot. Leafl.,* 1(8): 53 (1988)—Brazil.
lacteus *(S. Moore) P. Ravenna in Onira, Bot. Leafl.,* 1(8): 56 (1988): Zephyranthes lactea.
quilmesianus *P. Ravenna in Onira, Bot. Leafl.,* 1(8): 54 (1988)—Argentina.

HALIMOCNEMIS (Chenopodiac.).
subsect. Brachyphylla *U.P. Pratov in Aktualnye Voprosy Bot. v. SSSR: Tezisy Dokladov VIII S"ezda Vses. Bot. Obshch.:* 28 (1988), without latin descr.
subsect. Dolichophyllae *U.P. Pratov in Aktualnye Voprosy Bot. v. SSSR: Tezisy Dokladov VIII S"ezda Vses. Bot. Obshch.:* 28 (1988), without latin descr. or type.
sect. Pentamerae *U.P. Pratov in Aktualnye Voprosy Bot. v. SSSR: Tezisy Dokladov VIII S"ezda Vses. Bot. Obshch.:* 28 (1988), without latin descr. or type.

HALIMODENDRON (Legumin.).
halodendron *(Pall.) A. Voss in Vilmorin's Blumengärtnerei,* 1: (35) of Index (1896): Robinia halodendron.

HAMAMELIS (Hamamelidac.).
japonica
f. auriflora *N. Satomi in J. Phytogeogr. Taxon.*, 36(2): 64 (1988)—Japan.

HANABUSAYA (Campanulac.).
asiatica
f. alba *T.B. Lee in J. Korean Pl. Taxon.*, 6(1–2): 17 (1975)—Korea.

HAPLOPHYLLUM (Rutac.).
sanguineum *M. Thulin in Kew Bull.*, 44(1): 143 (1989)—Somali Republic.

HAPTANTHUS A. Goldberg & C. Nelson S. in Syst. Bot., 14(1): 16 (1989). *UNKNOWN.*
hazlettii *A. Goldberg & C. Nelson S. in Syst. Bot.*, 14(1): 16 (1989)—Honduras.

HARFORDIA (Polygonac.).
macroptera
var. fruticosa *(E. Greene) J.L. Reveal in Phytologia*, 66(3): 225 (1989): Pterostegia fruticosa.
var. galioides *(E. Greene) J.L. Reveal in Phytologia*, 66(3): 224 (1989): Pterostegia galioides.

HARVEYA (Scrophulariac.).
coccinea *F.N. Hepper in Kew Bull.*, 44(1): 163 (1989), illegit. later homonym of (Harv.) Schlechter (1899)—Tanzania.
coccinea *(Harv.) R. Schlechter in Engl. Bot. Jahrb.*, 27: 184 (1899): Aulaya coccinea.
kenyensis *F.N. Hepper in Kew Bull.*, 44(1): 163 (1989)—Kenya.

HASTINGSIA (Hyacinthac.).
serpentinicola *R.W. Becking in Madroño*, 36(3): 208 (1989)—U.S.A. (Oregon).

HECKELDORA (Meliac.).
mangenotiana *L. Aké Assi & J.G. Lorougnon in Bull. Soc. Bot. France, Lett. Bot.*, 136(2): 167 (1989)—Ivory Coast.

HEDYCHIUM (Zingiberac.).
glabrum *S.Q. Tong in Acta Phytotax. Sin.*, 27(4): 288 (1989)—China.

HEDYOTIS (Rubiac.).
sect. Gouldia *(A. Gray) W.L. Wagner & D.R. Herbst in Bishop Mus. Occas. Pap.*, 29: 111 (1989): gen. Gouldia.
sect. Phyllozygia *W.L. Wagner & D.R. Herbst in Bishop Mus. Occas. Pap.*, 29: 113 (1989).
fosbergii *W.L. Wagner & D.R. Herbst in Bishop Mus. Occas. Pap.*, 29: 111 (1989), nom. nov.: Gouldia st.-johnii *Fosb.*
galeottii *(Martens) E.E. Terrell & D.H. Lorence in Phytologia*, 66(1): 1 (1989): Declieuxia galeottii.
hillebrandii *(Fosb.) W.L. Wagner & D.R. Herbst in Bishop Mus. Occas. Pap.*, 29: 112 (1989): Gouldia hillebrandii.

HEDYOTIS :-
nagporensis *(Brace ex Haines) K.K. Sharma & S. Sharma in J. Econ. Taxon. Bot., Addit. Ser. 5:* 56 (1989), isonym; *see Rolla Rao & Hemadri in Indian Forester*, 99(6): 377 (June 1973); *et in Proc. Indian Sci. Congr.*, 60(3): 338 (1973): Oldenlandia nagporensis.
terminalis *(Hook. & Arnott) W.L. Wagner & D.R. Herbst in Bishop Mus. Occas. Pap.*, 29: 112 (1989): Petesia? terminalis.
tryblium *D.R. Herbst & W.L. Wagner in Bishop Mus. Occas. Pap.*, 29: 113 (1989)—Hawaiian Is.

HEDYSARUM (Legumin.).
austrokurilense *(N.S. Pavlova) N.S. Pavlova in Sosud. Rast. Sovet. Dal'nego Vostoka*, 4: 289 (1989): H. sacchalinense subsp. austrokurilense.
confertum *(N.S. Pavlova) N.S. Pavlova in Sosud. Rast. Sovet. Dal'nego Vostoka*, 4: 288 (1989): H. sachalinense subsp. confertum.
dshambulicum *N.V. Pavlov in Vestn. Akad. Nauk Kaz. SSR*, 1949(1): 33 (1949)—U.S.S.R. (Middle Asia).
falconeri
var. cachemirianum *(Benth. ex Baker) A. Pramanik & K. Thothathri in J. Jap. Bot.*, 64(3): 76 (1989): H. cachemirianum.
latibracteatum *N.S. Pavlova in Bot. Zhurn.*, 74(6): 908 (1989)—U.S.S.R. (Soviet Far East).
pallidiflorum *N.V. Pavlov in Vestn. Akad. Nauk Kaz. SSR*, 1949(1): 34 (1949)—U.S.S.R. (Middle Asia).
silicii *R. Lakušić in Glasn. Zem. Muz. Bosne Herceg. Prir. Nauke*, 21: 65 (1982)—Jugoslavia.

HELIANTHELLA (Compos.).
durangensis *B.L. Turner in Phytologia*, 65(6): 493 (1989)—Mexico.

HELIANTHEMUM (Cistac.).
leptophyllum
subsp. alypoides *(Losa España & Rivas Goday) C. Raynaud in Soc. Ech. Pl. Vasc. Eur. Bassin Médit. Bull.*, 22: 18 (1986–87 publ. 1988): H. alypoides.
raynaudii *A. Ortega Olivencia, A.T. Romero García & C. Morales Torres in Candollea*, 44(1): 233 (1989)—Spain.

HELICHRYSUM (Compos.).
albertense *O.M. Hilliard in Fl. S. Afr.*, 33(7:2): 177 (1983)—South Africa (Cape Province).
asperum
var. albidulum *(DC.) O.M. Hilliard in Fl. S. Afr.*, 33(7:2): 127 (1983): H. ericifolium var. albidulum.
var. appressifolium *(Moeser) O.M. Hilliard in Fl. S. Afr.*, 33(7:2): 128 (1983): H. ericifolium var. appressifolium.
var. comosum *(Sch. Bip.) O.M. Hilliard in Fl. S. Afr.*, 33(7:2): 128 (1983): H. comosum.

HELICHRYSUM :-
var. glabrum *O.M. Hilliard in Fl. S. Afr.*, 33(7:2): 128 (1983)—South Africa (Cape Province).
aureofolium *O.M. Hilliard in Fl. S. Afr.*, 33(7:2): 158 (1983)—South Africa (Cape Province).
deserticola *O.M. Hilliard in Fl. S. Afr.*, 33(7:2): 168 (1983)—Namibia.
dunense *O.M. Hilliard in Fl. S. Afr.*, 33(7:2): 143 (1983)—South Africa (Cape Province).
erubescens *O.M. Hilliard in Fl. S. Afr.*, 33(7:2): 174 (1983)—Namibia.
fourcadei *O.M. Hilliard in Fl. S. Afr.*, 33(7:2): 131 (1983)—South Africa (Cape Province).
jubilatum *O.M. Hilliard in Fl. S. Afr.*, 33(7:2): 179 (1983)—South Africa (Cape Province).
kitianum *B. Yildiz in Notes Roy. Bot. Gard. Edinburgh*, 45(3): 445 (1988 publ. 1989)—Turkey.
mixtum
var. grandiceps *O.M. Hilliard in Fl. S. Afr.*, 33(7:2): 253 (1983)—South Africa (Transvaal), Swaziland.
mutabile *O.M. Hilliard in Fl. S. Afr.*, 33(7:2): 302 (1983)—South Africa (Transvaal), Swaziland.
outeniquense *O.M. Hilliard in Fl. S. Afr.*, 33(7:2): 192 (1983)—South Africa (Cape Province).
pumilio
subsp. fleckii *(S. Moore) O.M. Hilliard in Fl. S. Afr.*, 33(7:2): 166 (1983): H. fleckii.
refractum *O.M. Hilliard in Fl. S. Afr.*, 33(7:2): 133 (1983)—South Africa (Cape Province).
rosum
var. arcuatum *O.M. Hilliard in Fl. S. Afr.*, 33(7:2): 107 (1983)—South Africa (Cape Province, Orange Free State).
saxicola *O.M. Hilliard in Fl. S. Afr.*, 33(7:2): 190 (1983)—South Africa (Cape Province).
sivasicum *Kit Tan & B. Yildiz in Notes Roy. Bot. Gard. Edinburgh*, 45(3): 447 (1988 publ. 1989)—Turkey.
solitarium *O.M. Hilliard in Fl. S. Afr.*, 33(7:2): 149 (1983)—South Africa (Cape Province).

HELICIA (Proteac.).
cochinchinensis
var. rengetiensis *(Masamune) S.S. Ying, Colour. Illustr. Pl. Taiwan, Pt. 3; Colour. Illustr. Ligneous Pl. Taiwan*, 1: 403 (1987): H. rengetiensis.

HELICONIA (Heliconiac.).
cucullata *W.J. Kress & L. Andersson in Selbyana*, 11: 49 (1989)—Panama.
faunorum *W.J. Kress & L. Andersson in Selbyana*, 11: 52 (1989)—Panama.
monteverdensis
var. vulcanicola *(Stiles) W.J. Kress in Selbyana*, 11: 53 (1989): H. vulcanicola.
mooreana *R.R. Smith in R. McVaugh, Fl. Novo-Galiciana*, 15: 83 (1989)—Mexico.

HELIETTA (Rutac.).
trifoliata *(Baillon) D.J. Mabberley, Plant-book:* 707 (1989 reprint): Picrella trifoliata.

×**HELIOCHIA** (Cactac.).
hybrida *(Géel) P.V. Heath in Taxon*, 38(1): 126 (1989): Cactus hybridus.

HELIOPHILA (Crucif.).
cornellsbergia *B.J. Pienaar & A. Nicholas in Bothalia*, 18(2): 183 (1988)—South Africa (Cape Province).

HELLEBORUS (Ranunculac.).
bocconei
subsp. intermedius *(Guss.) W. Greuter & Burdet in Willdenowia*, 19(1): 44 (1989): H. intermedius.
multifidus
subsp. bocconei *(Tenore) B. Mathew, Hellebores:* 126 (1989): H. bocconei.
subsp. hercegovinus *(Martinis) B. Mathew, Hellebores:* 125 (1989): H. hercegovinus.
orientalis
subsp. abchasicus *(A. Braun) B. Mathew, Hellebores:* 90 (1989): H. abchasicus.
subsp. guttatus *(A. Braun & Sauer) B. Mathew, Hellebores:* 90 (1989): H. guttatus.

HEMISTEPTA (Compos.).
lyrata
f. alba *M.H. Lee in Korean J. Pl. Taxon.*, 13(2): 110 (1983)—Korea.

HENRIETTEA (Melastomatac.).
lateriflora *(M. Vahl) R.A. Howard in Phytologia*, 65(4): 286 (1988): Melastoma lateriflora.
mucronata *(Gleason) S.S. Renner in Mem. New York Bot. Gard.*, 50: 52 (1989): Loreya mucronata.

HERACLEUM (Umbellif.).
marashicum *Kit Tan & B. Yildiz in Notes Roy. Bot. Gard. Edinburgh*, 45(3): 445 (1988 publ. 1989)—Turkey.

HERBERTIA (Iridac.).
quareimana *P. Ravenna in Nordic J. Bot.*, 9(1): 55 (1989)—Brazil, Uruguay.

HERITIERA (Sterculiac.).
magnifica *A.J.G.H. Kostermans in Reinwardtia*, 10(5): 477 (1988)—Borneo (Sarawak).

HERNIARIA (Illecebrac.).
amoena *T. Çelebioğlu & C. Favarger in Candollea*, 44(1): 332 (1989)—Turkey.

×**HESPEROTSUGA** C.N. Page in Notes Roy. Bot. Gard. Edinburgh, 45(2): 389 (1988 publ. 1989). *PINACEAE*. (Hesperopeuce × Tsuga).

HESPEROTSUGA :–
jeffreyi *(Henry) C.N. Page in Notes Roy. Bot. Gard. Edinburgh,* 45(2): 389 (1988 publ. 1989): Tsuga pattoniana var. jeffreyi.

HESSEA (Amaryllidac.).
incana *D. Snijman in S. Afr. J. Bot.,* 55(3): 356 (1989)—South Africa (Cape Province).
monticola *D. Snijman in S. Afr. J. Bot.,* 55(3): 349 (1989)—South Africa (Cape Province).
pusilla *D. Snijman in S. Afr. J. Bot.,* 55(3): 353 (1989)—South Africa (Cape Province).
undosa *D. Snijman in S. Afr. J. Bot.,* 55(3): 351 (1989)—South Africa (Cape Province).

HETEROCONDYLUS (Compos.).
didymus *(Klatt) R.M. King & H. Robinson in Phytologia,* 65(6): 487 (1989): Eupatorium didymum.

HETEROSTEMMA (Asclepiadac.).
sect. Oianthus *K. Swarupanandan & N. Sasidharan in Bot. J. Linn. Soc.,* 101(2): 253 (1989): gen. Oianthus.
beddomei *(Hook. f.) K. Swarupanandan & J.K. Mangaly in Bot. J. Linn. Soc.,* 101(2): 254 (1989): Oianthus beddomei.
decanense *(Talb.) K. Swarupanandan & J.K. Mangaly in Bot. J. Linn. Soc.,* 101(2): 255 (1989): Oianthus decanensis.
disciflorum *(Hook. f.) K. Swarupanandan & J.K. Mangaly in Bot. J. Linn. Soc.,* 101(2): 256 (1989): Oianthus disciflorus.
vasudevanii *K. Swarupanandan & J.K. Mangaly in Bot. J. Linn. Soc.,* 101(2): 257 (1989), as 'vasudevani'—India.

HETEROTHALAMUS (Compos.).
densus *N.E. Brown in Trans. Linn. Soc., Ser. 2,* 6: 39 (1901), listed in I.K. Supp. 3 as 'derisus'—Guyana.†

HETEROTROPA (Aristolochiac.).
kurosawae *(Sugimoto) F. Maekawa ex Y. Maekawa in J. Jap. Bot.,* 64(2): 59 (1989): Asarum kurosawae.
minamitaniana *(Hatusima) F. Maekawa ex Y. Maekawa in J. Jap. Bot.,* 64(2): 59 (1989): Asarum minamitanianum.
okinawensis *(Hatusima) F. Maekawa ex Y. Maekawa in J. Jap. Bot.,* 64(2): 60 (1989): Asarum okinawense.
senkakuinsularis *(Hatusima) F. Maekawa ex Y. Maekawa in J. Jap. Bot.,* 64(2): 60 (1989): Asarum senkakuinsulare.
yaeyamensis *(Hatusima) F. Maekawa ex Y. Maekawa in J. Jap. Bot.,* 64(2): 60 (1989): Asarum yaeyamense.

HEXACHLAMYS (Myrtac.).
sect. Macrohexachlamys *J.R. Mattos in Loefgrenia,* 94: 12 (1989).
excelsa *(Camb.) J.R. Mattos in Loefgrenia,* 94: 12 (1989): Myrtus excelsa.

HIBBERTIA (Dilleniac.).
hooglandii *J.R. Wheeler in Nuytsia,* 7(1): 69 (1989)—Australia (W. Australia).

HIBISCUS (Malvac.).
sect. Clypeati *O.J. Blanchard in P.A. Fryxell, Malvac. of Mexico (Syst. Bot. Monog., 25):* 470 (1988).
sect. Muenchhusia *(Heister ex Fabricius) O.J. Blanchard in P.A. Fryxell, Malvac. of Mexico (Syst. Bot. Monog., 25):* 471 (1988): gen. Muenchhusia.
sect. Striati *O.J. Blanchard in P.A. Fryxell, Malvac. of Mexico (Syst. Bot. Monog., 25):* 471 (1988).
arnottianus
subsp. immaculatus *(M. Roe) D.M. Bates in Bishop Mus. Occas. Pap.,* 29: 104 (1989): H. immaculatus.
subsp. punaluuensis *(Skottsb.) D.M. Bates in Bishop Mus. Occas. Pap.,* 29: 104 (1989): H. arnottianus var. punaluuensis.
brackenridgei
subsp. mokuleianus *(M. Roe) D.M. Bates in Bishop Mus. Occas. Pap.,* 29: 105 (1989): H. brackenridgei var. mokuleiana.
kokio
subsp. saintjohnianus *(M. Roe) D.M. Bates in Bishop Mus. Occas. Pap.,* 29: 105 (1989): H. saintjohnianus.
waimeae
subsp. hannerae *(Degener & I. Degener) D.M. Bates in Bishop Mus. Occas. Pap.,* 29: 106 (1989): H. waimeae var. hannerae.

HICORIA (Juglandac.).
tetraptera *(Liebm.) J. Rzedowski et al. in C. Conzatti, Fl. Taxon. Mexic., ed. 3,* 1: 673 (1988), with incorrect basionym ref.: Carya tetraptera.

HIDALGOA (Compos.).
uspanapa *B.L. Turner in Phytologia,* 65(5): 379 (1988)—Mexico.

HIERACIUM (Compos.).
laniferum
subsp. texedense *(Pau) G. Blanca, M. Cueto & M.C. Quesada in Saussurea,* 19: 26 (1988 publ. 1989): H. texedense.

HIEROCHLOE (Gramin.).
wrangelica *B.A. Yurtsev & Probat. in Bot. Zhurn.,* 74(1): 111 (1989)—U.S.S.R. (Soviet Far East).

HILLIA (Rubiac.).
subgen. Ravnia *(Oersted) C.M. Taylor in Selbyana,* 11: 30 (1989): gen. Ravnia.
allenii *C.M. Taylor in Selbyana,* 11: 32 (1989), nom. nov.: Ravnia panamensis *Steyerm.*
longifilamentosa *(Steyerm.) C.M. Taylor in Selbyana,* 11: 32 (1989): Ravnia longifilamentosa.
triflora *(Oersted) C.M. Taylor in Selbyana,* 11: 30 (1989): Ravnia triflora.

HILLIA :–
var. pittieri *(Standley) C.M. Taylor in Selbyana,* 11: 31 (1989): Ravnia pittieri.

HIMANTOGLOSSUM (Orchidac.).
samariensis *C. & A. Alibertis in Orchidophile,* 87: 110 (1989)—Crete.

HIPPOCREPIS (Legumin.).
balearica
var. minoricensis *P. Montserrat ex M. Laínz in Fontqueria,* 24: 2 (1989)—Balearic Is.
carpetana *P. Lassen in Willdenowia,* 19(1): 61 (1989)—Spain.
emerus *(L.) P. Lassen in Svensk Bot. Tidskr.,* 83(2): 86 (1989): Coronilla emerus.
subsp. emeroides *(Boiss. & Spruner) Greuter & Burdet ex P. Lassen in Svensk Bot. Tidskr.,* 83(2): 86 (1989): Coronilla emeroides.
monticola *Durieu ex P. Lassen in Willdenowia,* 19(1): 61 (1989)—Algeria, Morocco, Tunisia, Libya.
neglecta *P. Lassen in Willdenowia,* 19(1): 61 (1989)—Morocco.
unisiliquosa
subsp. armena *P. Lassen in Willdenowia,* 19(1): 61 (1989)—U.S.S.R. (Caucasus), Iran.

HIRTELLA (Chrysobalanac.).
liesneri *G.T. Prance in Fl. Neotrop.,* 9S: 89 (1989)—Venezuela.

HOFFMANNSEGGIA (Legumin.).
arequipensis *E.A. Ulibarri in Darwiniana,* 28(1–4): 232 (1987 publ. 1988)—Peru.

HOLOGRAPHIS (Acanthac.).
caput-medusae *T.F. Daniel in Proc. Calif. Acad. Sci.,* Ser. 4, 46(2): 76 (1988)—Mexico.
leticiana *T.F. Daniel in Proc. Calif. Acad. Sci.,* Ser. 4, 46(2): 79 (1988)—Mexico.
tolantongensis *T.F. Daniel in Proc. Calif. Acad. Sci.,* Ser. 4, 46(2): 75 (1988)—Mexico.

HOMOGNAPHALIUM (Compos.).
pulvinatum *(Del.) A.A. Fayed & M. Zareh in Willdenowia,* 18(2): 451 (1989): Gnaphalium pulvinatum.

HORTIA (Rutac.).
neblinensis *B. Maguire & B.M. Boom in Mem. New York Bot. Gard.,* 51: 121 (1989)—Brazil.

HOSTA (Hostac.).
jonesii *M.G. Chung in Ann. Missouri Bot. Gard.,* 76(3): 920 (1989)—Korea.
yingeri *S.B. Jones in Ann. Missouri Bot. Gard.,* 76(2): 604 (1989)—Korea.

HOVEA (Legumin.).
arnhemica *J.H. Ross in Muelleria,* 7(1): 135 (1989)—Australia (N. Terr.).

HOVEA :–
planifolia *(Domin) J.H. Ross in Muelleria,* 7(1): 138 (1989): H. longifolia subvar. planifolia.

HOYA (Asclepiadac.).
acuta
var. citrina *(Ridley) D.H. Kent in Hoyan,* 11(1:2): i (1989): H. citrina.
carnosa
var. gushanica *W. Xu in Guihaia,* 9(3): 199 (1989)—China.
heuschkeliana *Kloppenburg in Hoyan,* 11(1:2): i (1989)—Philippine Is.

HUERNIA (Asclepiadac.).
ser. Ampligibbae *L.C. Leach in Excelsa, Taxon. Ser.,* 4: 11 (1988).
ser. Annulatae *L.C. Leach in Excelsa, Taxon. Ser.,* 4: 9 (1988).
ser. Asperae *L.C. Leach in Excelsa, Taxon. Ser.,* 4: 11 (1988).
subser. Australes *L.C. Leach in Excelsa, Taxon. Ser.,* 4: 10 (1988).
subser. Boreales *L.C. Leach in Excelsa, Taxon. Ser.,* 4: 10 (1988).
subsect. Calostelmae *(Berger) L.C. Leach in Excelsa, Taxon. Ser.,* 4: 9 (1988): H. sect. Calostelma.
ser. Cleistostelmae *L.C. Leach in Excelsa, Taxon. Ser.,* 4: 12 (1988).
subser. Convexae *L.C. Leach in Excelsa, Taxon. Ser.,* 4: 11 (1988).
sect. Fallacistelma *L.C. Leach in Excelsa, Taxon. Ser.,* 4: 12 (1988).
ser. Humiles *L.C. Leach in Excelsa, Taxon. Ser.,* 4: 11 (1988).
ser. Multangulares *L.C. Leach in Excelsa, Taxon. Ser.,* 4: 9 (1988).
ser. Pauciangulosae *L.C. Leach in Excelsa, Taxon. Ser.,* 4: 10 (1988).
subser. Podostelmae *(Berger) L.C. Leach in Excelsa, Taxon. Ser.,* 4: 10 (1988): H. sect. Podostelma.
subser. Procumbentes *L.C. Leach in Excelsa, Taxon. Ser.,* 4: 12 (1988).
subser. Urceolatae *L.C. Leach in Excelsa, Taxon. Ser.,* 4: 11 (1988).
subser. Verekerae *L.C. Leach in Excelsa, Taxon. Ser.,* 4: 11 (1988).
archeri *L.C. Leach in Excelsa, Taxon. Ser.,* 4: 88 (1988)—Kenya.
bayeri *L.C. Leach in Excelsa, Taxon. Ser.,* 4: 187 (1988)—South Africa (Cape Province).
brevirostris
subsp. baviaana *L.C. Leach in Excelsa, Taxon. Ser.,* 4: 166 (1988)—South Africa (Cape Province).
subsp. intermedia *(N.E. Br.) L.C. Leach in Excelsa, Taxon. Ser.,* 4: 164 (1988): H. brevirostris var. intermedia.
echidnopsioides *(Leach) L.C. Leach in Excelsa, Taxon. Ser.,* 4: 51 (1988): H. pillansii subsp. echidnopsioides.

HUERNIA :–
guttata
subsp. calitzdorpensis *L.C. Leach in Excelsa, Taxon. Ser.*, 4: 38 (1988)—South Africa (Cape Province).
plowesii *L.C. Leach in Excelsa, Taxon. Ser.*, 4: 134 (1988)—Namibia.
quinta
var. blyderiverensis *L.C. Leach in Excelsa, Taxon. Ser.*, 4: 178 (1988)—South Africa (Transvaal).
thudichumii *L.C. Leach in Excelsa, Taxon. Ser.*, 4: 132 (1988)—South Africa (Cape Province).
thuretii
var. primulina *(N.E. Br.) L.C. Leach in Excelsa, Taxon. Ser.*, 4: 185 (1988): H. primulina.
zebrina
subsp. magniflora *(Phillips) L.C. Leach in Excelsa, Taxon. Ser.*, 4: 139 (1988): H. zebrina var. magniflora.

HUMBERTIOTURRAEA (Meliac.).
baronii *(C.DC.) M. Cheek in Kew Bull.*, 44(2): 370 (1989): Turraea baronii.
decaryana *(Danguy) M. Cheek in Kew Bull.*, 44(2): 370 (1989): Turraea decaryana.
grandidieri *(Baillon) M. Cheek in Kew Bull.*, 44(2): 370 (1989): Turraea grandidieri.
maculata *(Smith) M. Cheek in Kew Bull.*, 44(2): 369 (1989): Turraea maculata.
malifolia *(Baker) M. Cheek in Kew Bull.*, 44(2): 369 (1989): Turraea malifolia.
rhamnifolia *(Baker) M. Cheek in Kew Bull.*, 44(2): 369 (1989): Turraea rhamnifolia.
ripicola *(C.DC.) M. Cheek in Kew Bull.*, 44(2): 370 (1989): Turraea ripicola.

HYALOPOA (Gramin.).
czirahica *Sh.A. Guseĭnov in Bot. Zhurn.*, 73(12): 1743 (1988)—U.S.S.R. (Caucasus).

HYALOSPERMA (Compos.).
cotula *(Benth.) Paul G. Wilson in Nuytsia,* 7(1): 93 (1989): Helichrysum cotula.
demissum *(A. Gray) Paul G. Wilson in Nuytsia,* 7(1): 85 (1989): Pteropogon demissus.
glutinosum
subsp. venustum *(S. Moore) Paul G. Wilson in Nuytsia,* 7(1): 91 (1989): Helipterum venustum.
praecox *(F. Muell.) Paul G. Wilson in Nuytsia,* 7(1): 96 (1989): Helipterum praecox.
pusillum *(Turcz.) Paul G. Wilson in Nuytsia,* 7(1): 97 (1989): Helipterum pusillum.
semisterile *(F. Muell.) Paul G. Wilson in Nuytsia,* 7(1): 92 (1989): Helipterum semisterile.
simplex *(Steetz) Paul G. Wilson in Nuytsia,* 7(1): 94 (1989): Helipterum simplex.
subsp. graniticola *Paul G. Wilson in Nuytsia,* 7(1): 95 (1989)—Australia (W. Australia).
stoveae *(D.A. Cooke) Paul G. Wilson in Nuytsia,* 7(1): 87 (1989): Helipterum stoveae.

HYALOSPERMA :–
zacchaeus *(S. Moore) Paul G. Wilson in Nuytsia,* 7(1): 85 (1989): Helipterum zacchaeus.

HYBANTHUS (Violac.).
jefensis *C.A. Todzia in Ann. Missouri Bot. Gard.*, 76(1): 360 (1989)—Panama.

HYDRANGEA (Hydrangeac.).
serrata
var. minamitanii *H. Ohba in J. Jap. Bot.*, 64(7): 199 (1989)—Japan.

HYDROCOTYLE (Umbellif.).
calcicola *Y.H. Li in Guihaia,* 9(1): 25 (1989)—China.

HYDROLEA (Hydrophyllac.).
brevistyla *B. Verdcourt in Fl. Trop. E. Afr., Hydrophyllac.:* 4 (1989)—Zambia, Tanzania.

HYGROCHILUS (Orchidac.).
parishii
var. mariottianus *(Rchb. f.) K. Senghas in Schlechter Orchideen,* 1(22): 1357 (1989): Vanda parishii var. mariottiana.

HYMENOCALLIS (Amaryllidac.).
howardii *J.A. Bauml in R. McVaugh, Fl. Novo-Galiciana,* 15: 212 (1989)—Mexico.
leavenworthii *(Standl. & Steyerm.) J.A. Bauml in R. McVaugh, Fl. Novo-Galiciana,* 15: 215 (1989): Pancratium leavenworthii.
phalangidis *J.A. Bauml in R. McVaugh, Fl. Novo-Galiciana,* 15: 216 (1989)—Mexico.
proterantha *J.A. Bauml in R. McVaugh, Fl. Novo-Galiciana,* 15: 216 (1989)—Mexico.
pumila *J.A. Bauml in R. McVaugh, Fl. Novo-Galiciana,* 15: 218 (1989)—Mexico.

HYMENOLOBIUM (Legumin.).
mesoamericanum *H.C. de Lima in Ann. Missouri Bot. Gard.*, 75(3): 1145 (1988)—Costa Rica, Nicaragua, Panama.

HYMENOPAPPUS (Compos.).
carrizoanus *B.L. Turner in Phytologia,* 67(4): 294 (1989)—U.S.A. (Texas).

HYOSERIS (Compos.).
frutescens *S. Brullo & P. Pavone in Candollea,* 43(2): 717 (1988)—Malta.

HYPARRHENIA (Gramin.).
rufa
subsp. altissima *(Stapf) B.K. Simon in Austrobaileya,* 3(1): 86 (1989): H. altissima.

HYPECOUM (Papaverac.).
angustilobum *A.E. Dahl in Pl. Syst. Evol.*, 163(3–4): 260 (1989)—U.S.S.R. (Caucasus, Middle Asia), Iraq.

HYPECOUM :-
procumbens
subsp. atropunctatum *A.E. Dahl in Pl. Syst. Evol.,* 163(3–4): 255 (1989)—Turkey, East Aegean Is.
subsp. fragrantissimum *A.E. Dahl in Pl. Syst. Evol.,* 163(3–4): 254 (1989)—Greece.
torulosum *A.E. Dahl in Pl. Syst. Evol.,* 163(3–4): 268 (1989)—Greece, East Aegean Is., Crete, Algeria, Bulgaria, Cyprus, Corsica, France, Italy, Sicily, Romania, Turkey.
trullatum *A.E. Dahl in Pl. Syst. Evol.,* 163(3–4): 257 (1989)—Turkey.

HYPERBAENA (Menispermac.).
jalcomulcensis *E. Pérez Cueto & G. Castillo-Campos in Acta Bot. Mex.,* 4: 15 (1988)—Mexico.

HYPERICUM (Guttif.).
cerastoides
subsp. meuselianum *I. Hagemann in Flora (Germany),* 183(3–4): 272 (1989)—Bulgaria/Greece border.

HYPOBATHRUM (Rubiac.).
longifolium *(DC.) K.M. Wong in Tree Fl. Malaya,* 4: 354 (1989): Petunga longifolia.
var. hirta *(Ridl.) K.M. Wong in Tree Fl. Malaya,* 4: 354 (1989): Petunga hirta.
venulosum *(Hk. f.) K.M. Wong in Tree Fl. Malaya,* 4: 355 (1989): Petunga venulosa.

HYPOXIS (Hypoxidac.).
limicola *B.L. Burtt in Notes Roy. Bot. Gard. Edinburgh,* 45(2): 188 (1988 publ. 1989)—South Africa (Transvaal), Swaziland.
lucens *R. McVaugh, Fl. Novo-Galiciana,* 15: 225 (1989)—Mexico.
parvula
var. albiflora *B.L. Burtt in Notes Roy. Bot. Gard. Edinburgh,* 45(2): 190 (1988 publ. 1989)—South Africa (Natal).

I

ILEX (Aquifoliac.).
abscondita *J.A. Steyermark in Ann. Missouri Bot. Gard.,* 75(1): 320 (1988)—Venezuela.
acutidenticulata *J.A. Steyermark in Ann. Missouri Bot. Gard.,* 75(1): 321 (1988)—Venezuela.
altiplana *J.A. Steyermark in Ann. Missouri Bot. Gard.,* 75(1): 321 (1988)—Venezuela.
attenuata *J.A. Steyermark in Ann. Missouri Bot. Gard.,* 75(1): 322 (1988)—Brazil.
brevipedicellata *J.A. Steyermark in Ann. Missouri Bot. Gard.,* 75(1): 322 (1988)—Venezuela.

ILEX :-
cardonae *J.A. Steyermark in Ann. Missouri Bot. Gard.,* 75(1): 323 (1988)—Venezuela.
cheniana *T.R. Dudley in Holly Soc. J.,* 6(4): 15 (1988), nom. nov.: I. congesta *H.W. Li ex Y.R. Li.*
ciliolata *J.A. Steyermark in Ann. Missouri Bot. Gard.,* 75(1): 323 (1988)—Venezuela.
davidsei *J.A. Steyermark in Ann. Missouri Bot. Gard.,* 75(1): 324 (1988)—Venezuela.
glabella *J.A. Steyermark in Ann. Missouri Bot. Gard.,* 75(1): 324 (1988)—Venezuela.
gransabanensis *J.A. Steyermark in Ann. Missouri Bot. Gard.,* 75(1): 325 (1988)—Venezuela.
guaiquinimae *J.A. Steyermark in Ann. Missouri Bot. Gard.,* 75(1): 325 (1988)—Venezuela.
holstii *J.A. Steyermark in Ann. Missouri Bot. Gard.,* 75(1): 326 (1988)—Venezuela.
ignicola *J.A. Steyermark in Ann. Missouri Bot. Gard.,* 75(1): 326 (1988)—Venezuela.
jauaensis *J.A. Steyermark in Ann. Missouri Bot. Gard.,* 75(1): 327 (1988)—Venezuela.
longipilosa *J.A. Steyermark in Ann. Missouri Bot. Gard.,* 75(1): 327 (1988)—Venezuela.
magnifructa
var. minor *J.A. Steyermark in Ann. Missouri Bot. Gard.,* 75(1): 328 (1988)—Venezuela.
marahuacae *J.A. Steyermark in Ann. Missouri Bot. Gard.,* 75(1): 328 (1988)—Venezuela.
paruensis *J.A. Steyermark in Ann. Missouri Bot. Gard.,* 75(1): 330 (1988)—Venezuela.
paujiensis *J.A. Steyermark in Ann. Missouri Bot. Gard.,* 75(1): 330 (1988)—Venezuela.
pingheensis *C.J. Tseng in Bull. Bot. Res. North-East. Forest. Inst.,* 9(4): 29 (1989)—China.
polita *J.A. Steyermark in Ann. Missouri Bot. Gard.,* 75(1): 331 (1988)—Venezuela.
spathulata *J.A. Steyermark in Ann. Missouri Bot. Gard.,* 75(1): 331 (1988)—Venezuela.
summa *J.A. Steyermark in Ann. Missouri Bot. Gard.,* 75(1): 332 (1988)—Venezuela.
tetramera
var. glabra *(C.Y. Wu ex Y.R. Li) T.R. Dudley in Holly Soc. J.,* 6(4): 28 (1988): I. tephrophylla var. glabra.
wuiana *T.R. Dudley in Holly Soc. J.,* 6(4): 14 (1988), nom. nov.: I. mamillata *C.Y. Wu ex Y.R. Li.*
wurdackiana *J.A. Steyermark in Ann. Missouri Bot. Gard.,* 75(1): 332 (1988)—Venezuela.

IMPATIENS (Balsaminac.).
acehensis *C. Grey-Wilson in Kew Bull.,* 44(1): 78 (1989)—Sumatra.
arriensii *(Zoll.) T. Shimizu in Acta Phytotax. Geobot.,* 38: 53 (1987), as 'alliensii': Semeiocardium arriensii.†
subsp. harmandii *(Hook. f.) C. Grey-Wilson in Kew Bull.,* 44(1): 112 (1989): I. harmandii.
arunensis *C. Grey-Wilson in Kew Bull.,* 44(1): 65 (1989)—Nepal.
buennemeijeri *C. Grey-Wilson in Kew Bull.,* 44(1): 92 (1989)—Sumatra.

IMPATIENS :–
calendulina *C. Grey-Wilson in Kew Bull.*, 44(1): 104 (1989)—Sumatra.
dewildeana *C. Grey-Wilson in Kew Bull.*, 44(1): 87 (1989)—Sumatra.
var. parviflora *C. Grey-Wilson in Kew Bull.*, 44(1): 88 (1989)—Sumatra.
diepenhorstii
var. brevicornu *C. Grey-Wilson in Kew Bull.*, 44(1): 84 (1989)—Sumatra.
var. intermedia *C. Grey-Wilson in Kew Bull.*, 44(1): 84 (1989)—Sumatra.
elephanticeps *C. Grey-Wilson in Kew Bull.*, 44(1): 95 (1989)—Sumatra.
fenghwaiana *Y.L. Chen in Acta Phytotax. Sin.*, 27(5): 395 (1989)—China.
floretii *N. Hallé & A.M. Louis in Bull. Mus. Nation. Hist. Nat., B, Adansonia,* Ser. 4, 11(1): 12 (1989)—Gabon.
furfuracea *C. Grey-Wilson in Kew Bull.*, 44(1): 90 (1989)—Sumatra.
hunanensis *Y.L. Chen in Acta Phytotax. Sin.*, 27(5): 402 (1989)—China.
jinggangensis *Y.L. Chen in Acta Phytotax. Sin.*, 27(5): 399 (1989)—China.
kaliensis *C. Grey-Wilson in Kew Bull.*, 44(1): 65 (1989)—India.
kathmanduensis *C. Grey-Wilson in Kew Bull.*, 44(1): 119 (1989)—Nepal.
lampungensis *C. Grey-Wilson in Kew Bull.*, 44(1): 73 (1989)—Sumatra.
pilosivittata *C. Grey-Wilson in Kew Bull.*, 44(1): 85 (1989)—Sumatra.
pseudoperezii *C. Grey-Wilson in Kew Bull.*, 44(1): 95 (1989)—Sumatra.
subsp. bengkulensis *C. Grey-Wilson in Kew Bull.*, 44(1): 97 (1989)—Sumatra.
subsp. ophirensis *C. Grey-Wilson in Kew Bull.*, 44(1): 97 (1989)—Sumatra.
pyrrhotricha
var. grandis *C. Grey-Wilson in Kew Bull.*, 44(1): 78 (1989)—Sumatra.
rubriflora *C. Grey-Wilson in Kew Bull.*, 44(1): 105 (1989)—Sumatra.
sidikalangensis *C. Grey-Wilson in Kew Bull.*, 44(1): 89 (1989)—Sumatra.
singgalangensis *C. Grey-Wilson in Kew Bull.*, 44(1): 79 (1989)—Sumatra.
steenisii *C. Grey-Wilson in Kew Bull.*, 44(1): 103 (1989)—Sumatra.
tapanuliensis *C. Grey-Wilson in Kew Bull.*, 44(1): 75 (1989)—Sumatra.
vitellina *C. Grey-Wilson in Kew Bull.*, 44(1): 73 (1989)—Sumatra.
wuyuanensis *Y.L. Chen in Acta Phytotax. Sin.*, 27(5): 397 (1989)—China.

INDIGOFERA (Legumin.).
subsect. Bracteatae *Y.Y. Fang & C.Z. Zheng in Acta Phytotax. Sin.*, 27(3): 165 (1989).
subsect. Chaetodontae *Y.Y. Fang & C.Z. Zheng in Acta Phytotax. Sin.*, 27(3): 166 (1989).
subsect. Decorae *Y.Y. Fang & C.Z. Zheng in Acta Phytotax. Sin.*, 27(3): 164 (1989).
subsect. Hancockianae *Y.Y. Fang & C.Z. Zheng in Acta Phytotax. Sin.*, 27(3): 166 (1989).

INDIGOFERA :–
subsect. Henryanae *Y.Y. Fang & C.Z. Zheng in Acta Phytotax. Sin.*, 27(3): 165 (1989).
subsect. Pendulae *Y.Y. Fang & C.Z. Zheng in Acta Phytotax. Sin.*, 27(3): 163 (1989).
subsect. Pseudotinctoria *Y.Y. Fang & C.Z. Zheng in Acta Phytotax. Sin.*, 27(3): 167 (1989).
subsect. Scabridae *Y.Y. Fang & C.Z. Zheng in Acta Phytotax. Sin.*, 27(3): 167 (1989).
subsect. Simplicifoliae *(Harv.) Y.Y. Fang & C.Z. Zheng in Acta Phytotax. Sin.*, 27(3): 168 (1989): I. ser. Simplicifoliae.
subsect. Squamatae *Y.Y. Fang & C.Z. Zheng in Acta Phytotax. Sin.*, 27(3): 165 (1989).
subsect. Trifoliolatae *(Harv.) Y.Y. Fang & C.Z. Zheng in Acta Phytotax. Sin.*, 27(3): 168 (1989): I. ser. Trifoliolatae.
ancistrocarpa *M. Thulin in Nordic J. Bot.*, 8(5): 466 (1989)—Kenya.
curvirostrata *M. Thulin in Nordic J. Bot.*, 8(5): 467 (1989)—Ethiopia.
decora
var. chalara *(Craib) Y.Y. Fang & C.Z. Zheng in Acta Phytotax. Sin.*, 27(3): 164 (1989): I. chalara.
var. cooperi *(Craib) Y.Y. Fang & C.Z. Zheng in Acta Phytotax. Sin.*, 27(3): 164 (1989): I. cooperi.
var. ichangensis *(Craib) Y.Y. Fang & C.Z. Zheng in Acta Phytotax. Sin.*, 27(3): 164 (1989): I. ichangensis.
hiranensis *M. Thulin in Nordic J. Bot.*, 8(5): 469 (1989)—Somali Republic.
karinensis *M. Thulin in Nordic J. Bot.*, 8(5): 469 (1989)—Somali Republic.
kucharii *M. Thulin in Nordic J. Bot.*, 8(5): 471 (1989)—Somali Republic.
neosericopetala *P.C. Li in Acta Bot. Yunnanica*, 11(3): 285 (1989)—China.
nugalensis *M. Thulin in Nordic J. Bot.*, 8(5): 471 (1989)—Somali Republic.
pellucida *Gillett & M. Thulin in Nordic J. Bot.*, 8(5): 473 (1989)—Somali Republic/Kenya border.
sabulosa *M. Thulin in Nordic J. Bot.*, 8(5): 474 (1989)—Somali Republic, Ethiopia.
sessilis *M. Thulin in Nordic J. Bot.*, 8(5): 475 (1989)—Somali Republic.
tenuirostris *M. Thulin in Nordic J. Bot.*, 8(5): 476 (1989)—Somali Republic.

INDOSASA (Gramin.).
laotica *(A. Camus) C.S. Chao & S.A. Renvoize in Kew Bull.*, 44(2): 365 (1989): Arundinaria laotica.

INEZIA (Compos.).
speciosa *F. Brusse in Bothalia,* 19(1): 27 (1989)—South Africa (Transvaal).

INGA (Legumin.).
amazonica *L. Cárdenas in Ernstia,* 51: 1 (1989), nom. nov.: I. speciosa *Benth.*

INGA :–
var. bracteifera *(Ducke) L. Cárdenas in Ernstia,* 51: 1 (1989): I. speciosa var. bracteifera.
var. lomatophylla *(Benth.) L. Cárdenas in Ernstia,* 51: 1 (1989): I. speciosa var. lomatophylla.
var. membranacea *(Ducke) L. Cárdenas in Ernstia,* 51: 1 (1989): I. speciosa var. membranacea.
jefensis *R.L. Liesner & W.G. D'Arcy in Ann. Missouri Bot. Gard.,* 75(1): 385 (1988)—Panama.
spiralis *R.L. Liesner & W.G. D'Arcy in Ann. Missouri Bot. Gard.,* 75(1): 385 (1988)—Panama.

IPOMOEA (Convolvulac.).
bracteata
var. viridibracta *A. McDonald in Brenesia,* 28: 60 (1987 publ. 1989)—Mexico.
fissifolia *(McPherson) J.E. Eckenwalder in Brittonia,* 41(1): 79 (1989): Quamoclit fissifolia.
praematura *J.E. Eckenwalder in Brittonia,* 41(1): 75 (1989)—Windward Is., Trinidad, Guyana, Colombia.
schaijesii *J. Lejoly & S. Lisowski in Bull. Jard. Bot. Nation. Belg.,* 58(3–4): 554 (1988)—Zaïre.

IPOMOPSIS (Polemoniac.).
tenuituba
subsp. latiloba *V. Grant & D.H. Wilken in Bot. Gaz.,* 149(4): 446 (1988)—U.S.A. (Arizona, Utah).
subsp. macrosiphon *(Kearney & Peebles) V. Grant & D.H. Wilken in Bot. Gaz.,* 149(4): 447 (1988): Gilia aggregata var. macrosiphon.

IRANECIO B. Nordenstam in Fl. Iran., 164: 53 (1989). *COMPOSITAE.*
sect. Synotoides *B. Nordenstam in Fl. Iran.,* 164: 57 (1989).
elbrusensis *(Boiss.) B. Nordenstam in Fl. Iran.,* 164: 55 (1989): Senecio elbrusensis.
oligolepis *(Boiss.) B. Nordenstam in Fl. Iran.,* 164: 54 (1989): Senecio oligolepis.
othonnae *(M.B.) B. Nordenstam in Fl. Iran.,* 164: 58 (1989): Senecio othonnae.
paucilobus *(DC.) B. Nordenstam in Fl. Iran.,* 164: 56 (1989): Senecio paucilobus.

IRIS (Iridac.).
sect. Brevituba *B. Mathew in Davis & Hedge Festschrift (ed. Kit Tan):* 84 (1989).
sect. Monolepis *(Rodionenko) B. Mathew in Davis & Hedge Festschrift (ed. Kit Tan):* 87 (1989): Iridodictyum sect. Monolepis.
histrio
subsp. aintabensis *(G.P. Baker) B. Mathew in Davis & Hedge Festschrift (ed. Kit Tan):* 97 (1989): A. histrio var. aintabensis.

IRIS :–
pseudonotha *A.I. Galushko in A.I. Galushko (ed.), Fl. Severn. Kavkaza i Voprosў ee istorii:* 9 (1983)—U.S.S.R. (Caucasus).

ISCHAEMUM (Gramin.).
albovillosum *B.K. Simon in Austrobaileya,* 3(1): 86 (1989)—Australia (W. Australia).
tropicum *B.K. Simon in Austrobaileya,* 3(1): 88 (1989)—Australia (N. Terr., Queensland), New Guinea (Papua New Guinea).

ISOMETRUM (Gesneriac.).
sect. Chorianthera *W.T. Wang & K.Y. Pan in Acta Phytotax. Sin.,* 26(6): 432 (1988).
eximium *Chun ex K.Y. Pan in Acta Phytotax. Sin.,* 26(6): 432 (1988)—China.

IVESIA (Rosac.).
aperta
var. canina *B. Ertter in Brittonia,* 40(4): 398 (1988)—U.S.A. (California).
arizonica *(Eastw. ex J.T. Howell) B. Ertter in Syst. Bot.,* 14(2): 233 (1989): Purpusia arizonica.
var. saxosa *(T.S. Brandegee) B. Ertter in Syst. Bot.,* 14(2): 233 (1989): Purpusia saxosa.
baileyi
var. beneolens *(Nelson & J.F. Macbr.) B. Ertter in Syst. Bot.,* 14(2): 236 (1989): Horkelia beneolens.
kingii
var. eremica *(Cov.) B. Ertter in Syst. Bot.,* 14(2): 244 (1989): Potentilla eremica.
longibracteata *B. Ertter in Syst. Bot.,* 14(2): 233 (1989)—U.S.A. (California).
patellifera *(J.T. Howell) B. Ertter in Syst. Bot.,* 14(2): 232 (1989): Potentilla patellifera.
pityocharis *B. Ertter in Syst. Bot.,* 14(2): 241 (1989)—U.S.A. (Nevada).
rhypara
var. shellyi *B. Ertter in Syst. Bot.,* 14(2): 239 (1989)—U.S.A. (Oregon).
saxosa *(Lemmon ex E. Greene) B. Ertter in Syst. Bot.,* 14(2): 232 (1989): Potentilla saxosa.
shockleyi
var. ostleri *B. Ertter in Syst. Bot.,* 14(2): 237 (1989)—U.S.A. (Utah).

IXIA (Iridac.).
esterhuyseniae *M.P. de Vos in S. Afr. J. Bot.,* 54(6): 600 (1988)—South Africa (Cape Province).
frederickii *M.P. de Vos in S. Afr. J. Bot.,* 54(6): 596 (1988)—South Africa (Cape Province).
mostertii *M.P. de Vos in S. Afr. J. Bot.,* 54(6): 598 (1988)—South Africa (Cape Province).

IXORA (Rubiac.).
casei
var. medialoba *F.R. Fosberg & H.H. Sachet in Baileya,* 23(2): 85 (1989)—Caroline Is.
coccinea
var. hermannii *F.R. Fosberg & H.H. Sachet in Baileya,* 23(2): 79 (1989)—Sri Lanka.

IXORA :–
var. intermedia *F.R. Fosberg & H.H. Sachet in Baileya,* 23(2): 80 (1989)—Cultivation.
f. lutea *(Hutch.) F.R. Fosberg & H.H. Sachet in Baileya,* 23(2): 79 (1989): I. lutea.
pendula
var. montana *(Ridl.) K.M. Wong in Tree Fl. Malaya,* 4: 359 (1989): I. montana.

J

JACQUINIA (Theophrastac.).
macrocarpa
subsp. panamensis *(Lundell) B. Ståhl in Nordic J. Bot.,* 9(1): 23 (1989): J. panamensis.
subsp. pungens *(A. Gray) B. Ståhl in Nordic J. Bot.,* 9(1): 23 (1989): J. pungens.
montana *B. Ståhl in Nordic J. Bot.,* 9(1): 27 (1989)—Nicaragua.
nitida *B. Ståhl in Nordic J. Bot.,* 9(1): 24 (1989)—Honduras, Guatemala, Nicaragua.

×**JANKAEBERLEA** J. Halda in Preslia, 61(2): 125 (1989). *GESNERIACEAE.* (Haberlea × Jankaea).
panayotii *J. Halda in Preslia,* 61(2): 125 (1989)—Cultivation. (Haberlea rhodopensis × Jankaea heldreichii).

×**JANKAENDRON** J. Halda in Preslia, 61(2): 125 (1989). *GESNERIACEAE.* (Conandron × Jankaea).
stevensiae *J. Halda in Preslia,* 61(2): 125 (1989)—Cultivation. (Conandron ramondioides × Jankaea heldreichii).

×**JANKAESSANDRA** J. Halda in Preslia, 61(2): 127 (1989). *GESNERIACEAE.* (Didissandra × Jankaea).
bluemelii *J. Halda in Preslia,* 61(2): 127 (1989)—Cultivation. (Didissandra lanuginosa × Jankaea heldreichii).

JOSEANTHUS H. Robinson in Rev. Acad. Colomb. Cienc. Exact. Fis. Nat., 65: 210 (1989). *COMPOSITAE.*
chimborazensis *(Hieron.) H. Robinson in Rev. Acad. Colomb. Cienc. Exact. Fis. Nat.,* 65: 211 (1989): Vernonia chimborazensis.
crassilanatus *(Cuatr.) H. Robinson in Rev. Acad. Colomb. Cienc. Exact. Fis. Nat.,* 65: 211 (1989): Vernonia crassilanata.
cuatrecasasii *H. Robinson in Rev. Acad. Colomb. Cienc. Exact. Fis. Nat.,* 65: 211 (1989)—Ecuador.
sparrei *(H. Robinson) H. Robinson in Rev. Acad. Colomb. Cienc. Exact. Fis. Nat.,* 65: 211 (1989): Basionym not stated.

JOSEANTHUS :–
trichotomus *(Gleason) H. Robinson in Rev. Acad. Colomb. Cienc. Exact. Fis. Nat.,* 65: 211 (1989): Basionym not stated.

JOVIBARBA (Crassulac.).
hirta
subsp. arenaria *(Koch) J. Parnell in Acta Bot. Hung.,* 34(1–2): 222 (1988): Sempervivum arenarium.
subsp. borealis *(H. Huber) J. Parnell in Acta Bot. Hung.,* 34(1–2): 222 (1988), isonym; *see R. Soo in Acta Bot. Acad. Sci. Hung.,* 23(3–4): 380 (1977 publ. 1978): Diopogon hirtus subsp. borealis.

JUNCUS (Juncac.).
chiapasensis *H. Balslev in Ann. Missouri Bot. Gard.,* 75(1): 379 (1988)—Mexico.
exsertus
subsp. lesuticus *B.L. Burtt in Notes Roy. Bot. Gard. Edinburgh,* 45(2): 191 (1988 publ. 1989)—South Africa (Cape Province), Lesotho.
liebmannii
var. polycephalus *H. Balslev in Ann. Missouri Bot. Gard.,* 75(1): 381 (1988)—Mexico.
sosnowskyi *V.S. Novikov in Byull. Mosk. Obshch. Ispŷt. Prir., Biol.,* 94(1): 110 (1989)—U.S.S.R. (Caucasus).

JURINEA (Compos.).
cartilaginea *V. Mozaffarian in Iranian J. Bot.,* 4(1): 62 (1988)—Iran.

JURINELLA (Compos.).
macrocephala *(Royle) B.S. Aswal & A.K. Goel in Indian J. Forest.,* 11(4): 339 (1988 publ. 1989): Dolomiaea macrocephala.

JUSTICIA (Acanthac.).
brevipedunculata *Ensermu Kelbessa in Pl. Syst. Evol.,* 163(1–2): 121 (1989)—Tanzania.
cuatrecasasii *D.C. Wasshausen in Rev. Acad. Colomb. Cienc. Exact. Fis. Nat.,* 65: 257 (1989)—Colombia.
diclipteroides
subsp. aethiopica *M. Hedrén in Bull. Mus. Nation. Hist. Nat., B, Adansonia,* Ser. 4, 10(4): 354 (1988 publ. 1989)—Ethiopia.
subsp. kibarianiensis *M. Hedrén in Bull. Mus. Nation. Hist. Nat., B, Adansonia,* Ser. 4, 10(4): 362 (1988 publ. 1989)—Tanzania.
subsp. megaensis *M. Hedrén in Bull. Mus. Nation. Hist. Nat., B, Adansonia,* Ser. 4, 10(4): 359 (1988 publ. 1989)—Ethiopia.
subsp. nierensis *(Mildbr.) M. Hedrén in Bull. Mus. Nation. Hist. Nat., B, Adansonia,* Ser. 4, 10(4): 363 (1988 publ. 1989): J. nierensis.
subsp. praetervisa *(Lindau) M. Hedrén in Bull. Mus. Nation. Hist. Nat., B, Adansonia,* Ser. 4, 10(4): 356 (1988 publ. 1989): J. praetervisa.

JUSTICIA :-
subsp. usambarica *M. Hedrén in Bull. Mus. Nation. Hist. Nat., B, Adansonia,* Ser. 4, 10(4): 359 (1988 publ. 1989)—Tanzania.
hansenii *V.V. Sivarajan & P. Mathew in Taxon,* 38(3): 486 (1989), nom. nov.: J. simplex var. serpyllifolia *Benth. ex C.B. Clarke.*
heterocarpa
subsp. dinteri *(S. Moore) M. Hedrén in Symb. Bot. Upsal.,* 29(1): 106 (1989): J. dinteri.
imlayae *B. Hansen in Nordic J. Bot.,* 9(2): 210 (1989), nom. nov.: J. vasculosa var. laxa *Imlay.*
internodialis *B. Hansen in Nordic J. Bot.,* 9(2): 210 (1989), nom. nov.: Rhyticalymma longipetiolatum *Brem.*
kirkbridei *D.C. Wasshausen in Rev. Acad. Colomb. Cienc. Exact. Fis. Nat.,* 65: 255 (1989)—Colombia.
leucophloea *(Nees) D. Wasshausen in Rhodora,* 71: 551 (1969): Rhytiglossa leucophloea.
luteocinerea *M. Hedrén in Symb. Bot. Upsal.,* 29(1): 79 (1989)—Kenya, Ethiopia.
stolonifera *(C.B. Clarke) B. Hansen in Nordic J. Bot.,* 9(2): 210 (1989): Rungia stolonifera.
striata
var. dyschoristeoides *(C.B. Clarke) M. Hedrén in Symb. Bot. Upsal.,* 29(1): 111 (1989): J. dyschoristeoides.
unyorensis
var. keniensis *(Rendle) M. Hedrén in Symb. Bot. Upsal.,* 29(1): 108 (1989): J. keniensis.
upembensis *M. Hedrén in Symb. Bot. Upsal.,* 29(1): 112 (1989)—Zaïre.

K

KAEMPFERIA (Zingiberac.).
larsenii *P. Sirirugsa in Nordic J. Bot.,* 9(3): 257 (1989)—Thailand.
siamensis *P. Sirirugsa in Nordic J. Bot.,* 9(3): 257 (1989)—Thailand.
spoliata *P. Sirirugsa in Nordic J. Bot.,* 9(3): 259 (1989)—Thailand.

KAILARSENIA (Rubiac.).
jardinei *(F. Muell. ex Benth.) C.F. Puttock in Austrobaileya,* 3(1): 53 (1989): Gardenia jardinei.
ochreata *(F. Muell.) C.F. Puttock in Austrobaileya,* 3(1): 54 (1989): Gardenia ochreata.
suffruticosa *(R. Br. ex Benth.) C.F. Puttock in Austrobaileya,* 3(1): 59 (1989): Gardenia suffruticosa.

KALANCHOE (Crassulac.).
deficiens
var. glabra *E. Raadts in Willdenowia,* 18(2): 423 (1989)—Yemen, South Yemen.

KEFERSTEINIA (Orchidac.).
andreettae *G. Gerlach, T. Neudecker & H.G. Seeger in Orchidee,* 40(4): 133 (1989)—Ecuador.

KICKXIA (Scrophulariac.).
abhaica *D.A. Sutton, Revis. tribe Antirrhineae:* 241 (1988)—Saudi Arabia.
adpressa *D.A. Sutton, Revis. tribe Antirrhineae:* 200 (1988)—Kenya.
cabulica *(Bentham) D.A. Sutton, Revis. tribe Antirrhineae:* 196 (1988): Linaria cabulica.
collenetteana *D.A. Sutton, Revis. tribe Antirrhineae:* 208 (1988)—Saudi Arabia.
corallicola *D.A. Sutton, Revis. tribe Antirrhineae:* 225 (1988)—Saudi Arabia.
dentata *(Vahl) D.A. Sutton, Revis. tribe Antirrhineae:* 184 (1988): Antirrhinum dentatum.
elegans *(G. Forster) D.A. Sutton, Revis. tribe Antirrhineae:* 211 (1988): Antirrhinum elegans.
glaberrima *(J.A. Schmidt) D.A. Sutton, Revis. tribe Antirrhineae:* 215 (1988): Linaria brunneri var. glaberrima.
gracilis *D.A. Sutton, Revis. tribe Antirrhineae:* 222 (1988), nom. nov.: Linaria gracilis *R. Br. ex Bentham.*
heterophylla
subsp. canariensis *V.W. Smith in D.A. Sutton, Revis. tribe Antirrhineae:* 220 (1988)—Canary Is.
kasalensis *V.W. Smith in D.A. Sutton, Revis. tribe Antirrhineae:* 223 (1988)—Sudan.
membranacea *D.A. Sutton, Revis. tribe Antirrhineae:* 191 (1988)—Bhutan.
monodiana *(Maire) G. Alziar in Biocosme Mésogéen,* 5(3–4): 86 (1988): Linaria monodiana.
petiolata *D.A. Sutton, Revis. tribe Antirrhineae:* 244 (1988)—Yemen.
pseudoscoparia *V.W. Smith & D.A. Sutton in D.A. Sutton, Revis. tribe Antirrhineae:* 231 (1988)—Saudi Arabia.
qaraica *D.A. Sutton, Revis. tribe Antirrhineae:* 227 (1988)—Oman.
sabarum *V.W. Smith & D.A. Sutton in D.A. Sutton, Revis. tribe Antirrhineae:* 233 (1988)—Yemen.
saccata *D.A. Sutton, Revis. tribe Antirrhineae:* 246 (1988)—Yemen.
scalarum *D.A. Sutton, Revis. tribe Antirrhineae:* 237 (1988)—Yemen.
spiniflora *(Schwartz) D.A. Sutton, Revis. tribe Antirrhineae:* 229 (1988): Linaria spiniflora.
woodii *D.A. Sutton, Revis. tribe Antirrhineae:* 239 (1988)—Yemen, ?South Yemen.

KLEINIA (Compos.).
grandiflora
var. major *(Ramasw.) B.P. Uniyal in R.R. Rao et al., Fl. Ind. Enumerat. - Asterac.:* 47 (1988): Notonia grandiflora var. major.
longifolia *A.H. Haworth, Revis. Pl. Succ.:* 193 (1821)—Locality not stated.
shevaroyensis *(Fyson) B.P. Uniyal in R.R. Rao et al., Fl. Ind. Enumerat. - Asterac.:* 47 (1988): Notonia shevaroyensis.
walkeri *(Clarke) B.P. Uniyal in R.R. Rao et al., Fl. Ind. Enumerat. - Asterac.:* 48 (1988): Notonia walkeri.

KNOXIA (Rubiac.).
hedyotoidea *(K. Schum.) C. Puff & E. Robbrecht in Bot. Jahrb.,* 110(4): 548 (1989): Baumannia hedyotoidea.
manika *(Verdcourt) C. Puff & E. Robbrecht in Bot. Jahrb.,* 110(4): 548 (1989): Afroknoxia manika.

KOANOPHYLLON (Compos.).
monanthum *(Schultz-Bip.) T.J. Ayers & B.L. Turner in Sida,* 13(3): 341 (1989): Eupatorium monanthum.

KOHAUTIA (Rubiac.).
caespitosa
subsp. brachyloba *(Sonder) D. Mantell in Fl. Zambes.,* 5(1): 88 (1989): Hedyotis brachyloba.
microflora *D. Mantell in Fl. Zambes.,* 5(1): 90 (1989)—Botswana, Namibia, South Africa (Cape Province, Transvaal).
subverticillata *(K. Schum.) D. Mantell in Fl. Zambes.,* 5(1): 89 (1989): Oldenlandia subverticillata.

KOYAMAEA W.W. Thomas & G. Davidse in Syst. Bot., 14(2): 189 (1989). *CYPERACEAE.*
neblinensis *W.W. Thomas & G. Davidse in Syst. Bot.,* 14(2): 189 (1989)—Venezuela, Brazil.

L

LABLAB (Legumin.).
purpureus
var. lignosus *(Prain) Kumari ex M. Chandrabose & N.C. Nair, Fl. Coimbatore:* 85 (1988): Dolichos lablab var. lignosus.

LABORDIA (Loganiac.).
hosakana *(Sherff) W.L. Wagner, D.R. Herbst & S.H. Sohmer in Bishop Mus. Occas. Pap.,* 29: 102 (1989): L. hedyosmifolia var. hosakana.

LACANDONIA E. Martínez & C.H. Ramos in Ann. Missouri Bot. Gard., 76(1): 128 (1989). *LACANDONIACEAE.*
schismatica *E. Martínez & C.H. Ramos in Ann. Missouri Bot. Gard.,* 76(1): 130 (1989)—Mexico.

LAELIA (Orchidac.).
duveenii *J.A. Fowlie in Orchid Dig.,* 52(4): 180 (1988)—Brazil.

LAGOCHILUS (Labiat.).
aucheri
var. elegans *Z. Jamzad in Iranian J. Bot.,* 4(1): 96 (1988)—Iran.
subsp. heterophyllus *Z. Jamzad in Iranian J. Bot.,* 4(1): 97 (1988)—Iran.
var. tomentosus *Z. Jamzad in Iranian J. Bot.,* 4(1): 96 (1988)—Iran.
lasiocalyx *(Stapf) Z. Jamzad in Iranian J. Bot.,* 4(1): 97 (1988): L. aucheri var. lasiocalyx.
quadridentatus *Z. Jamzad in Iranian J. Bot.,* 4(1): 100 (1988)—Iran.

LAGURUS (Gramin.).
ovatus
subsp. vestitus *(Messeri) S. Brullo in Colloqu. Phytosociolog.,* 12: 81 (1983); *see Pubbl. Ist. Bot. Univ. Catania* (1985): L. ovatus var. vestitus.

LAMIASTRUM (Labiat.).
galeobdolon
subsp. argentatum *(Smejkal) C.A. Stace in Watsonia,* 17(4): 443 (1989): Galeobdolon argentatum.

LAMIUM (Labiat.).
sect. Amplexicaule *J. Mennema, Taxon. Revis. Lamium (Leiden Bot. Ser., 11):* 48 (1989).
album
subsp. barbatum *(Siebold & Zucc.) J. Mennema, Taxon. Revis. Lamium (Leiden Bot. Ser., 11):* 70 (1989): L. barbatum.
amplexicaule
var. orientale *(Pacz.) J. Mennema, Taxon. Revis. Lamium (Leiden Bot. Ser., 11):* 57 (1989): L. amplexicaule f. orientale.
galeobdolon
f. argentatum *(Smejkal) J. Mennema, Taxon. Revis. Lamium (Leiden Bot. Ser., 11):* 46 (1989): Galeobdolon argentatum.
garganicum
var. armenum *(Boiss.) J. Mennema, Taxon. Revis. Lamium (Leiden Bot. Ser., 11):* 109 (1989): L. armenum.
subsp. corsicum *(Gren. & Godron) J. Mennema, Taxon. Revis. Lamium (Leiden Bot. Ser., 11):* 104 (1989): L. corsicum.
var. microphyllum *(Boiss.) J. Mennema, Taxon. Revis. Lamium (Leiden Bot. Ser., 11):* 110 (1989): L. microphyllum.

LAMIUM :–
moschatum
subsp. micranthum *(Boiss.) J. Mennema, Taxon. Revis. Lamium (Leiden Bot. Ser., 11):* 85 (1989): L. moschatum var. micranthum.
purpureum
subsp. caucasicum *(Grossh.) Yu.L. Menitskiĭ in Fl. Armenii,* 8: 80 (1987): L. caucasicum.
var. ehrenbergii *(Boiss. & Reuter ex Boiss.) J. Mennema, Taxon. Revis. Lamium (Leiden Bot. Ser., 11):* 120 (1989): L. ehrenbergii.
sempervirens *A.P. Khokhryakov in Byull. Glavn. Bot. Sada (Moscow),* 152: 61 (1989)—U.S.S.R. (Caucasus), Turkey.

LAMPROCOCCUS (Bromeliac.).
brevicollis *(L.B. Smith) L.B. Smith & W.J. Kress in Phytologia,* 66(1): 70 (1989): Aechmea brevicollis.
campanulatus *(L.B. Smith) L.B. Smith & W.J. Kress in Phytologia,* 66(1): 70 (1989): Aechmea campanulata.
carvalhoi *(E. Pereira & Leme) L.B. Smith & W.J. Kress in Phytologia,* 66(1): 71 (1989): Aechmea carvalhoi.
cathcartii *(C.F. Reed & R.W. Read) L.B. Smith & W.J. Kress in Phytologia,* 66(1): 71 (1989): Aechmea cathcartii.
farinosus
var. discolor *(Beer ex Baker) L.B. Smith & W.J. Kress in Phytologia,* 66(1): 71 (1989): Aechmea conglomerata var. discolor.
var. glomeratus *(Beer) L.B. Smith & W.J. Kress in Phytologia,* 66(1): 71 (1989): L. glomeratus.
pedicellatus *(E.M.C. Leme & H.E. Luther) L.B. Smith & W.J. Kress in Phytologia,* 66(1): 71 (1989): Aechmea pedicellata.
podanthus *(L.B. Smith) L.B. Smith & W.J. Kress in Phytologia,* 66(1): 71 (1989): Aechmea podantha.
racinae *(L.B. Smith) L.B. Smith & W.J. Kress in Phytologia,* 66(1): 71 (1989): Aechmea racinae.
var. erectus *(L.B. Smith) L.B. Smith & W.J. Kress in Phytologia,* 66(1): 71 (1989): Aechmea racinae var. erecta.
var. tubiformis *(E. Pereira) L.B. Smith & W.J. Kress in Phytologia,* 66(1): 71 (1989): Aechmea racinae var. tubiformis.
victorianus *(L.B. Smith) L.B. Smith & W.J. Kress in Phytologia,* 66(1): 71 (1989): Aechmea victoriana.
var. discolor *(M.B. Foster) L.B. Smith & W.J. Kress in Phytologia,* 66(1): 71 (1989): Aechmea victoriana var. discolor.
warasii *(E. Pereira) L.B. Smith & W.J. Kress in Phytologia,* 66(1): 71 (1989): Aechmea warasii.

LANTANA (Verbenac.).
camara
f. glandulosa *R.B. Fernandes in Bol. Soc. Brot.,* 61: 132 (1988)—Angola, Mexico, Guatemala, Costa Rica, Venezuela, Brazil, Uruguay, Argentina, Chile.
machadoi *R.B. Fernandes in Bol. Soc. Brot.,* 61: 142 (1988)—Angola.
moldenkei *R.B. Fernandes in Bol. Soc. Brot.,* 61: 147 (1988)—Zimbabwe.
ruiz-teranii *S. López-Palacios & J.A. Steyermark in Acta Bot. Venezuel.,* 14(1): 163 (1983)—Venezuela.
tiliifolia
f. glandulosa *(Schau.) R.B. Fernandes in Bol. Soc. Brot.,* 61: 179 (1988): L. tiliifolia var. glandulosa.
urticifolia
subsp. moldenkei *R.W. Sanders in Moscosoa,* 5: 202 (1989)—Hispaniola (Dominican Republic, Haiti), Puerto Rico.
subsp. zanonii *R.W. Sanders in Moscosoa,* 5: 206 (1989)—Hispaniola (Dominican Republic).
viburnoides
subsp. richardii *R.B. Fernandes in Bol. Soc. Brot.,* 61: 200 (1988)—Ethiopia, Uganda, Kenya, Tanzania, Zambia, Zimbabwe, Malawi, Mozambique, Angola.
var. richardii *R.B. Fernandes in Bol. Soc. Brot.,* 61: 201 (1988)—Ethiopia, Uganda, Kenya, Tanzania, Zambia, Zimbabwe, Malawi, Mozambique, Angola.

LAPORTEA (Urticac.).
glandulosa *(Wedd.) V.C. Lima in Prog. Resum. Soc. Bot. Brasil, XXXVI Congr. Nac. Bot., Curitiba:* 362 (1985), without basionym ref.: Fleurya glandulosa.

LASIANTHAEA (Compos.).
beckeri *B.L. Turner in Phytologia,* 65(5): 361 (1988)—Mexico.
belizeana *B.L. Turner in Phytologia,* 65(5): 363 (1988)—Belize.
breedlovei *B.L. Turner in Phytologia,* 65(5): 364 (1988)—Mexico.
calva *(Greenm.) B.L. Turner in Phytologia,* 65(5): 365 (1988): Perymenium (?) calvum.
gentryi *B.L. Turner in Phytologia,* 66(6): 496 (1989)—Mexico.
guatemalensis *(J.D. Smith) B.L. Turner in Phytologia,* 65(5): 366 (1988): Zexmenia guatemalensis.
kingii *(H. Rob.) B.L. Turner in Phytologia,* 65(5): 366 (1988): Zexmenia kingii.
lundellii *(H. Rob.) B.L. Turner in Phytologia,* 65(5): 367 (1988): Oyedaea lundellii.
salvinii *(Hemsl.) B.L. Turner in Phytologia,* 65(5): 367 (1988): Zexmenia salvinii.
steyermarkii *(Blake) B.L. Turner in Phytologia,* 65(5): 367 (1988): Oyedaea steyermarkii.

LASIANTHUS (Rubiac.).
bahorucanus *T.A. Zanoni in Moscosoa,* 5: 101 (1989)—Hispaniola (Dominican Republic).

LASTARRIAEA (Polygonac.).
ptilota *J.L. Reveal in Phytologia,* 66(3): 217 (1989)—Mexico.

LATHYRUS (Legumin.).
plitmannii *W. Greuter & Burdet in Willdenowia,* 19(1): 32 (1989), nom. nov.: L. gorgoni var. lineatus *Plitmann.*
pseudocicera
subsp. negevensis *(Plitmann) Plitmann & Heyn ex W. Greuter in Willdenowia,* 19(1): 32 (1989): L. cicera var. negevensis.

LAUNAEA (Compos.).
fragilis
var. lucentica *O. de Bolòs & J. Vigo in Folia Bot. Misc.,* 6: 86 (1989)—Spain.
subsp. pumila *(Cav.) O. de Bolòs & J. Vigo in Folia Bot. Misc.,* 6: 86 (1989): Scorzonera pumila.
var. pumila *(Cav.) O. de Bolòs & J. Vigo in Folia Bot. Misc.,* 6: 86 (1989): Scorzonera pumila.
var. viminea *(Lange) O. de Bolòs & J. Vigo in Folia Bot. Misc.,* 6: 86 (1989): Zollikoferia resedifolia var. viminea.
obtusus *(DC.) Mundhe & Almeida in J. Bombay Nat. Hist. Soc.,* 85(3): 523 (1988 publ. 1989): Brachyramphus obtusus.

LAURUS (Laurac.).
magnoliaefolia *A.J. Cavanilles in Anales Ci. Nat.,* 3(7): 53 (1801)—Canary Is., Madeira.

LAWIELLA (Podostemac.).
chinensis *H.C. Chao in Contrib. Inst. Bot. Nation. Acad. Peiping,* 6(1): 9 (1948 publ. 1949)—China.
fukienensis *H.C. Chao in Contrib. Inst. Bot. Nation. Acad. Peiping,* 6(1): 6 (1948 publ. 1949)—China.

LECOINTEA (Legumin.).
tango *(Standl.) L. Emygdio de Mello Filho & A. Grave de Andrade in Atas Simpósio sôbre Biota Amazônica (Bot.),* 4: 156, 166 for bas. date (1967): Zollernia tango.

LEDEBOURIA (Hyacinthac.).
violacea *(Hutchinson) W.L. Tjaden in Brit. Cact. Succ. J.,* 7(1): 11 (1989): Scilla violacea.

LEIOTHRIX (Eriocaulac.).
cipoensis *A.M. Giulietti in Prog. Resum. Soc. Bot. Brasil, XXXVI Congr. Nac. Bot., Curitiba:* 346 (1985), without latin descr. or type—Locality not stated.

LEIOTHRIX :-
curvifolia
var. mucronata *(Bong.) A.M. Giulietti in Prog. Resum. Soc. Bot. Brasil, XXXVI Congr. Nac. Bot., Curitiba:* 346 (1985), without basionym ref.: Basionym not stated.
flagellaris
var. pedunculosa *(Ruhl.) A.M. Giulietti in Prog. Resum. Soc. Bot. Brasil, XXXVI Congr. Nac. Bot., Curitiba:* 346 (1985), without basionym ref.: Basionym not stated.
flavescens
var. disticophylla *(Silv.) A.M. Giulietti in Prog. Resum. Soc. Bot. Brasil, XXXVI Congr. Nac. Bot., Curitiba:* 346 (1985), without basionym ref.: Basionym not stated.
fulgida
var. milho-verdensis *(Silv.) A.M. Giulietti in Prog. Resum. Soc. Bot. Brasil, XXXVI Congr. Nac. Bot., Curitiba:* 346 (1985), without basionym ref.: Basionym not stated.
plantago *(Mart.) A.M. Giulietti in Prog. Resum. Soc. Bot. Brasil, XXXVI Congr. Nac. Bot., Curitiba:* 346 (1985), without basionym ref.: Basionym not stated.
rupestris *A.M. Giulietti in Prog. Resum. Soc. Bot. Brasil, XXXVI Congr. Nac. Bot., Curitiba:* 346 (1985), without latin descr. or type—Locality not stated.
sinuosa *A.M. Giulietti in Prog. Resum. Soc. Bot. Brasil, XXXVI Congr. Nac. Bot., Curitiba:* 346 (1985), without latin descr. or type—Locality not stated.

LEMBOGLOSSUM (Orchidac.).
bictoniense *(Batem.) Halbinger ex M.A. Soto Arenas in Orquídea (México),* 11: 271 (1988): Cyrtochilum bictoniense.
bictoniense *(Bateman) Halb. ex E.A. Christenson in Lindleyana,* 3(4): 222 (1988 publ. 1989), isonym see M.A. Soto Arenas *in Orquídea (México),* 11: 271 (1988): Cyrtochilum bictoniense.

LEMOORIA P.S. Short in Muelleria, 7(1): 112 (1989). *COMPOSITAE.*
burkittii *(Benth.) P.S. Short in Muelleria,* 7(1): 113 (1989): Gnephosis burkittii.

LEMURODENDRON J.F. Villiers & P. Guinet in Bull. Mus. Nation. Hist. Nat., B, Adansonia, Ser. 4, 11(1): 3 (1989). *LEGUMINOSAE.*
capuronii *J.F. Villiers & P. Guinet in Bull. Mus. Nation. Hist. Nat., B, Adansonia,* Ser. 4, 11(1): 6 (1989)—Madagascar.

LEONTOPODIUM (Compos.).
villosulum *A.P. Khokhryakov in Byull. Mosk. Obshch. Ispÿt. Prir., Biol.,* 93(6): 65 (1988)—U.S.S.R. (Soviet Far East).

LEONURUS (Labiat.).
ser. Glaucescentes *V. Krecz. & Kuprian. ex T.V. Krestovskaya in Bot. Zhurn.,* 73(12): 1751 (1988).
subsect. Panzerioidei *V. Krecz. & Kuprian. ex T.V. Krestovskaya in Bot. Zhurn.,* 73(12): 1754 (1988).
subsect. Pubescentes *T.V. Krestovskaya in Bot. Zhurn.,* 73(12): 1754 (1988).
ser. Tatarici *V. Krecz. & Kuprian. ex T.V. Krestovskaya in Bot. Zhurn.,* 73(12): 1748 (1988).
deminutus
var. orientalis *T.V. Krestovskaya in Bot. Zhurn.,* 73(12): 1748 (1988)—U.S.S.R. (Siberia).
glaucescens
f. latifolius *(Bunge) T.V. Krestovskaya in Bot. Zhurn.,* 73(12): 1752 (1988): L. glaucescens var. latifolius.
kuprijanoviae *T.V. Krestovskaya in Bot. Zhurn.,* 73(12): 1753 (1988)—Pakistan.
macranthus
var. villosissimus *T.V. Krestovskaya in Novosti Sist. Vŷssh. Rast.,* 25: 134 (1988)—Japan, China.
pseudopanzerioides *T.V. Krestovskaya in Bot. Zhurn.,* 73(12): 1749 (1988)—Mongolia, Xinjiang.
quinquelobatus
var. caucasicus *T.V. Krestovskaya in Bot. Zhurn.,* 73(12): 1751 (1988)—U.S.S.R. (Caucasus).
sibiricus
var. glaber *T.V. Krestovskaya in Novosti Sist. Vŷssh. Rast.,* 25: 133 (1988)—U.S.S.R. (Siberia), Mongolia, China.
turkestanicus
var. songoricus *T.V. Krestovskaya in Bot. Zhurn.,* 73(12): 1752 (1988)—U.S.S.R. (Middle Asia).

LEPANTHES (Orchidac.)
aprica *P.M. Catling & V.R. Catling in Canad. J. Bot.,* 66(11): 2130 (1988)—Mexico.
bahorucana *H.A. Hespenheide & D.D. Dod in Moscosoa,* 5: 252 (1989), as 'barahucana'—Hispaniola (Dominican Republic, Haiti).
caudatisepala *H.A. Hespenheide & D.D. Dod in Moscosoa,* 5: 254 (1989)—Hispaniola (Dominican Republic).
domingensis *H.A. Hespenheide & D.D. Dod in Moscosoa,* 5: 256 (1989)—Hispaniola (Dominican Republic, Haiti).
fiskei *C.A. Luer in Amer. Orchid Soc. Bull.,* 58(2): 135 (1989)—Ecuador.
hagsateri *G.A. Salazar Chávez & M.A. Soto Arenas in Orquídea (México),* 11: 16 (1988)—Mexico.
nutanticaulis *H.A. Hespenheide & D.D. Dod in Moscosoa,* 5: 258 (1989)—Hispaniola (Dominican Republic, Haiti).
oreophila *P.M. Catling & V.R. Catling in Canad. J. Bot.,* 66(11): 2130 (1988)—Mexico.

LEPANTHES :-
palatoflora *H.A. Hespenheide & D.D. Dod in Moscosoa,* 5: 260 (1989)—Hispaniola (Dominican Republic).

LEPIDOCORDIA (Boraginac.).
williamsii *(I.M. Johnston) J.S. Miller in Phytologia,* 66(6): 476 (1989): Antrophora williamsii.

LEPTAULUS (Icacinac.).
madagascariensis *J.F. Villiers in Bull. Mus. Nation. Hist. Nat., B, Adansonia,* Ser. 4, 10(4): 369 (1988 publ. 1989)—Madagascar.

LESPEDEZA (Legumin.).
maruyamana *G. Murata in Acta Phytotax. Geobot.,* 39(4–6): 149 (1988)—Japan.

LEUCANTHEMOPSIS (Compos.).
pallida
subsp. alpina *(Boiss. & Reuter) S. Rivas-Martínez, F. Fernández González & D. Sánchez-Mata in Opusc. Bot. Pharm. Complutensis,* 2: 112 (1986): Pyrethrum sulphureum var. alpinum.

LEUCANTHEMUM (Compos.).
catalaunicum *R. Vogt in Mitt. Bot. Staatssamml. München,* 27: 21 (1988)—Spain, France.

LEUCAS (Labiat.).
flaccida
var. sebastiana *(Subbarao & Kumari) N. Rama Rao in J. Econ. Taxon. Bot.,* 12(2): 378 (1988 publ. 1989): L. mollissima var. sebastiana.
indica
var. nagalapuramiana *(Chandr. & Srin.) N. Rama Rao in J. Econ. Taxon. Bot.,* 12(2): 378 (1988 publ. 1989): L. lavandulifolia var. nagalapuramiana.

LEWISIA (Portulacac.).
sect. Brachycalyx *B. Mathew, Genus Lewisia:* 139 (1989).
sect. Cotyledon *J.E. Hohn ex B. Mathew, Genus Lewisia:* 139 (1989).
sect. Erocallis *B. Mathew, Genus Lewisia:* 139 (1989).
sect. Oppositifolia *B. Mathew, Genus Lewisia:* 139 (1989).
sect. Pygmaea *B. Mathew, Genus Lewisia:* 139 (1989).
subgen. Strophiolum *B. Mathew, Genus Lewisia:* 139 (1989), as '(J.E. Hohn) ex B. Mathew'.
columbiana
subsp. wallowensis *(C.L. Hitchcock) J.E. Hohn ex B. Mathew, Genus Lewisia:* 101 (1989): L. columbiana var. wallowensis.

LIBANOTIS (Umbellif.).
jinanensis *L.C. Xu & M.D. Xu in Bull. Bot. Res. North-East. Forest. Inst.,* 9(1): 37 (1989)—China.

LICANIA (Chrysobalanac.).
arianeae *G.T. Prance in Fl. Neotrop.*, 9S: 44 (1989)—Brazil.
dodsonii *G.T. Prance in Fl. Neotrop.*, 9S: 27 (1989)—Ecuador.
filomenoi *G.T. Prance in Fl. Neotrop.*, 9S: 20 (1989)—Peru.
gentryi *G.T. Prance in Fl. Neotrop.*, 9S: 24 (1989)—Colombia.
grandibracteata *G.T. Prance in Fl. Neotrop.*, 9S: 22 (1989)—Ecuador.
hispida *G.T. Prance in Fl. Neotrop.*, 9S: 37 (1989)—Venezuela.
laevigata *G.T. Prance in Fl. Neotrop.*, 9S: 41 (1989)—Brazil, Surinam.
lamentanda *G.T. Prance in Fl. Neotrop.*, 9S: 51 (1989)—Brazil.
nelsonii *G.T. Prance in Fl. Neotrop.*, 9S: 53 (1989)—Brazil.
occultans *G.T. Prance in Fl. Neotrop.*, 9S: 42 (1989)—Brazil.
tachirensis *G.T. Prance in Fl. Neotrop.*, 9S: 30 (1989)—Venezuela.
tambopatensis *G.T. Prance in Fl. Neotrop.*, 9S: 36 (1989)—Peru.
teixeirae *G.T. Prance in Fl. Neotrop.*, 9S: 47 (1989)—Brazil.

LICARIA (Laurac.).
tomentosa *H. van der Werff in Ann. Missouri Bot. Gard.*, 76(2): 462 (1989)—Venezuela.

LIGULARIA (Compos.).
dux *(Clarke) R. Mathur in R.R. Rao et al., Fl. Ind. Enumerat. - Asterac.:* 53 (1988): Senecio dux.
hookeri
f. angustifolium *(R. Good) R. Mathur in R.R. Rao et al., Fl. Ind. Enumerat. - Asterac.:* 53 (1988): Cremanthodium hookeri f. angustifolium.
f. latiligulatum *(R. Good) R. Mathur in R.R. Rao et al., Fl. Ind. Enumerat. - Asterac.:* 53 (1988): Cremanthodium hookeri f. latiligulatum.
kingiana *(W.W. Smith) R. Mathur in R.R. Rao, Fl. Ind. Enumerat. - Asterac.:* 53 (1988): Senecio kingianus.

LILIUM (Liliac.).
megalanthum *(Wang & Tang) S.Q. Sun in Bull. Bot. Res. North-East. Forest. Inst.*, 9(3): 135 (1989): L. concolor var. megalanthum.

LIMNANTHES (Limnanthac.).
alba
subsp. versicolor *(E. Greene) C. Mason in Madroño,* 36(1): 51 (1989): Floerkea versicolor.
douglasii
subsp. nivea *(C. Mason) C. Mason in Madroño,* 36(1): 50 (1989): L. douglasii var. nivea.
subsp. rosea *(Benth.) C. Mason in Madroño,* 36(1): 50 (1989): L. rosea.

LIMNANTHES :-
subsp. sulphurea *(C. Mason) C. Mason in Madroño,* 36(1): 50 (1989): L. douglasii var. sulphurea.
gracilis
subsp. parishii *(Jepson) C. Mason in Madroño,* 36(1): 51 (1989): L. versicolor var. parishii.

LIMNIBOZA (Labiat.).
lisowskiana *P. Bamps in Bull. Jard. Bot. Nation. Belg.*, 58(3–4): 547 (1988)—Zaïre.

LIMONIUM (Plumbaginac.).
aucheri *(Girard) W. Greuter & Burdet in Willdenowia,* 19(1): 39 (1989): Statice aucheri.
carpetanicum *M. Erben in Mitt. Bot. Staatssamml. München,* 27: 388 (1988)—Spain.
hibericum *M. Erben in Mitt. Bot. Staatssamml. München,* 27: 385 (1988), nom. nov.: Statice duriuscula var. procera *Willk.*
ilerdense *M. Erben in Mitt. Bot. Staatssamml. München,* 27: 386 (1988), illegit. superfl. nom. nov., as 'sp. nov.': Statice tournefortii *Girard.*
longibracteatum *M. Erben in Mitt. Bot. Staatssamml. München,* 27: 393 (1988)—Spain.
minus *(Boiss.) M. Erben in Mitt. Bot. Staatssamml. München,* 27: 382 (1988): Statice globulariaefolia var. minor.
mucronulatum *(H. Lindb.) W. Greuter & Burdet in Willdenowia,* 19(1): 40 (1989): Statice mucronulata.
pericotii *(O. Bolòs & Vigo) W. Greuter & Burdet in Willdenowia,* 19(1): 40 (1989): L. minutum subsp. pericotii.
pseudominutum *M. Erben in Mitt. Bot. Staatssamml. München,* 27: 397 (1988)—France.
quesadense *M. Erben in Mitt. Bot. Staatssamml. München,* 27: 390 (1988)—Spain.
santapolense *M. Erben in Mitt. Bot. Staatssamml. München,* 27: 392 (1988)—Spain.
tournefortii *(Pignatti) M. Erben in Mitt. Bot. Staatssamml. München,* 14: 514 (1978), as '(Girard)' but Statice tournefortii Girard is illegit.: Limonium delicatulum subsp. tournefortii.†

LINARIA (Scrophulariac.).
subsect. Elegantes *(Viano) D.A. Sutton, Revis. tribe Antirrhineae:* 451 (1988): L. sect. Elegantes.
subsect. Trimerocalyx *(Murb.) D.A. Sutton, Revis. tribe Antirrhineae:* 420 (1988): L. sect. Trimerocalyx.
aeruginea
subsp. nevadensis *(Boiss.) D.A. Sutton, Revis. tribe Antirrhineae:* 380 (1988): L. supina var. nevadensis.

LINARIA :–
amethystea
subsp. ignescens *(G. Kunze) D.A. Sutton, Revis. tribe Antirrhineae:* 401 (1988): L. ignescens.
bipunctata
subsp. glutinosa *(Hoffmanns. & Link) D.A. Sutton, Revis. tribe Antirrhineae:* 415 (1988): L. glutinosa.
bordiana
subsp. kralikiana *(Maire) D.A. Sutton, Revis. tribe Antirrhineae:* 440 (1988): L. elegans var. kralikiana.
dalmatica
subsp. macedonica *(Griseb.) D.A. Sutton, Revis. tribe Antirrhineae:* 321 (1988): L. macedonica.
genistifolia
subsp. euxina *(Velen.) D.A. Sutton, Revis. tribe Antirrhineae:* 317 (1988): L. euxina.
jaxartica *I.G. Levichev in Opred. Rast. Sred. Azii,* 9: 363 (1987)—U.S.S.R. (Middle Asia).
kurdica
subsp. sintenisii *(P.H. Davis) D.A. Sutton, Revis. tribe Antirrhineae:* 308 (1988): L. sintenisii.
multicaulis
subsp. aurasiaca *(Pomel) D.A. Sutton, Revis. tribe Antirrhineae:* 449 (1988): L. aurasiaca.
subsp. galioides *(Ball) D.A. Sutton, Revis. tribe Antirrhineae:* 448 (1988): L. galioides.
subsp. heterophylla *(Desf.) D.A. Sutton, Revis. tribe Antirrhineae:* 448 (1988): L. heterophylla.
oblongifolia
subsp. aragonensis *(Lange) D.A. Sutton, Revis. tribe Antirrhineae:* 391 (1988): L. diffusa var. aragonensis.
oligantha
subsp. valentina *D.A. Sutton, Revis. tribe Antirrhineae:* 412 (1988)—Spain.
parviracemosa *D.A. Sutton, Revis. tribe Antirrhineae:* 355 (1988)—Algeria.
polygalifolia
subsp. lamarckii *(Rouy) D.A. Sutton, Revis. tribe Antirrhineae:* 373 (1988): L. lamarckii.
pyramidalis
subsp. kopetdaghensis *(Kuprian.) D.A. Sutton, Revis. tribe Antirrhineae:* 299 (1988): L. kopetdaghensis.
subsp. lenkoranica *(Kuprian.) D.A. Sutton, Revis. tribe Antirrhineae:* 297 (1988): L. lenkoranica.
quasisessilis *I.G. Levichev in Opred. Rast. Sred. Azii,* 9: 363 (1987)—U.S.S.R. (Middle Asia).
reflexa
subsp. agglutinans *(Pomel) D.A. Sutton, Revis. tribe Antirrhineae:* 344 (1988): L. agglutinans.

LINARIA :–
subsp. brevicalcarata *D.A. Sutton, Revis. tribe Antirrhineae:* 339 (1988)—Algeria, Tunisia.
subsp. doumetii *(Bonnet & G. Barratte) D.A. Sutton, Revis. tribe Antirrhineae:* 339 (1988): L. doumetii.
subsp. drummondiae *D.A. Sutton, Revis. tribe Antirrhineae:* 341 (1988)—Morocco.
subsp. puberula *(Doumergue) D.A. Sutton, Revis. tribe Antirrhineae:* 342 (1988): L. reflexa var. puberula.
supina
subsp. ajmasiana *(Pau) A. Dobignard in Saussurea,* 19: 110 (1988 publ. 1989): L. supina var. ajmasiana.
subsp. duriaeana *D.A. Sutton, Revis. tribe Antirrhineae:* 365 (1988)—Spain.
tristis
subsp. mesatlantica *D.A. Sutton, Revis. tribe Antirrhineae:* 377 (1988)—Morocco.
tuberculata *D.A. Sutton, Revis. tribe Antirrhineae:* 386 (1988)—Spain.
viscosa
subsp. crassifolia *(Coutinho) D.A. Sutton, Revis. tribe Antirrhineae:* 436 (1988): L. viscosa var. crassifolia.
subsp. spicata *(Coutinho) D.A. Sutton, Revis. tribe Antirrhineae:* 436 (1988): L. spicata.

LINDERA (Laurac.).
aggregata
f. playfairii *(Hemsl.) J.C. Liao in Quart. J. Chinese Forest.,* 21(3): 78 (1988): Litsea? playfairii.

LINUM (Linac.).
hellenicum *G.A. Iatrou in Willdenowia,* 19(1): 69 (1989)—Greece.
ockendonii *W. Greuter & Burdet in Willdenowia,* 19(1): 35 (1989)—France.
usitatissimum
subsp. bienne *(Mill.) A.K. Stankevich in Sborn. Nauch. Tr. Prikl. Bot. Genet. Selek.,* 113: 61 (1987): L. bienne.
subsp. humile *(Mill.) N.M. Chernomorskaya in Sborn. Nauch. Tr. Prikl. Bot. Genet. Selek.,* 113: 58 (1987), as '(L.)': L. humile.
subsp. intermedium *N.M. Chernomorskaya in Sborn. Nauch. Tr. Prikl. Bot. Genet. Selek.,* 113: 60 (1987)—Cultivation.
subsp. latifolium *(L.) A.K. Stankevich in Sborn. Nauch. Tr. Prikl. Bot. Genet. Selek.,* 113: 59 (1987), given basionym non-existent: L. usitatissimum ζ. latifolium.

LIPARIS (Orchidac.).
aaronii *P.J. Cribb & B. Lewis in Orchid Rev.,* 97(1150): 251 (1989)—Vanuatu.

LIPARIS :–
tigerhillensis *A.P. Das & S. Chanda in J. Econ. Taxon. Bot.*, 12(2): 403 (1988 publ. 1989)—India, Sikkim.

LISIANTHIUS (Gentianac.).
aurantiacus *K.J. Sytsma in Ann. Missouri Bot. Gard.*, 75(4): 1599 (1988 publ. 1989)—Panama.
habuensis *K.J. Sytsma in Ann. Missouri Bot. Gard.*, 75(4): 1597 (1988 publ. 1989)—Panama.
weaveri *K.J. Sytsma in Ann. Missouri Bot. Gard.*, 75(4): 1596 (1988 publ. 1989)—Panama.

LITOSANTHES (Rubiac.).
biermannii *(King ex Hook. f.) D.B. Deb & M. Gangopadhyay in Candollea*, 44(1): 211 (1989): Lasianthus biermannii.
capitulatus *(Wt.) D.B. Deb & M. Gangopadhyay in Candollea*, 44(1): 213 (1989): Lasianthus capitulatus.
cinereus *(Gamble) D.B. Deb & M. Gangopadhyay in Candollea*, 44(1): 215 (1989): Lasianthus cinereus.
dichotomus *(Wt.) D.B. Deb & M. Gangopadhyay in Candollea*, 44(1): 215 (1989): Lasianthus dichotomus.
filiformis *(King & Gamble) D.B. Deb & M. Gangopadhyay in Candollea*, 44(1): 222 (1989): Lasianthus filiformis.
var. bracteata *(King & Gamble) D.B. Deb & M. Gangopadhyay in Candollea*, 44(1): 222 (1989): Lasianthus filiformis var. bracteata.
longicauda *(Hook. f.) D.B. Deb & M. Gangopadhyay in Candollea*, 44(1): 218 (1989): Lasianthus longicauda.
micranthus *(Hook. f.) D.B. Deb & M. Gangopadhyay in Candollea*, 44(1): 220 (1989): Lasianthus micranthus.
pendulus *(Ridley) D.B. Deb & M. Gangopadhyay in Candollea*, 44(1): 222 (1989): Lasianthus pendulus.
pseudolucidus *(King) D.B. Deb & M. Gangopadhyay in Candollea*, 44(1): 222 (1989): Lasianthus pseudo-lucidus.
robinsonii *(Ridley) D.B. Deb & M. Gangopadhyay in Candollea*, 44(1): 222 (1989): Lasianthus robinsonii.
scalariformis *(King & Gamble) D.B. Deb & M. Gangopadhyay in Candollea*, 44(1): 222 (1989): Lasianthus scalariformis.
subspicatus *(King & Gamble) D.B. Deb & M. Gangopadhyay in Candollea*, 44(1): 223 (1989): Lasianthus subspicatus.
varians *(Hook. f.) D.B. Deb & M. Gangopadhyay in Candollea*, 44(1): 223 (1989): Lasianthus varians.
venulosus *(Wt. & Arn.) D.B. Deb & M. Gangopadhyay in Candollea*, 44(1): 220 (1989): Santia venulosa.
villosus *(Ridley) D.B. Deb & M. Gangopadhyay in Candollea*, 44(1): 223 (1989): Lasianthus villosus.

LITSEA (Laurac.).
subgen. Actinodaphne *(Nees) J.C. Liao, Taxon. Revis. Fam. Lauraceae in Taiwan:* vi (1988), without basionym ref.: Basionym not stated.
auricolor *A.J.G.H. Kostermans in Reinwardtia*, 10(5): 464 (1988), nom. nov.: L. aureo-sericea *Kostermans (1974)*.
boerlagei *A.J.G.H. Kostermans in Reinwardtia*, 10(5): 464 (1988), nom. nov.: L. ambigua *(Bl.) Boerlage*.
caulostachys *A.J.G.H. Kostermans in Reinwardtia*, 10(5): 464 (1988), nom. nov.: L. cauliflora *(Meissn.) Trimen*.
citronella *A.J.G.H. Kostermans in Reinwardtia*, 10(5): 465 (1988)—Vietnam, Hainan.
elongata
var. mushaensis *(Hay.) J.C. Liao, Taxon. Revis. Fam. Lauraceae in Taiwan:* 92 (1988): L. mushaensis.
f. postiglabra *J.C. Liao, Taxon. Revis. Fam. Lauraceae in Taiwan:* 94 (1988)—Taiwan.
elongatoides *A.J.G.H. Kostermans in Reinwardtia*, 10(5): 466 (1988), nom. nov., homotypic with L. pseudolongifolia Kostermans: L. pseudo-elongata *Kostermans*.
fluminensis *A.J.G.H. Kostermans in Reinwardtia*, 10(5): 466 (1988), nom. nov.: L. acuminata *(Teschn.) Kostermans*.
foveola *A.J.G.H. Kostermans in Reinwardtia*, 10(5): 466 (1988), nom. nov.: L. foveolata *Yang & P.H. Huang*.
pseudolongifolia *A.J.G.H. Kostermans in Reinwardtia*, 10(5): 467 (1988), nom. nov., homotypic with L. elongatoides Kostermans: L. pseudo-elongata *Kostermans*.
rotundata *(Bl.) A.J.G.H. Kostermans in Reinwardtia*, 10(5): 468 (1988): Tetranthera resinosa var. rotundata.

LLERASIA (Compos.).
pascoensis *A. Sagástegui-Alva & M.O. Dillon in Brittonia*, 40(4): 363 (1988)—Peru.
sanmartinensis *A. Sagástegui-Alva & M.O. Dillon in Brittonia*, 40(4): 365 (1988)—Peru.

LOBANILIA A. Radcliffe-Smith in Kew Bull., 44(2): 334 (1989). *EUPHORBIACEAE*.
asterothrix *A. Radcliffe-Smith in Kew Bull.*, 44(2): 335 (1989)—Madagascar.
bakeriana *(Baill.) A. Radcliffe-Smith in Kew Bull.*, 44(2): 335 (1989): Claoxylon bakerianum.
claoxyloides *A. Radcliffe-Smith in Kew Bull.*, 44(2): 337 (1989)—Madagascar.
crotonoides *A. Radcliffe-Smith in Kew Bull.*, 44(2): 337 (1989)—Madagascar.
hirtella *(Baill.) A. Radcliffe-Smith in Kew Bull.*, 44(2): 338 (1989): Claoxylon hirtellum.
luteobrunnea *(Bak.) A. Radcliffe-Smith in Kew Bull.*, 44(2): 338 (1989): Croton luteo-brunneus.
ovalis *(Baill.) A. Radcliffe-Smith in Kew Bull.*, 44(2): 338 (1989): Claoxylon ovale.

LOBELIA (Campanulac.).
gypsophila *T.J. Ayers in Sida,* 13(2): 144 (1988), nom. nov.: Heterotoma pringlei *B.L. Robinson.*

×LOCKOCHILETTIA hort. in Orchid Rev., 97(1145): centre page pull-out p. 9 (1989). *ORCHIDACEAE.* (Comparettia × Leochilus × Lockhartia).

LONCHOCARPUS (Legumin.).
balsensis *M. Sousa S. & J.C. Soto N. in An. Inst. Biol. Univ. Nac. Autón. Méx., Bot.,* 58: 70 (1987 publ. 1989)—Mexico.
epigaeus *M. Sousa S. in An. Inst. Biol. Univ. Nac. Autón. Méx., Bot.,* 58: 73 (1987 publ. 1989)—Mexico.
huetamoensis *M. Sousa S. & J.C. Soto N. in An. Inst. Biol. Univ. Nac. Autón. Méx., Bot.,* 58: 75 (1987 publ. 1989)—Mexico.
subsp. xochipalensis *M. Sousa S. & J.C. Soto N. in An. Inst. Biol. Univ. Nac. Autón. Méx., Bot.,* 58: 78 (1987 publ. 1989)—Mexico.
longipedunculatus *M. Sousa S. & J.C. Soto N. in An. Inst. Biol. Univ. Nac. Autón. Méx., Bot.,* 58: 79 (1987 publ. 1989)—Mexico.
schubertiae *M. Sousa S. in An. Inst. Biol. Univ. Nac. Autón. Méx., Bot.,* 58: 82 (1987 publ. 1989)—Mexico.
torrensis *N.F. Mattos in Loefgrenia,* 92: 2 (1988)—Brazil.

LONCOMELOS (Hyacinthac.).
purpureus *C.S. Rafinesque, Fl. Tellur.,* 3: 62 (1837)—Sicily.
pyrenaicus *(L.) Hrouda ex J. Holub in Folia Geobot. Phytotax.,* 23(4): 413 (1988): Ornithogalum pyrenaicum.
subsp. sphaerocarpus *(A. Kern.) J. Holub in Folia Geobot. Phytotax.,* 23(4): 413 (1988): Ornithogalum sphaerocarpum.

LONICERA (Caprifoliac.).
confusa
var. glabrocalyx *R.H. Miao & X.J. Liu in Acta Sci. Nat. Univ. Sunyatseni,* 28(4): 78 (1989)—China.
fengkaiensis *R.H. Miao & X.J. Liu in Acta Sci. Nat. Univ. Sunyatseni,* 28(4): 79 (1989)—China.

LOREYA (Melastomatac.).
gracilis *S.S. Renner in Mem. New York Bot. Gard.,* 50: 45 (1989)—Bolivia.
klugii *S.S. Renner in Mem. New York Bot. Gard.,* 50: 52 (1989)—Colombia, Peru, Bolivia.
riparia *S.S. Renner in Mem. New York Bot. Gard.,* 50: 46 (1989)—Brazil.
subrotundifolia *(Wurdack) S.S. Renner in Mem. New York Bot. Gard.,* 50: 46 (1989): Bellucia subrotundifolia.
wurdackiana *S.S. Renner in Mem. New York Bot. Gard.,* 50: 42 (1989)—Brazil.

LOTONONIS (Legumin.).
acocksii *B.E. van Wyk in S. Afr. J. Bot.,* 55(2): 178 (1989)—South Africa (Cape Province).
azureoides *B.E. van Wyk in Bothalia,* 19(1): 1 (1989)—South Africa (Cape Province).
difformis *B.E. Van Wyk in S. Afr. J. Bot.,* 55(5): 529 (1989)—South Africa (Transvaal).
gracilifolia *B.E. van Wyk in Bothalia,* 19(1): 3 (1989)—South Africa (Cape Province).
lamprifolia *B.E. van Wyk in Bothalia,* 19(1): 4 (1989)—South Africa (Cape Province).
minima *B.E. van Wyk in S. Afr. J. Bot.,* 54(6): 628 (1988)—South Africa (Cape Province).
nutans *B.E. van Wyk in S. Afr. J. Bot.,* 55(2): 180 (1989)—South Africa (Cape Province).
purpurescens *B.E. van Wyk in S. Afr. J. Bot.,* 55(2): 180 (1989)—South Africa (Cape Province).
racemiflora *B.E. van Wyk in Bothalia,* 19(1): 7 (1989)—South Africa (Cape Province).

LOZANIA (Lacistematac.).
glabrata *A.H. Gentry in Ann. Missouri Bot. Gard.,* 75(4): 1431 (1988 publ. 1989)—Colombia.

LUCILIA (Compos.).
breweri *(Steyerm.) V.M. Badillo in Ernstia,* 50: 11 (1988): Chionolaena breweri.
lycopodioides *(Less.) S.E. Freire in Taxon,* 38(2): 298 (1989): Oligandra lycopodioides.
plicatifolia *(Sagást. & Dillon) S.E. Freire in Darwiniana,* 28(1–4): 411 (1987 publ. 1988): Belloa plicatifolia.
saxatilis *V.M. Badillo in Ernstia,* 50: 9 (1988)—Venezuela.

LUNDELLIANTHUS (Compos.).
belizeanus *(B. Turner) J.L. Strother in Syst. Bot.,* 14(4): 547 (1989): Lasianthaea belizeana.
breedlovei *(B. Turner) J.L. Strother in Syst. Bot.,* 14(4): 547 (1989): Lasianthaea breedlovei.
guatemalensis *(J.D. Smith) J.L. Strother in Syst. Bot.,* 14(4): 544 (1989): Zexmenia guatemalensis.
harrimanii *J.L. Strother in Syst. Bot.,* 14(4): 545 (1989)—Nicaragua.
jaliscensis *(McVaugh) J.L. Strother in Syst. Bot.,* 14(4): 544 (1989): Otopappus jaliscensis.
kingii *(H. Robinson) J.L. Strother in Syst. Bot.,* 14(4): 545 (1989): Zexmenia kingii.
salvinii *(Hemsley) J.L. Strother in Syst. Bot.,* 14(4): 545 (1989): Zexmenia salvinii.
steyermarkii *(S.F. Blake) J.L. Strother in Syst. Bot.,* 14(4): 546 (1989): Oyedea steyermarkii.

LUPINUS (Legumin.).
magnistipulatus *A.M. Planchuelo & D.B. Dunn in Ann. Missouri Bot. Gard.,* 76(1): 306 (1989)—Brazil.

LUPINUS :-
setifolius *A.M. Planchuelo & D.B. Dunn in Ann. Missouri Bot. Gard.*, 76(1): 307 (1989)—Brazil.

LUZULA (Juncac.).
abchasica *V.S. Novikov in Byull. Mosk. Obshch. Ispȳt. Prir., Biol.*, 94(1): 111 (1989)—U.S.S.R. (Caucasus).

LYCASTE (Orchidac.).
colombiana *M. Ospina H. in Mutisia*, 71: 10 (1988)—Colombia.

LYCIUM (Solanac.).
grandicalyx *A.M. Joubert & H.J.T. Venter in S. Afr. J. Bot.*, 55(5): 516 (1989)—Namibia.

LYSIMACHIA (Primulac.).
vulgaris
f. koreana *W.T. Lee in Korean J. Pl. Taxon.*, 18(3): 197 (1988), without latin descr. or type—Korea.

LYTHRUM (Lythrac.).
castiliae *W. Greuter & Burdet in Willdenowia*, 19(1): 35 (1989)—Spain.

M

MABRYA (Scrophulariac.).
flaviflora *(I.M. Johnston) D.A. Sutton, Revis. tribe Antirrhineae:* 491 (1988): Maurandya flaviflora.

MACAIREA (Melastomatac.).
cuieirasii *S.S. Renner in Mem. New York Bot. Gard.*, 50: 82 (1989)—Brazil.
philipsonii *S.S. Renner in Mem. New York Bot. Gard.*, 50: 90 (1989)—Colombia.

MACHILUS (Laurac.).
glaucescens *(Nees) H.W. Li in Acta Bot. Yunnanica*, 10(4): 490 (1988): Ocotea glaucescens.
miaoshanensis *F.N. Wei & C.Q. Lin in Guihaia*, 8(4): 305 (1988)—China.
reticulata *S. Lee & F.N. Wei in Guihaia*, 8(4): 306 (1988)—China.

MACLURA (Morac.).
cochinchinensis
var. gerontogea *(Sieb. & Zucc.) S.S. Ying, Colour. Illustr. Pl. Taiwan, Pt. 3; Colour. Illustr. Ligneous Pl. Taiwan*, 1: 382 (1987): M. gerontogea.

MACROCENTRUM (Melastomatac.).
chimantense *J.J. Wurdack in Mem. New York Bot. Gard.*, 51: 103 (1989)—Venezuela.

MACROCHORDION (Bromeliac.).
alba *(Mez) L.B. Smith & W.J. Kress in Phytologia*, 66(1): 77 (1989): Aechmea alba.
chlorophylla *(L.B. Smith) L.B. Smith & W.J. Kress in Phytologia*, 66(1): 77 (1989): Aechmea chlorophylla.
kautskyana *(E. Pereira & L.B. Smith) L.B. Smith & W.J. Kress in Phytologia*, 66(1): 77 (1989): Aechmea kautskyana.
lamarchei *(Mez) L.B. Smith & W.J. Kress in Phytologia*, 66(1): 77 (1989): Aechmea lamarchei.
maculata *(L.B. Smith) L.B. Smith & W.J. Kress in Phytologia*, 66(1): 77 (1989): Aechmea maculata.
nervata *(L.B. Smith) L.B. Smith & W.J. Kress in Phytologia*, 66(1): 77 (1989): Aechmea nervata.
pabstii *(E. Pereira & Moutinho) L.B. Smith & W.J. Kress in Phytologia*, 66(1): 77 (1989): Aechmea pabstii.
triangularis *(L.B. Smith) L.B. Smith & W.J. Kress in Phytologia*, 66(1): 77 (1989): Aechmea triangularis.
turbinocalyx *(Mez) L.B. Smith & W.J. Kress in Phytologia*, 66(1): 77 (1989): Aechmea turbinocalyx.

MACROMERIA (Boraginac.).
alba *G.L. Nesom in Madroño*, 36(1): 28 (1989)—Mexico.

MACROPTILIUM (Legumin.).
sect. Microcochle *(Bentham) V.P. Barbosa-Fevereiro in Arq. Jard. Bot. Rio de Janeiro*, 28: 144 (1986–87 publ. 1988): Phaseolus sect. Microcochle.
sabaraense *(Hoehne) V.P. Barbosa-Fevereiro in Arq. Jard. Bot. Rio de Janeiro*, 28: 150 (1986–87 publ. 1988): Phaseolus sabaraensis.

MACROSCIADIUM V.N. Tikhomirov & T.V. Lavrova in Byull. Mosk. Obshch. Ispȳt. Prir., Biol., 93(6): 63 (1988). *UMBELLIFERAE.*
alatum *(Bieb.) V.N. Tikhomirov & T.V. Lavrova in Byull. Mosk. Obshch. Ispȳt. Prir., Biol.*, 93(6): 63 (1988): Athamanta alata.
physospermifolium *(Albov) V.N. Tikhomirov & T.V. Lavrova in Byull. Mosk. Obshch. Ispȳt. Prir., Biol.*, 93(6): 64 (1988): Ligusticum physospermifolium.

MAEVIELLA R.A. Rossow in Arq. Jard. Bot. Rio de Janeiro, 27: 172 (1983–85 publ. 1985). *SCROPHULARIACEAE.*
cochlearia *(Huber) R.A. Rossow in Arq. Jard. Bot. Rio de Janeiro*, 27: 173 (1983–85 publ. 1985): Herpestis cochlearia.

MAGNOLIA (Magnoliac.).
andamanica *(King) D.C.S. Raju & M.P. Nayar in Indian J. Bot.*, 3(2): 171 (1980): Talauma andamanica.

MAGNOLIA :–
caveana *(Hook. f. & Thoms.) D.C.S. Raju & M.P. Nayar in Indian J. Bot.*, 3(2): 171 (1980): Manglietia caveana.
hookeri *(Cubitt & Smith) D.C.S. Raju & M.P. Nayar in Indian J. Bot.*, 3(2): 171 (1980): Manglietia hookeri.
macrophylla
var. ashei *(Weath.) D.L. Johnson in Baileya*, 23(1): 55 (1989): M. ashei.
var. dealbata *(Zucc.) D.L. Johnson in Baileya*, 23(1): 56 (1989): M. dealbata.
rabaniana *(Hook. f. & Thoms.) D.C.S. Raju & M.P. Nayar in Indian J. Bot.*, 3(2): 171 (1980): Talauma rabaniana.
sinostellata *P.L. Chiu & Z.H. Chen in Acta Phytotax. Sin.*, 27(1): 79 (1989)—China.
villariana *(Rolfe) D.C.S. Raju & M.P. Nayar in Indian J. Bot.*, 3(2): 171 (1980): Talauma villariana.

MAHONIA (Berberidac.).
russellii *N.P. Taylor in Kew Mag.*, 6(2): 58 (1989)—Mexico.

MALAXIS (Orchidac.).
sect. Oistochilus *(Schlechter) P.J. Kores in Allertonia*, 5(1): 46 (1989): Microstylis sect. Oistochilus.
reineckeana *(Kraenzl.) P.J. Kores in Allertonia*, 5(1): 51 (1989): Microstylis reineckeana.
tetraloba *(Schlechter) P.J. Kores in Allertonia*, 5(1): 50 (1989): Microstylis tetraloba.
whitmeei *(Rolfe) P.J. Kores in Allertonia*, 5(1): 52 (1989): Microstylis whitmeei.

MALPIGHIA (Malpighiac.).
pusillifolia *(Ekman & Niedenzu) F.K. Meyer in Wiss. Zeitschr. Friedrich-Schiller-Univ. Jena, Mat. Naturwiss.*, 38(2): 295 (1989): M. nummulariifolia var. pusillifolia.

MALUS (Rosac.).
daochengensis *C.L. Li in Acta Phytotax. Sin.*, 27(4): 301 (1989)—China.

MAMMEA (Guttif.).
decaryana *(Perrier) J.F. Leroy in Compt. R. Hebd. Séances Acad. Sci., D (Paris)*, 284(16): 1522 (1977), without basionym ref.: Ochrocarpos decaryanus.
perrieri *(Vig. & Humb.) J.F. Leroy in Compt. R. Hebd. Séances Acad. Sci., D (Paris)*, 284(16): 1521 (1977), without basionym ref.: Ochrocarpos perrieri.

MAMMILLARIA (Cactac.).
albrechtiana *M. Wohlschlager in Kakt. And. Sukk.*, 40(9): 219 (1989)—Mexico.
arroyensis *W. Reppenhagen in Kakt. And. Sukk.*, 40(12): 292, 290 (1989)—Mexico.

MANDEVILLA (Apocynac.).
aracamunensis *G. Morillo in Ernstia*, 53: 17 (1989)—Venezuela.
holstii *G. Morillo in Ernstia*, 53: 18 (1989)—Venezuela.

MANDEVILLA :–
huberi *G. Morillo in Ernstia*, 53: 19 (1989)—Venezuela.

MANFREDA (Agavac.).
involuta *R. McVaugh, Fl. Novo-Galiciana*, 15: 231 (1989)—Mexico.
scabra *(Ort.) R. McVaugh, Fl. Novo-Galiciana*, 15: 234 (1989): Agave scabra.

MANIHOT (Euphorbiac.).
tripartita
subsp. indivisa *A.C. Allem in Rev. Brasil. Biol.*, 49(1): 13 (1989)—Brazil.
subsp. xavantinensis *(Rogers & Appan) A.C. Allem in Rev. Brasil. Biol.*, 49(1): 13 (1989): M. xavantinensis.
violacea
subsp. jacobinensis *(Muell. Arg.) A.C. Allem in Rev. Brasil. Biol.*, 49(1): 25 (1989): M. jacobinensis.

MAPANIA (Cyperac.).
assimilis
subsp. guianensis *D.A. Simpson in Kew Bull.*, 44(1): 141 (1989)—French Guiana.
sylvatica
subsp. gabonica *(Chermezon) D.A. Simpson in Kew Bull.*, 44(1): 140 (1989): M. gabonica.

MARAHUACAEA B. Maguire in Acta Bot. Venezuel., 14(3): 16 (1984). *RAPATEACEAE.*
schomburgkii *(Maguire) B. Maguire in Acta Bot. Venezuel.*, 14(3): 16 (1984): Amphiphyllum schomburgkii.

MARILA (Guttif.).
parviflora *A.H. Gentry in Ann. Missouri Bot. Gard.*, 75(4): 1431 (1988 publ. 1989)—Colombia.

MARISCUS (Cyperac.).
diamantinus *D.A. Simpson in Kew Bull.*, 44(2): 283 (1989)—Brazil.
fauriei *(Kükenth.) T. Koyama in Bishop Mus. Occas. Pap.*, 29: 122 (1989): Cyperus fauriei.
hillebrandii *(Boeck.) T. Koyama in Bishop Mus. Occas. Pap.*, 29: 122 (1989): Cyperus hillebrandii.
subsp. decipiens *(Hillebr.) T. Koyama in Bishop Mus. Occas. Pap.*, 29: 122 (1989): Cyperus decipiens.
hypochlorus
subsp. brevior *(Kükenth.) T. Koyama in Bishop Mus. Occas. Pap.*, 29: 122 (1989): Cyperus hypochlorus var. brevior.
pennatiformis *(Kükenth.) T. Koyama in Bishop Mus. Occas. Pap.*, 29: 123 (1989): Cyperus pennatiformis.
subsp. bryanii' *(Kükenth.) T. Koyama in Bishop Mus. Occas. Pap.*, 29: 123 (1989): Cyperus pennatiformis var. bryanii.

MARISCUS :–
phleoides
subsp. hawaiiensis *(H. Mann) T. Koyama in Bishop Mus. Occas. Pap.*, 29: 124 (1989): Cyperus hawaiiensis.
pohlii *(Nees) D.A. Simpson in Kew Bull.*, 44(2): 285 (1989): Papyrus pohlii.
subsp. bahiensis *D.A. Simpson in Kew Bull.*, 44(2): 286 (1989)—Brazil.
rockii *(Kükenth.) T. Koyama in Bishop Mus. Occas. Pap.*, 29: 124 (1989): Cyperus rockii.
sandwicensis *(Kükenth.) T. Koyama in Bishop Mus. Occas. Pap.*, 29: 124 (1989): Cyperus sandwicensis.

MARLIEREA (Myrtac.).
cordata *J.R. Mattos in Loefgrenia*, 94: 7 (1989)—Brazil.

MARSDENIA (Asclepiadac.).
liisae *J.B. Williams in Austrobaileya*, 3(1): 45 (1989)—Australia (New South Wales).
lloydii *P.I. Forster in Austrobaileya*, 2(5): 455 (1988), nom. nov.: M. suberosa *S.T. Blake*.

MASDEVALLIA (Orchidac.).
berthae *C.A. Luer & Andreetta in Lindleyana*, 4(3): 105 (1989)—Ecuador.
cleistogama *C.A. Luer in Lindleyana*, 4(3): 105 (1989)—Peru.
cranion *C.A. Luer in Lindleyana*, 4(3): 108 (1989)—Peru.
cylix *C.A. Luer & Malo in Lindleyana*, 3(4): 195 (1988 publ. 1989)—Ecuador.
descendens *C.A. Luer & Andreetta in Lindleyana*, 3(4): 195 (1988 publ. 1989)—Ecuador.
dryada *C.A. Luer & Escobar in Lindleyana*, 4(3): 108 (1989)—Colombia.
eumeces *C.A. Luer in Lindleyana*, 4(3): 111 (1989)—Peru.
foeda *C.A. Luer & Vásquez in Lindleyana*, 3(4): 198 (1988 publ. 1989)—Bolivia.
glossacles *C.A. Luer in Lindleyana*, 4(3): 111 (1989)—Ecuador.
hercules *C.A. Luer & Andreetta in Lindleyana*, 3(4): 198 (1988 publ. 1989)—Ecuador.
hirtzii *C.A. Luer & Andreetta in Amer. Orchid Soc. Bull.*, 58(2): 137 (1989)—Ecuador.
manchinazae *C.A. Luer & Andreetta in Lindleyana*, 3(4): 201 (1988 publ. 1989)—Ecuador.
maxilimax *(Luer) C.A. Luer in Lindleyana*, 4(3): 114 (1989): M. limax subsp. maxilimax.
mayaycu *C.A. Luer & Andreetta in Lindleyana*, 3(4): 201 (1988 publ. 1989)—Ecuador.
melanoglossa *C.A. Luer in Lindleyana*, 4(3): 114 (1989)—Ecuador.
microptera *C.A. Luer & Würstle in Lindleyana*, 4(3): 116 (1989)—Peru.
morochoi *C.A. Luer & Andreetta in Lindleyana*, 4(3): 116 (1989)—Ecuador.
mystica *C.A. Luer in Lindleyana*, 4(3): 119 (1989)—Colombia.

MASDEVALLIA :–
nivea *(Luer & Escobar) C.A. Luer & Escobar in C.A. Luer, Thesaurus Masdevalliarum*, 13: t. 9 (1989): M. purpurella subsp. nivea.
os-viperae *C.A. Luer & Andreetta in Lindleyana*, 4(3): 119 (1989)—Ecuador.
paquishae *C.A. Luer & Hirtz in Lindleyana*, 3(4): 204 (1988 publ. 1989)—Ecuador.
rex *C.A. Luer & Hirtz in Lindleyana*, 4(3): 122 (1989)—Ecuador.
solomonii *C.A. Luer & Vásquez in Lindleyana*, 3(4): 204 (1988 publ. 1989)—Bolivia.
stirpis *C.A. Luer in Lindleyana*, 4(3): 122 (1989)—Venezuela.
tinekeae *C.A. Luer & Vásquez in Lindleyana*, 3(4): 207 (1988 publ. 1989)—Bolivia.
vidua *C.A. Luer & Andreetta in Lindleyana*, 3(4): 207 (1988 publ. 1989)—Ecuador.
vieirana *C.A. Luer & Escobar in Lindleyana*, 3(4): 211 (1988 publ. 1989)—Colombia.
zebracea *C.A. Luer in Lindleyana*, 4(3): 126 (1989)—Peru.

MATAYBA (Sapindac.).
affinis *J.A. Steyermark in Ann. Missouri Bot. Gard.*, 75(3): 1067 (1988)—Venezuela.
longipes
subsp. tepuiensis *J.A. Steyermark in Ann. Missouri Bot. Gard.*, 75(3): 1069 (1988)—Venezuela, Brazil.
oligandra
var. occidentalis *J.A. Steyermark in Ann. Missouri Bot. Gard.*, 75(3): 1069 (1988)—Venezuela.
yutajensis *J.A. Steyermark in Ann. Missouri Bot. Gard.*, 75(3): 1070 (1988)—Venezuela.

MATELEA (Asclepiadac.).
aenea *(Woodson) W.D. Stevens in Ann. Missouri Bot. Gard.*, 75(4): 1558 (1988 publ. 1989): Dictyanthus aeneus.
aymardii *G. Morillo in Ernstia*, 50: 19 (1988)—Venezuela.
camiloana *G. Morillo in Ernstia*, 50: 20 (1988)—Peru.
chimboracensis *G. Morillo in Ernstia*, 50: 20 (1988)—Ecuador.
dwyeri *G. Morillo in Ernstia*, 50: 21 (1988)—Panama.
eximia *W.D. Stevens in Ann. Missouri Bot. Gard.*, 75(4): 1552 (1988 publ. 1989)—Mexico.
fontellana *G. Morillo in Ernstia*, 50: 22 (1988)—Brazil.
hamata *W.D. Stevens in Ann. Missouri Bot. Gard.*, 75(4): 1540 (1988 publ. 1989)—Mexico.
hildegardiana *G. Morillo in Ernstia*, 50: 22 (1988)—Venezuela.
honorana *G. Morillo in Ernstia*, 50: 24 (1988)—Ecuador.
lauta *W.D. Stevens in Ann. Missouri Bot. Gard.*, 75(4): 1548 (1988 publ. 1989)—Mexico.
liesneri *G. Morillo in Ernstia*, 50: 26 (1988)—Venezuela.

MATELEA :–
macvaughiana *W.D. Stevens in Ann. Missouri Bot. Gard.*, 75(4): 1545 (1988 publ. 1989)—Mexico.
microphylla *G. Morillo in Ernstia,* 50: 26 (1988)—Brazil.
ochracea *G. Morillo in Ernstia,* 50: 28 (1988)—Colombia.
suffruticosa *W.D. Stevens in Ann. Missouri Bot. Gard.*, 75(4): 1558 (1988 publ. 1989)—Mexico.
vargasii *G. Morillo in Ernstia,* 50: 28 (1988)—Peru.
woytkowskii *G. Morillo in Ernstia,* 50: 29 (1988)—Peru.

MATISIA (Bombacac.).
dolichosiphon *(Robyns & Nilsson) W.S. Alverson in Taxon,* 38(3): 386 (1989): Quararibea dolichosiphon.
exalata *W.S. Alverson in Syst. Bot.*, 14(4): 473 (1989)—Panama.
intricata *(Robyns & Nilsson) W.S. Alverson in Taxon,* 38(3): 386 (1989): Quararibea intricata.
jefensis *(Robyns & Nilsson) W.S. Alverson in Taxon,* 38(3): 386 (1989): Quararibea jefensis.
malacocalyx *(Robyns & Nilsson) W.S. Alverson in Taxon,* 38(3): 386 (1989): Quararibea malacocalyx.
palenquiana *(Robyns) W.S. Alverson in Taxon,* 38(3): 386 (1989): Quararibea palenquiana.

MATTIASTRUM (Boraginac.).
karakoricum *D. Podlech & F. Sadat in Mitt. Bot. Staatssamml. München,* 27: 65 (1988)—Pakistan.

MATUCANA (Cactac.).
weberbaueri
f. flammea *(Don.) R. Bregman, A. Meerstadt, P. Melis & A.B. Pullen in Succulenta,* 68(6): 143 (1989): Borzicactus weberbaueri var. flammeus.

MAXILLARIA (Orchidac.).
auyantepuiensis
subsp. epiphytica *G. Carnevali & I. Ramírez in Ann. Missouri Bot. Gard.*, 76(2): 376 (1989)—Venezuela.
foldatsiana *G. Carnevali & I. Ramírez in Ann. Missouri Bot. Gard.*, 76(2): 376 (1989)—Venezuela.
rodrigueziana *J.T. Atwood & Mora-Retana in Ic. Pl. Trop.*, 14 (Orch. Costa Rica, pt. 1): t. 1361 (1989)—Panama, Costa Rica.
santanae *G. Carnevali & I. Ramírez in Ann. Missouri Bot. Gard.*, 76(2): 377 (1989)—Venezuela.
valenzuelana
subsp. angustata *J.T. Atwood in Ic. Pl. Trop.*, 14 (Orch. Costa Rica, pt. 1): t. 1371 (1989)—Nicaragua, Honduras.

MAXIMOVICZIA A.P. Khokhryakov, Fl. Magadansk. Obl.: 73 (1985), nom. nov.: Scirpus sect. Japonici T. Koyama (Cyperac.).
japonica *(Maxim.) A.P. Khokhryakov, Fl. Magadansk. Obl.*: 73 (1985): Eriophorum japonicum.

MAXIMOWICZIELLA A.P. Khokhryakov, Analiz Fl. Kolўmskogo Nagor'ya: 15 (1989), nom. nov.: gen. Maximoviczia A.P. Khokhryakov (Cyperac.).
japonica *(Maxim.) A.P. Khokhryakov, Analiz Fl. Kolўmskogo Nagor'ya:* 15 (1989), without basionym ref.: Basionym not stated.

MAYTENUS (Celastrac.).
harenensis *Sebsebe Demissew in Bull. Brit. Mus. (Nat. Hist.), Bot.*, 19: 1 (1989)—Ethiopia.
huberi *J.A. Steyermark in Ann. Missouri Bot. Gard.*, 75(3): 1062 (1988)—Venezuela.
insculpta *J.A. Steyermark in Ann. Missouri Bot. Gard.*, 75(3): 1063 (1988)—Venezuela.
longistipitata *J.A. Steyermark in Ann. Missouri Bot. Gard.*, 75(3): 1064 (1988)—Venezuela.
neblinae *J.A. Steyermark in Ann. Missouri Bot. Gard.*, 75(3): 1064 (1988)—Venezuela.
pustulata *J.A. Steyermark in Ann. Missouri Bot. Gard.*, 75(3): 1064 (1988)—Venezuela.

MECONOPSIS (Papaverac.).
rebeccae *H.S. Debnath & M.P. Nayar in J. Jap. Bot.*, 64(5): 157 (1989)—Nepal.

MEDICAGO (Legumin.).
falcata
subsp. glandulosa *(Koch) W. Greuter & Burdet in Willdenowia,* 19(1): 32 (1989): M. falcata var. glandulosa.
italica
subsp. striata *(Bast.) W. Greuter & Burdet in Willdenowia,* 19(1): 32 (1989): M. striata.
monantha
subsp. noeana *(Boiss.) W. Greuter & Burdet in Willdenowia,* 19(1): 32 (1989): Trigonella noeana.
polyceratia
subsp. amandiana *(Samp.) W. Greuter & Burdet in Willdenowia,* 19(1): 32 (1989): Trigonella amandiana.

MEDIOCALCAR (Orchidac.).
decoratum *A. Schuiteman in Blumea,* 34(1): 167 (1989)—New Guinea (Papua New Guinea, W. Irian).
subteres *A. Schuiteman in Blumea,* 34(1): 171 (1989)—New Guinea (Papua New Guinea, W. Irian).

MELALEUCA (Myrtac.).
quinquenerva
var. rubriflora *(Vieill. ex Brongn. & Gris) M.G. Baumann-Bodenheim, Syst. Fl. Neu-Caledonien,* 4: 102 (1988 publ. Jan. 1989), without basionym ref.; *et Syst. Fl. Neu-Caledonien,* 5: 97 (May 1989), with basionym ref., basionym as 'M. leucadendron auct. non L. var. rubriflora': M. viridiflora var. rubriflora.

MELAMPYRUM (Scrophulariac.).
laxum
f. edentatum *(Tuyama) T. Yamazaki in J. Jap. Bot.,* 64(4): 123 (1989): M. australe f. edentatum.
yezoense *T. Yamazaki in J. Jap. Bot.,* 64(4): 121 (1989)—Japan.

MELANDRIUM (Caryophyllac.).
auritipetalum *Y.Z. Zhao & P. Ma in Acta Phytotax. Sin.,* 27(3): 225 (1989)—China.
verrucosoalatum *Y.Z. Zhao & P. Ma in Acta Phytotax. Sin.,* 27(3): 227 (1989), as 'verrucoso-alatum'—China.

MELANOSPERMUM O.M. Hilliard in Notes Roy. Bot. Gard. Edinburgh, 45(3): 482 (1988 publ. 1989). *SCROPHULARIACEAE.*
foliosum *(Benth.) O.M. Hilliard in Notes Roy. Bot. Gard. Edinburgh,* 45(3): 482 (1988 publ. 1989): Polycarena foliosa.
italae *O.M. Hilliard in Notes Roy. Bot. Gard. Edinburgh,* 45(3): 482 (1988 publ. 1989)—South Africa (Natal), Swaziland.
rudolfii *O.M. Hilliard in Notes Roy. Bot. Gard. Edinburgh,* 45(3): 483 (1988 publ. 1989)—South Africa (Transvaal).
rupestre *(Hiern) O.M. Hilliard in Notes Roy. Bot. Gard. Edinburgh,* 45(3): 483 (1988 publ. 1989): Phyllopodium rupestre.
swazicum *O.M. Hilliard in Notes Roy. Bot. Gard. Edinburgh,* 45(3): 483 (1988 publ. 1989)—Swaziland.
transvaalense *(Hiern) O.M. Hilliard in Notes Roy. Bot. Gard. Edinburgh,* 45(3): 483 (1988 publ. 1989): Polycarena transvaalensis.

MELICA (Gramin.).
glabrescens *(M.A. Torres) M.A. Torres in Op. Lilloana,* 29: 61 (1980): M. violacea var. glabrescens.
montana *Lamarck, Tabl. Encycl.,* 1: 179 (1791)—Europe.
sarmentosa
var. monticola *M.A. Torres in Op. Lilloana,* 29: 21 (1980)—Argentina.

MELICOPE (Rutac.).
adscendens *(St. John & Hume) T.G. Hartley & B.C. Stone in Taxon,* 38(1): 120 (1989): Pelea adscendens.
anisata *(H. Mann) T.G. Hartley & B.C. Stone in Taxon,* 38(1): 122 (1989): Pelea anisata.
balloui *(Rock) T.G. Hartley & B.C. Stone in Taxon,* 38(1): 121 (1989): Pelea balloui.

MELICOPE :–
christophersenii *(St. John) T.G. Hartley & B.C. Stone in Taxon,* 38(1): 121 (1989): Pelea christophersenii.
cinereops *(St. John) T.G. Hartley & B.C. Stone in Taxon,* 38(1): 120 (1989): Pelea cinereops.
clusiifolia *(A. Gray) T.G. Hartley & B.C. Stone in Taxon,* 38(1): 122 (1989): Pelea clusiaefolia.
cruciata *(Heller) T.G. Hartley & B.C. Stone in Taxon,* 38(1): 121 (1989): Pelea cruciata.
degeneri *(B.C. Stone) T.G. Hartley & B.C. Stone in Taxon,* 38(1): 122 (1989): Pelea degeneri.
fatuhivensis *(F. Brown) T.G. Hartley & B.C. Stone in Taxon,* 38(1): 123 (1989): Pelea fatuhivensis.
feddei *(Léveillé) T.G. Hartley & B.C. Stone in Taxon,* 38(1): 121 (1989): Pelea feddei.
haleakalae *(B.C. Stone) T.G. Hartley & B.C. Stone in Taxon,* 38(1): 122 (1989): Pelea haleakalae.
haupuensis *(St. John) T.G. Hartley & B.C. Stone in Taxon,* 38(1): 120 (1989): Pelea haupuensis.
hawaiensis *(Wawra) T.G. Hartley & B.C. Stone in Taxon,* 38(1): 121 (1989): Pelea hawaiensis.
hiiakae *(B.C. Stone) T.G. Hartley & B.C. Stone in Taxon,* 38(1): 122 (1989): Pelea hiiakae.
honoluluensis *(St. John) T.G. Hartley & B.C. Stone in Taxon,* 38(1): 121 (1989): Pelea honoluluensis.
kaalaensis *(St. John) T.G. Hartley & B.C. Stone in Taxon,* 38(1): 121 (1989): Pelea kaalaensis.
kavaiensis *(H. Mann) T.G. Hartley & B.C. Stone in Taxon,* 38(1): 121 (1989): Pelea kavaiensis.
knudsenii *(Hillebr.) T.G. Hartley & B.C. Stone in Taxon,* 38(1): 121 (1989): Pelea knudsenii.
lakae *(B.C. Stone) T.G. Hartley & B.C. Stone in Taxon,* 38(1): 122 (1989): Pelea lakae.
lohiauana *(B.C. Stone) T.G. Hartley & B.C. Stone in Taxon,* 38(1): 122 (1989): Pelea lohiauana.
lydgatei *(Hillebr.) T.G. Hartley & B.C. Stone in Taxon,* 38(1): 121 (1989): Pelea lydgatei.
macropus *(Hillebr.) T.G. Hartley & B.C. Stone in Taxon,* 38(1): 121 (1989): Pelea macropus.
makahae *(B.C. Stone) T.G. Hartley & B.C. Stone in Taxon,* 38(1): 121 (1989): Pelea makahae.
mauii *T.G. Hartley & B.C. Stone in Taxon,* 38(1): 121 (1989), nom. nov.: Pelea parvifolia *Hillebr.*
molokaiensis *(Hillebr.) T.G. Hartley & B.C. Stone in Taxon,* 38(1): 121 (1989): Pelea molokaiensis.
mucronulata *(St. John) T.G. Hartley & B.C. Stone in Taxon,* 38(1): 121 (1989): Pelea mucronulata.

MELICOPE :–
munroi *(St. John) T.G. Hartley & B.C. Stone in Taxon,* 38(1): 121 (1989): Pelea munroi.
nealae *(B.C. Stone) T.G. Hartley & B.C. Stone in Taxon,* 38(1): 121 (1989): Pelea nealae.
nukuhivensis *(F. Brown) T.G. Hartley & B.C. Stone in Taxon,* 38(1): 123 (1989): Pelea nukuhivensis.
oahuensis *(Léveillé) T.G. Hartley & B.C. Stone in Taxon,* 38(1): 122 (1989): Pelea oahuensis.
obovata *(St. John) T.G. Hartley & B.C. Stone in Taxon,* 38(1): 121 (1989): Pelea obovata.
orbicularis *(Hillebr.) T.G. Hartley & B.C. Stone in Taxon,* 38(1): 121 (1989): Pelea orbicularis.
ovalis *(St. John) T.G. Hartley & B.C. Stone in Taxon,* 38(1): 122 (1989): Pelea ovalis.
ovata *(St. John & Hume) T.G. Hartley & B.C. Stone in Taxon,* 38(1): 121 (1989): Pelea ovata.
pallida *(Hillebr.) T.G. Hartley & B.C. Stone in Taxon,* 38(1): 121 (1989): Pelea pallida.
paniculata *(St. John) T.G. Hartley & B.C. Stone in Taxon,* 38(1): 122 (1989): Pelea paniculata.
peduncularis *(Léveillé) T.G. Hartley & B.C. Stone in Taxon,* 38(1): 122 (1989): Pelea peduncularis.
pseudoanisata *(Rock) T.G. Hartley & B.C. Stone in Taxon,* 38(1): 121 (1989): Pelea pseudoanisata.
puauluensis *(St. John) T.G. Hartley & B.C. Stone in Taxon,* 38(1): 123 (1989): Pelea puauluensis.
puberula *(St. John) T.G. Hartley & B.C. Stone in Taxon,* 38(1): 121 (1989): Pelea puberula.
quadrangularis *(St. John & Hume) T.G. Hartley & B.C. Stone in Taxon,* 38(1): 122 (1989): Pelea quadrangularis.
radiata *(St. John) T.G. Hartley & B.C. Stone in Taxon,* 38(1): 121 (1989): Pelea radiata.
reflexa *(St. John) T.G. Hartley & B.C. Stone in Taxon,* 38(1): 122 (1989): Pelea reflexa.
rotundifolia *(A. Gray) T.G. Hartley & B.C. Stone in Taxon,* 38(1): 122 (1989): Pelea rotundifolia.
saint-johnii *(Hume) T.G. Hartley & B.C. Stone in Taxon,* 38(1): 121 (1989): Pelea saint-johnii.
sandwicensis *(Hook. & Arn.) T.G. Hartley & B.C. Stone in Taxon,* 38(1): 121 (1989): Brunellia sandwicensis.
volcanica *(A. Gray) T.G. Hartley & B.C. Stone in Taxon,* 38(1): 122 (1989): Pelea volcanica.
waialealae *(Wawra) T.G. Hartley & B.C. Stone in Taxon,* 38(1): 122 (1989): Pelea waialealae.
wailauensis *(St. John) T.G. Hartley & B.C. Stone in Taxon,* 38(1): 122 (1989): Pelea wailauensis.
wawraeana *(Rock) T.G. Hartley & B.C. Stone in Taxon,* 38(1): 122 (1989): Pelea wawraeana.

MELICOPE :–
zahlbruckneri *(Rock) T.G. Hartley & B.C. Stone in Taxon,* 38(1): 122 (1989): Pelea zahlbruckneri.

MELIENTHA (Opiliac.).
longistaminea *W.Z. Li in Acta Bot. Yunnanica,* 11(4): 407 (1989)—China.

MELIOSMA (Meliosmac.).
caldasii *J.M. Idrobo in Mutisia,* 72: 2 (1988)—Colombia.
cuneifolia
var. multinervia *(Beus.) Law ex C.Y. Chang in Fl. Sichuanica,* 4: 172 (1988): M. dilleniifolia var. multinervia.
yunnanensis
var. fischeriana *(Rehd. & Wils.) C.Y. Chang in Fl. Sichuanica,* 4: 167 (1988): M. fischeriana.

MELISSITUS (Legumin.).
liaosiensis *(P.Y. Fu & Y.A. Chen) P.Y. Fu & Y.A. Chen in Fl. Liaoningica, chief ed. S.X. Li,* 1: 901 (1988), isonym; *see G.P. Yakovlev in Novosti Sist. Vÿssh. Rast.,* 16: 138 (1979): Pocockia liaosiensis.

MELOCACTUS (Cactac.).
azureus
var. krainzianus *(Buining & Brederoo) P.J. Braun in Bradleya,* 6: 94 (1988): M. krainzianus.
deinacanthus
subsp. florschuetzianus *(Buining & Brederoo) P.J. Braun in Bradleya,* 6: 94 (1988): M. florschuetzianus.
subsp. longicarpus *(Buining & Brederoo) P.J. Braun in Bradleya,* 6: 94 (1988): M. longicarpus.
f. mulequensis *(Buining & Brederoo) P.J. Braun in Bradleya,* 6: 94 (1988): M. mulequensis.
diersianus
f. rubrispinus *(Ritter) P.J. Braun in Bradleya,* 6: 94 (1988): M. rubrispinus.
oreas
f. azulensis *(Buining et al.) P.J. Braun in Bradleya,* 6: 95 (1988): M. azulensis.
subsp. cremnophilus *(Buining & Brederoo) P.J. Braun in Bradleya,* 6: 95 (1988): M. cremnophilus.
subsp. ernestii *(Vaupel) P.J. Braun in Bradleya,* 6: 95 (1988): Melocactus ernestii.
f. erythracanthus *(Buining & Brederoo) P.J. Braun in Bradleya,* 6: 95 (1988): M. erythracanthus.
var. longispinus *(Buining et al.) P.J. Braun in Bradleya,* 6: 95 (1988): M. longispinus.
subsp. rubrisaetosus *(Buining et al.) P.J. Braun in Bradleya,* 6: 95 (1988): M. rubrisaetosus.
pruinosus
var. concinnus *(Buining & Brederoo) P.J. Braun in Bradleya,* 6: 95 (1988): M. concinnus.

MELOCACTUS :–
zehntneri
subsp. canescens *(Ritter) P.J. Braun in Bradleya,* 6: 98 (1988): M. canescens.
var. curvicornis *(Buining & Brederoo) P.J. Braun in Bradleya,* 6: 98 (1988): M. curvicornis.
subsp. robustispinus *(Buining et al.) P.J. Braun in Bradleya,* 6: 98 (1988): M. robustispinus.

MELOCALAMUS (Gramin.).
arrectus *T.P. Yi in Acta Bot. Yunnanica,* 10(4): 440 (1988)—China.

MELOCHIA (Sterculiac.).
spicata *(L.) P.A. Fryxell, Malvac. of Mexico (Syst. Bot. Monog., 25):* 457 (1988): Malva spicata.

MEMECYLON (Melastomatac.).
randeriana *S.M. Almeida & M.R. Almeida in J. Bombay Nat. Hist. Soc.,* 85(3): 521 (1988 publ. 1989), nom. nov.: M. malabaricum *Cogn.*

MENTHA (Labiat.).
×piperita
var. ouweneelii *J. Lebeau & J. Lambinon in Soc. Ech. Pl. Vasc. Eur. Bassin Médit. Bull.,* 22: 21 (1986–87 publ. 1988)—Belgium.

MENZIESIA (Ericac.).
katsumatae
f. duplex *(Satomi & M. Hashimoto) N. Satomi in J. Phytogeogr. Taxon.,* 36(2): 112 (1988): M. multiflora f. duplex.

MERREMIA (Convolvulac.).
eberhardtii *(Gagnep.) N.T. Nhan in L.V. Aver'yanov, N.T. Ban & E.P. Kudryavtseva, Mater. Fl. Rast. Ostrov. V'etnama:* 43 (1988): Ipomoea eberhardtii.
gracilis *E.J.F. Campbell & G. Argent in Notes Roy. Bot. Gard. Edinburgh,* 45(2): 345 (1988 publ. 1989)—Borneo (Sabah).

METADACRYDIUM M.G. Baumann-Bodenheim, Syst. Fl. Neu-Caledonien, 5: 76 (1989), nom. nov., without replaced synonym ref.: Dacrydium author not stated (Podocarpac.).
araucarioides *(Brongn. & Gris) M.G. Baumann-Bodenheim, Syst. Fl. Neu-Caledonien,* 5: 76 (1989), generic name not validly publ.: Dacrydium araucarioides.
balansae *(Brongn. & Gris) M.G. Baumann-Bodenheim, Syst. Fl. Neu-Caledonien,* 5: 76 (1989), generic name not validly publ.: Dacrydium balansae.

METROSIDEROS (Myrtac.).
polymorpha
var. dieteri *J. Wyndham Dawson & L. Stemmermann in Bishop Mus. Occas. Pap.,* 29: 108 (1989)—Hawaiian Is.

METROSIDEROS :–
var. pumila *(A. Heller) J. Wyndham Dawson & L. Stemmermann in Bishop Mus. Occas. Pap.,* 29: 108 (1989): Nania pumila.
waialealae
var. fauriei *(H. Lév.) J. Wyndham Dawson & L. Stemmermann in Bishop Mus. Occas. Pap.,* 29: 109 (1989): Nania × fauriei.

METTENIUSA (Icacinac.).
cundinamarcensis *G. Lozano-C. in Fl. Colombia,* 11: 32 (1988)—Colombia.
huilensis *G. Lozano-C. in Fl. Colombia,* 11: 36 (1988)—Colombia.
santanderensis *G. Lozano-C. in Fl. Colombia,* 11: 38 (1988)—Colombia.

MICONIA (Melastomatac.).
cacatin *(Aublet) S.S. Renner in Mem. New York Bot. Gard.,* 50: 29 (1989): Melastoma cacatin.
calocoma *F. Almeda in Proc. Calif. Acad. Sci.,* Ser. 4, 46(5): 144 (1989)—Costa Rica.
dissitiflora *F. Almeda in Proc. Calif. Acad. Sci.,* Ser. 4, 46(5): 146 (1989)—Costa Rica.
huberi *J.J. Wurdack in Mem. New York Bot. Gard.,* 51: 103 (1989)—Venezuela.
jimenezii *W.S. Judd & R.S. Beaman in Brittonia,* 40(4): 386 (1988)—Hispaniola (Dominican Republic, Haiti).
septentrionalis *W.S. Judd & R.S. Beaman in Brittonia,* 40(4): 383 (1988)—Hispaniola (Dominican Republic).

MICRODESMIS (Pandac.).
afrodecandra *J.J. Floret, A.M. Louis & J.M. Reitsma in Bull. Mus. Nation. Hist. Nat., B, Adansonia,* Ser. 4, 11(2): 104 (1989)—Gabon.

MICROLOMA (Asclepiadac.).
calycinum
subsp. flavescens *(E. Mey.) H.E. Wanntorp in Op. Bot.,* 98: 36 (1988 publ. 1989): M. calycinum var. flavescens.
campanulatum *H.E. Wanntorp in Op. Bot.,* 98: 63 (1988 publ. 1989)—South Africa (Cape Province).
dregei *(E. Mey.) H.E. Wanntorp in Op. Bot.,* 98: 62 (1988 publ. 1989): Haemax dregei.
glabratum
subsp. subglabratum *H.E. Wanntorp in Op. Bot.,* 98: 46 (1988 publ. 1989)—South Africa (Cape Province).
hystricoides *H.E. Wanntorp in Op. Bot.,* 98: 64 (1988 publ. 1989)—South Africa (Orange Free State, Cape Province).
lanatum *H.E. Wanntorp in Op. Bot.,* 98: 53 (1988 publ. 1989)—Namibia.
sagittatum
subsp. pillansii *H.E. Wanntorp in Op. Bot.,* 98: 41 (1988 publ. 1989)—South Africa (Cape Province).
spinosum
subsp. dinteri *(Schlecht.) H.E. Wanntorp in Op. Bot.,* 98: 62 (1988 publ. 1989): M. dinteri.

MICROLOMA :–
subsp. velutinum *H.E. Wanntorp in Op. Bot.*, 98: 62 (1988 publ. 1989)—Namibia.

MICROSIDEROS M.G. Baumann-Bodenheim, Syst. Fl. Neu-Caledonien, 5: 77 (1989), nom. nov., without replaced synonym ref.: Metrosideros author not stated (Myrtac.).
tetrasticha *(Guill.) M.G. Baumann-Bodenheim, Syst. Fl. Neu-Caledonien,* 5: 77 (1989), without basionym page: Metrosideros tetrasticha.

MICROSISYMBRIUM (Crucif.).
qingshuiheense *Y.Q. Ma & Z.Y. Zhu in Acta Sci. Nat. Univ. Intramongolicae,* 20(4): 538 (1989)—China.

MICROTATORCHIS (Orchidac.).
smithii *P.J. Kores in Allertonia,* 5(1): 205 (1989)—Fiji.

MIKANIA (Compos.).
cuatrecasasii *W.C. Holmes in Phytologia,* 65(3): 242 (1988)—Colombia.
daspitii *W.C. Holmes & S. McDaniel in Candollea,* 44(1): 32 (1989)—Argentina.
longicarpa *W.C. Holmes in Phytologia,* 65(3): 243 (1988)—Colombia.
macdanielii *W.C. Holmes in Syst. Bot.,* 14(3): 377 (1989)—Peru.
marahuacensis *J. Steyermark & B. Maguire in Acta Bot. Venezuel.,* 14(3): 28 (1984)—Venezuela.
pseudorimachii *W.C. Holmes & S. McDaniel in Candollea,* 44(1): 32 (1989)—Paraguay.
tristachya *W.C. Holmes in Phytologia,* 65(3): 244 (1988)—Colombia.

MILIUSA (Annonac.).
traceyi *L.W. Jessup in Austrobaileya,* 2(5): 522 (1988)—Australia (Queensland, N. Terr.).

MIMETES (Proteac.).
chrysanthus *J.P. Rourke in S. Afr. J. Bot.,* 54(6): 636 (1988)—South Africa (Cape Province).

MIMOSA (Legumin.).
barnebiana *R.H. Fortunato & S.G. Tressens in Candollea,* 44(1): 35 (1989)—Argentina, Paraguay.
pauperoides *(Burk.) R.H. Fortunato in Ann. Missouri Bot. Gard.,* 76(2): 381 (1989): M. balansae var. pauperoides.

MIMULUS (Scrophulariac.).
cupriphilus *M.R. Macnair in Bot. J. Linn. Soc.,* 100(1): 3 (1989)—U.S.A. (California).

MINUARTIA (Caryophyllac.).
recurva
subsp. bigerrensis *(Pau) S. Rivas-Martínez, F. Fernández González & D. Sánchez-Mata in Opusc. Bot. Pharm. Complutensis,* 2: 113 (1986): Alsine recurva f. bigerrensis.

MISCANTHUS (Gramin.).
nudipes
subsp. taylorii *(Bor) Y.N. Lee in J. Korean Pl. Taxon.,* 3(1–2): 18 (1971): M. taylorii.†
subsp. wardii *(Bor) Y.N. Lee in J. Korean Pl. Taxon.,* 3(1–2): 18 (1971), as 'wordii', without basionym ref.: M. wardii.†
sacchariflorus
f. purpurascens *Y.N. Lee in J. Korean Pl. Taxon.,* 3(1–2): 18 (1971)—Great Britain.†

MISOPATES (Scrophulariac.).
marraicum *D.A. Sutton, Revis. tribe Antirrhineae:* 153 (1988)—Sudan.
microcarpum *(Pomel) D.A. Sutton, Revis. tribe Antirrhineae:* 151 (1988): Antirrhinum microcarpum.
oranense *(Faure) D.A. Sutton, Revis. tribe Antirrhineae:* 148 (1988): Antirrhinum orontium var. oranense.
orontium
subsp. gibbosum *(Wall.) D.A. Sutton, Revis. tribe Antirrhineae:* 147 (1988): Antirrhinum gibbosum.
salvagense *D.A. Sutton, Revis. tribe Antirrhineae:* 151 (1988)—Salvage Is.

MITRACARPUS (Rubiac.).
bahorucanus *T.A. Zanoni in Moscosoa,* 5: 103 (1989)—Hispaniola (Dominican Republic).

MNESITHEA (Gramin.).
pilosa *B.K. Simon in Austrobaileya,* 3(1): 90 (1989)—Australia (Queensland).

MOLLINEDIA (Monimiac.).
foreroi *A.L. Peixoto in Arq. Jard. Bot. Rio de Janeiro,* 27: 195 (1983–85 publ. 1985)—Colombia.
gentrii *A.L. Peixoto in Arq. Jard. Bot. Rio de Janeiro,* 27: 196 (1983–85 publ. 1985)—Mexico.
glabricaulis *B. Maguire & J.A. Steyermark in Mem. New York Bot. Gard.,* 51: 115 (1989)—Brazil.
jorgeorum *A.L. Peixoto in Arq. Jard. Bot. Rio de Janeiro,* 27: 194 (1983–85 publ. 1985), as 'jorgearum'—Brazil.
luizae *A.L. Peixoto in Arq. Jard. Bot. Rio de Janeiro,* 27: 194 (1983–85 publ. 1985)—Brazil.
neblinensis *B. Maguire & J.A. Steyermark in Mem. New York Bot. Gard.,* 51: 113 (1989)—Venezuela.

MONANTHOCITRUS (Rutac.).
oblanceolata *B.C. Stone & D.T. Jones in Proc. Acad. Nat. Sci. Philadelphia,* 140(2): 267 (1988)—Borneo (Sabah).

MONOPSIS (Campanulac.).
unidentata
subsp. intermedia *P.B. Phillipson in Bot. J. Linn. Soc.,* 99(3): 268 (1989)—South Africa (Cape Province).

MONOPSIS :–
subsp. laevicaulis *(C. Presl) P.B. Phillipson in Bot. J. Linn. Soc.,* 99(3): 270 (1989): Dobrowskya laevicaulis.

MONOTES (Dipterocarpac.).
lutambensis *B. Verdcourt in Fl. Trop. E. Afr., Dipterocarpac.:* 9 (1989)—Tanzania.

MONTANOA (Compos.).
leucantha
var. arborescens *(DC.) B.L. Turner in Phytologia,* 65(5): 382 (1988): M. arborescens.
tomentosa
var. microcephala *(Sch.-Bip.) B.L. Turner in Phytologia,* 65(5): 382 (1988): M. microcephala.
var. xanthifolia *(Sch.-Bip.) B.L. Turner in Phytologia,* 65(5): 382 (1988): M. xanthifolia.

MONTROUZIERA (Guttif.).
sphaeroidea
var. rhodoneura *(Schltr.) M.G. Baumann-Bodenheim, Syst. Fl. Neu-Caledonien,* 4: 102 (1988 publ. Jan. 1989), without basionym ref.; *et Syst. Fl. Neu-Caledonien,* 5: 97 (May 1989), with basionym ref.: M. rhodoneura.

MORINDA (Rubiac.).
coreia
var. stenophylla *(Spreng.) M. Chandrabose in M. Chandrabose & N.C. Nair, Fl. Coimbatore:* 142 (1988): M. stenophylla.
longipedunculata *J.A. Steyermark in Ann. Missouri Bot. Gard.,* 75(1): 340 (1988)—Venezuela.

MORMODES (Orchidac.).
auriculata *F. Miranda in Lindleyana,* 4(2): 95 (1989)—Brazil.
densiflora *F. Miranda in Lindleyana,* 4(2): 95 (1989)—Brazil.
elegans *F. Miranda in Lindleyana,* 4(2): 94 (1989)—Brazil.
hirsutissima *F. Miranda in Lindleyana,* 4(2): 98 (1989)—Brazil.
tuxtlensis *G.A. Salazar Chávez in Orquídea (México),* 11: 52 (1988)—Mexico.

MORUS (Morac.).
australis
var. incisa *C.Y. Wu in Acta Bot. Yunnanica,* 11(1): 26 (1989)—China.
var. inusitata *(Lévl.) C.Y. Wu in Acta Bot. Yunnanica,* 11(1): 25 (1989): M. inusitata.
mongolica
var. hopeiensis *S.S. Chang in Acta Bot. Yunnanica,* 11(1): 24 (1989)—China.

MOURIRI (Melastomatac.).
gleasoniana
var. coibensis *(Morley) T. Morley in Ann. Missouri Bot. Gard.,* 76(2): 437 (1989): M. coibensis.
latihila *T. Morley in Ann. Missouri Bot. Gard.,* 76(2): 434 (1989)—Venezuela.
osaensis *T. Morley in Ann. Missouri Bot. Gard.,* 76(2): 430 (1989)—Costa Rica.
papillosa *T. Morley in Ann. Missouri Bot. Gard.,* 76(2): 432 (1989)—Venezuela.
retentipetala *T. Morley in Ann. Missouri Bot. Gard.,* 76(2): 434 (1989)—Brazil, Peru.

MUCUNA (Legumin.).
killipiana *J.I. Hernández Camacho & C. Barbosa in Trianea, Acta Cient. Tecnol. Inderena,* 1: 38 (1988)—Colombia.

MUHLENBERGIA (Gramin.).
majalcensis *P.M. Peterson in Syst. Bot.,* 14(3): 316 (1989)—Mexico.
palmirensis *I. Grignon & S. Laegaard in Nordic J. Bot.,* 9(1): 47 (1989)—Ecuador.

MULGEDIUM (Compos.).
bracteatum *(Hook. f. & Thoms. ex C.B. Clarke) C. Shih in Acta Phytotax. Sin.,* 26(5): 390 (1988): Lactuca bracteata.
meridionale *C. Shih in Acta Phytotax. Sin.,* 26(5): 392 (1988)—China.
monocephalum *(Chang) C. Shih in Acta Phytotax. Sin.,* 26(5): 391 (1988): Lactuca monocephala.
polypodifolium *(Franch.) C. Shih in Acta Phytotax. Sin.,* 26(5): 392 (1988): Lactuca polypodifolia.
umbrosum *(Dunn) C. Shih in Acta Phytotax. Sin.,* 26(5): 391 (1988): Lactuca umbrosa.

MURALTIA (Polygalac.).
bondii *J.H.J. Vlok in S. Afr. J. Bot.,* 55(3): 365 (1989)—South Africa (Cape Province).

MUSA (Musac.).
lushanensis *J.L. Liu in Acta Bot. Yunnanica,* 11(2): 171 (1989)—China.

MUSCARI (Hyacinthac.).
macbeathianum *Kit Tan in Herbertia,* 44(1): 26 (1988), as 'mcbeathianum'—Turkey.
mirum *F. Speta in Phyton (Austria),* 29(1): 107 (1989)—Turkey.

MYCETIA (Rubiac.).
membranacea *N. Fukuoka in Acta Phytotax. Geobot.,* 40(1–4): 112 (1989)—Thailand.
ovatistipulata *N. Fukuoka in Acta Phytotax. Geobot.,* 40(1–4): 110 (1989)—Thailand.
paniculiformis *N. Fukuoka in Acta Phytotax. Geobot.,* 40(1–4): 114 (1989)—Thailand.
siamensis *N. Fukuoka in Acta Phytotax. Geobot.,* 40(1–4): 114 (1989)—Thailand.

MYOSOTIS (Boraginac.).
radix-palaris *A.P. Khokhryakov in Byull. Glavn. Bot. Sada (Moscow),* 152: 60 (1989)—U.S.S.R. (Caucasus).
superalpina *A.P. Khokhryakov in Byull. Glavn. Bot. Sada (Moscow),* 152: 59 (1989)—U.S.S.R. (Caucasus).

MYRCEUGENIA (Myrtac.).
decussata *J.R. Mattos in Loefgrenia,* 94: 8 (1989)—Brazil.

MYRCIA (Myrtac.).
botris *(Berg) N.J.E. Silveira in Loefgrenia,* 91: 1 (1987): Aulomyrcia botris.
rufipes
var. dives *(Berg) N.J.E. Silveira in Loefgrenia,* 91: 2 (1987): Aulomyrcia rufipes var. dives.
selloi *(Spreng.) N.J.E. Silveira in Loefgrenia,* 89: 5 (1986): Myrtus selloi.
var. acutata *(Bg.) N.J.E. Silveira in Loefgrenia,* 89: 6 (1986): Aulomyrcia ramulosa var. acutata.
var. australis *(Berg) N.J.E. Silveira in Loefgrenia,* 89: 7 (1986): Aulomyrcia ramulosa var. australis.
var. megapotamica *(Legr.) N.J.E. Silveira in Loefgrenia,* 92: 3 (1988): M. ramulosa var. megapotamica.
var. microsiphonata *(Legr.) N.J.E. Silveira in Loefgrenia,* 89: 7 (1986): M. microsiphonata.
velhensis *(Berg) N.J.E. Silveira in Loefgrenia,* 91: 1 (1987): Aulomyrcia velhensis.

MYRCIARIA (Myrtac.).
sulcata *J.R. Mattos in Loefgrenia,* 90: 2 (1986)—Brazil.

MYRSINE (Myrsinac.).
rolletii *R.A. Howard in Phytologia,* 65(4): 287 (1988), nom. nov.: Rapanea rotundifolia *Mez.*
vaccinioides *W.L. Wagner, D.R. Herbst & S.H. Sohmer in Bishop Mus. Occas. Pap.,* 29: 107 (1989)—Hawaiian Is.

N

NAGEIA (Podocarpac.).
formosensis *(Dummer) C.N. Page in Notes Roy. Bot. Gard. Edinburgh,* 45(2): 382 (1988 publ. 1989): Podocarpus formosensis.

NAMA (Hydrophyllac.).
californicum *(A. Gray) J.D. Bacon in Aliso,* 12(2): 332 (1989): Lemmonia californica.

NARCISSUS (Amaryllidac.).
albimarginatus *D. & U. Müller-Doblies in Flow. Pl. Afr.,* 50(2): t. 1986 (1989)—Morocco.
×montsiccianus *Á.M. Romo in Fontqueria,* 24: 15 (1989)—Spain. (N. assoanus × palearensis).
palearensis *A.M. Romo in Fontqueria,* 24: 15 (1989)—Spain.

NAUTILOCALYX (Gesneriac.).
pemphidius *L.E. Skog in Syst. Bot.,* 14(3): 281 (1989)—Venezuela.

NAVIA (Bromeliac.).
albiflora *L.B. Smith, J. Steyermark & Robinson in Acta Bot. Venezuel.,* 14(3): 11 (1984)—Venezuela.
aliciae *L.B. Smith, J. Steyermark & Robinson in Acta Bot. Venezuel.,* 14(3): 13 (1984)—Venezuela.
geaster *L.B. Smith, J. Steyermark & Robinson in Acta Bot. Venezuel.,* 14(3): 14 (1984)—Venezuela.
lactea *L.B. Smith, J. Steyermark & Robinson in Acta Bot. Venezuel.,* 14(3): 14 (1984)—Venezuela.

NAYARIOPHYTON T.K. Paul in Fasc. Fl. India, 19: 183 (30 March 1988); et in Bot. Jahrb., 110(1): 43 (20 Oct. 1988). *MALVACEAE.* †
jujubifolium *(Griff.) T.K. Paul in Fasc. Fl. India,* 19: 185 (30 March 1988); *et in Bot. Jahrb.,* 110(1): 43 (20 Oct. 1988): Kydia jujubifolia.†

NECTANDRA (Laurac.).
aurea *J.G. Rohwer in Ann. Missouri Bot. Gard.,* 75(4): 1525 (1988 publ. 1989)—Venezuela.
fulva *J.G. Rohwer in Ann. Missouri Bot. Gard.,* 75(4): 1527 (1988 publ. 1989)—Venezuela.
ruforamula *J.G. Rohwer in Ann. Missouri Bot. Gard.,* 75(4): 1529 (1988 publ. 1989)—Venezuela.

NEEA (Nyctaginac.).
grandis *B. Maguire & J.A. Steyermark in Mem. New York Bot. Gard.,* 51: 109 (1989)—Brazil.
neblinensis *B. Maguire & J.A. Steyermark in Mem. New York Bot. Gard.,* 51: 112 (1989)—Venezuela.

NEMESIA (Scrophulariac.).
acornis *K.E. Steiner in S. Afr. J. Bot.,* 55(4): 405 (1989)—South Africa (Cape Province).

NEMUM (Cyperac.).
parviflorum *K.A. Lye in Lidia,* 2(1): 33 (1989)—Guinea.

NEOCINNAMOMUM (Laurac.).
complanifructum *S. Lee & F.N. Wei in Guihaia,* 8(4): 307 (1988)—Vietnam, China.

×NEOGLOSSUM hort. in Orchid Rev., 97(1145): centre page pull-out p. 9 (1989). *ORCHIDACEAE.* (Ascoglossum × Neofinetia).

NEOLITSEA (Laurac.).
subsect. Nonsecundinervatae *J.C. Liao, Taxon. Revis. Fam. Lauraceae in Taiwan:* vii (1988), without type.
subsect. Parallelinervatae *J.C. Liao, Taxon. Revis. Fam. Lauraceae in Taiwan:* vii (1988), without type.
subsect. Parumelevatae *J.C. Liao, Taxon. Revis. Fam. Lauraceae in Taiwan:* vii (1988), without type.
sect. Retisupraelevata *J.C. Liao, Taxon. Revis. Fam. Lauraceae in Taiwan:* vii (1988), without type.
sect. Retisupraplana *J.C. Liao, Taxon. Revis. Fam. Lauraceae in Taiwan:* vii (1988), without type.
subsect. Valdelevatae *J.C. Liao, Taxon. Revis. Fam. Lauraceae in Taiwan:* vii (1988), without type.

NEOMITRANTHES (Myrtac.).
sect. Dichotomae *J.R. Mattos in Loefgrenia,* 95: 1 (1989).
sect. Multibracteatae *J.R. Mattos in Loefgrenia,* 95: 1 (1989).
sect. Pedunculatae *J.R. Mattos in Loefgrenia,* 95: 1 (1989).
wilsoniana *J.R. Mattos in Loefgrenia,* 94: 7 (1989)—Brazil.

NEONAUCLEA (Rubiac.).
anthraciticus *C.E. Ridsdale in Blumea,* 34(1): 242 (1989)—New Guinea (W. Irian).
artocarpoides *C.E. Ridsdale in Blumea,* 34(1): 207 (1989)—Borneo (Brunei, Sabah, Kalimantan).
bartlingii
var. cumingiana *(Vidal) C.E. Ridsdale in Blumea,* 34(1): 219 (1989): Nauclea cumingiana.
bomberaiensis *C.E. Ridsdale in Blumea,* 34(1): 247 (1989)—New Guinea (W. Irian).
borneensis *C.E. Ridsdale in Blumea,* 34(1): 212 (1989)—Borneo (Brunei, Kalimantan, Sarawak).
butonensis *C.E. Ridsdale in Blumea,* 34(1): 217 (1989)—Celebes.
calcarea *C.E. Ridsdale in Blumea,* 34(1): 212 (1989)—Borneo (Sarawak).
ceramensis *C.E. Ridsdale in Blumea,* 34(1): 238 (1989)—Moluccas.
circumscissa *C.E. Ridsdale in Blumea,* 34(1): 217 (1989)—Philippine Is.
colla *C.E. Ridsdale in Blumea,* 34(1): 249 (1989)—New Guinea (W. Irian).
coronata *C.E. Ridsdale in Blumea,* 34(1): 249 (1989)—New Guinea (W. Irian, Papua New Guinea).
endertii *C.E. Ridsdale in Blumea,* 34(1): 201 (1989)—Borneo (Kalimantan).

NEONAUCLEA :–
excelsioides *C.E. Ridsdale in Blumea,* 34(1): 202 (1989)—Borneo (Sabah, Kalimantan).
glabra *(Roxb.) Bakh. f. & C.E. Ridsdale in Blumea,* 34(1): 240 (1989): Nauclea glabra.
glandulifera *C.E. Ridsdale in Blumea,* 34(1): 237 (1989)—Moluccas.
havilandii *Koord. ex C.E. Ridsdale in Blumea,* 34(1): 221 (1989)—Celebes.
intercontinentalis *Bakh. f. & C.E. Ridsdale in Blumea,* 34(1): 240 (1989)—Celebes, Moluccas.
kraboensis *C.E. Ridsdale in Blumea,* 34(1): 246 (1989)—New Guinea (W. Irian).
lanceolata
subsp. gracilis *(Vidal) C.E. Ridsdale in Blumea,* 34(1): 225 (1989): Nauclea gracilis.
montana *C.E. Ridsdale in Blumea,* 34(1): 226 (1989)—Borneo (Sarawak).
morotaiensis *C.E. Ridsdale in Blumea,* 34(1): 234 (1989)—Moluccas.
pallida
subsp. malaccensis *(Gand.) C.E. Ridsdale in Blumea,* 34(1): 199 (1989): Nauclea malaccensis.
paracyrtopoda *Bakh. f. & C.E. Ridsdale in Blumea,* 34(1): 206 (1989)—Borneo (Sarawak).
parviflora *(Koord. & Valeton) C.E. Ridsdale in Blumea,* 34(1): 204 (1989): Nauclea purpurascens var. parviflora.
pseudocalycina *C.E. Ridsdale in Blumea,* 34(1): 204 (1989)—Borneo (Sabah, Kalimantan), Philippine Is.
pseudopeduncularis *C.E. Ridsdale in Blumea,* 34(1): 222 (1989)—Celebes.
rupestris *Bakh. f. & C.E. Ridsdale in Blumea,* 34(1): 235 (1989)—Celebes.
sericea *C.E. Ridsdale in Blumea,* 34(1): 230 (1989)—Moluccas.
subulifera *C.E. Ridsdale in Blumea,* 34(1): 245 (1989)—Moluccas, New Guinea (W. Irian).
tricephala *C.E. Ridsdale in Blumea,* 34(1): 247 (1989)—New Guinea (W. Irian).
unicapitulifera *C.E. Ridsdale in Blumea,* 34(1): 239 (1989)—Celebes.
ventricosa *C.E. Ridsdale in Blumea,* 34(1): 207 (1989)—Celebes.
vinkiorum *C.E. Ridsdale in Blumea,* 34(1): 244 (1989)—New Guinea (W. Irian).

NEONOTONIA (Legumin.).
wightii
var. longicauda *(Schweinf.) J.A. Lackey in Iselya,* 2(1): 11 (1981): Glycine longicauda.
var. mearnsii *(De Wild.) J.A. Lackey in Iselya,* 2(1): 11 (1981): Glycine mearnsii.
subsp. petitiana *(A. Rich.) J.A. Lackey in Iselya,* 2(1): 11 (1981): Johnia petitiana.
var. petitiana *(A. Rich) J.A. Lackey in Iselya,* 2(1): 11 (1981): Johnia petitiana.
subsp. pseudojavanica *(Taub.) J.A. Lackey in Iselya,* 2(1): 11 (1981): Glycine pseudojavanica.

NEOPLATYTAENIA A.M. Gel'dikhanov in V.V. Nikitin & A.M. Gel'dikhanov, Opred. Rast. Turkmenistana: 455 (1988), nom. nov.: Platytaenia Nevski & Vved. (Umbellif.).
dichotoma *(Boiss.) A.M. Gel'dikhanov in V.V. Nikitin & A.M. Gel'dikhanov, Opred. Rast. Turkmenistana:* 455 (1988): Zosima dichotoma.
pimpinelloides *(Nevski) A.M. Gel'dikhanov in V.V. Nikitin & A.M. Gel'dikhanov, Opred. Rast. Turkmenistana:* 455 (1988): Platytaenia pimpinelloides.

NEOREGELIA (Bromeliac.).
johnsoniae *H.E. Luther in J. Bromeliad Soc.*, 39(2): 70 (1989)—Peru.

NEOSASAMORPHA (Gramin:).
kagamiana
subsp. yoshinoi *(Koidz.) S. Suzuki in J. Jap. Bot.*, 64(2): 46 (1989): Sasa kashidensis var. yoshinoi.
magnifica *(Nakai) S. Suzuki in J. Jap. Bot.*, 64(2): 44 (1989): Arundinaria magnifica.
subsp. fujitae *(S. Suzuki) S. Suzuki in J. Jap. Bot.*, 64(2): 45 (1989): Sasa magnifica subsp. fujitae.
oshidensis
subsp. glabra *(Koidz.) S. Suzuki in J. Jap. Bot.*, 64(2): 45 (1989): Sasa kashidensis var. glabra.
pubiculmis *(Makino) S. Suzuki in J. Jap. Bot.*, 64(9): 273 (1989): Sasa pubiculmis.
subsp. sugimotoi *(Nakai) S. Suzuki in J. Jap. Bot.*, 64(9): 273 (1989): Sasa sugimotoi.
shimidzuana
subsp. kashidensis *(Makino & Koidz.) S. Suzuki in J. Jap. Bot.*, 64(2): 46 (1989): Sasa kashidensis.
stenophylla *(Koidz.) S. Suzuki in J. Jap. Bot.*, 64(2): 43 (1989): Sasa stenophylla.
subsp. tobagenzoana *(Koidz.) S. Suzuki in J. Jap. Bot.*, 64(2): 44 (1989): Sasa tobagenzoana.
takizawana
subsp. nakashimana *(Koidz.) S. Suzuki in J. Jap. Bot.*, 64(2): 47 (1989): Sasamorpha nakashimana.
tsukubensis *(Nakai) S. Suzuki in J. Jap. Bot.*, 64(2): 44 (1989): Sasa tsukubensis.
subsp. pubifolia *(Koidz.) S. Suzuki in J. Jap. Bot.*, 64(2): 44 (1989): Sasa kashidensis var. pubifolia.

NEPETA (Labiat.).
alii *A. Jehan in Willdenowia*, 18(2): 427 (1989)—Pakistan.
betonicifolia
subsp. somkhetica *(Kapeller) Yu.L. Menitskiĭ in Fl. Armenii*, 8: 47 (1987): N. somkhetica.
subsp. strictifolia *(Pojark.) Yu.L. Menitskiĭ in Fl. Armenii*, 8: 47 (1987): N. strictifolia.

NEPETA :-
supina
subsp. buschii *(Sosn. & Manden.) Yu.L. Menitskiĭ in Fl. Armenii*, 8: 42 (1987): N. buschii.

NEWTONIA (Legumin.).
grandifolia *J.F. Villiers in Bull. Jard. Bot. Nation. Belg.*, 58(3–4): 544 (1988)—Gabon, Cameroun, Congo-Brazzaville.

NIDULARIUM (Bromeliac.).
altimontanum *E.M.C. Leme in Bradea*, 5(16): 167 (1989)—Brazil.
ambiguum *M. das Graças L. Wanderley & E.M.C. Leme in Bradea*, 5(16): 168 (1989)—Brazil.
bocainensis *E.M.C. Leme in Bradea*, 5(16): 169 (1989)—Brazil.

NIGELLA (Ranunculac.).
arvensis
subsp. glaucescens *(Guss.) W. Greuter & Burdet in Willdenowia*, 19(1): 45 (1989): N. arvensis var. glaucescens.
subsp. latilabris *(Zohary) W. Greuter & Burdet in Willdenowia*, 19(1): 45 (1989): N. deserti var. latilabris.
subsp. negevensis *(Zohary) W. Greuter & Burdet in Willdenowia*, 19(1): 45 (1989): N. arvensis var. negevensis.
subsp. palaestina *(Zohary) W. Greuter & Burdet in Willdenowia*, 19(1): 45 (1989): N. arvensis var. palaestina.

NOTHODORITIS Z.H. Tsi in Acta Phytotax. Sin., 27(1): 58 (1989). *ORCHIDACEAE.*
zhejiangensis *Z.H. Tsi in Acta Phytotax. Sin.*, 27(1): 59 (1989)—China.

NOTHOSCORDUM (Alliac.).
catharinense *P. Ravenna in Onira, Bot. Leafl.*, 1(6): 46 (1988)—Brazil.
gracilipes *P. Ravenna in Onira, Bot. Leafl.*, 1(6): 45 (1988)—Brazil.

NOTHOTSUGA H.H. Hu ex C.N. Page in Notes Roy. Bot. Gard. Edinburgh, 45(2): 390 (1988 publ. 1989). *PINACEAE.*
longibracteata *(Cheng) H.H. Hu ex C.N. Page in Notes Roy. Bot. Gard. Edinburgh*, 45(2): 390 (1988 publ. 1989): Tsuga longibracteata.

NOTOCACTUS (Cactac.).
concinnus
var. aceguensis *N. Gerloff in Internoto*, 10(4): 105 (1989)—Brazil.
var. eremiticus *(F. Ritter) N. Gerloff in Internoto*, 9(4): 112 (1988): N. eremiticus.
var. gibberulus *(Prestlé) N. Gerloff in Internoto*, 9(4): 112 (1988): N. gibberulus.
var. rubrigemmatus *(Abraham) N. Gerloff in Internoto*, 9(4): 112 (1988): N. rubrigemmatus.

NOTOCACTUS :-
glomeratus *R. Menges in Internoto,* 10(1): 8 (1989), without latin descr. or type—Brazil.
grandiensis *S. Bergner in Internoto,* 10(2): 43 (1989), without latin descr. or type—Brazil.
herteri
f. pseudoherteri *(Buining) K. Herm in Internoto,* 10(2): 39 (1989): N. pseudoherteri.
linkii
var. albispinus *N. Gerloff & G. Schäfer in Internoto,* 10(1): 11 (1989), without latin descr. or type—Brazil.
muegelianus *T. Engel in Internoto,* 9(4): 121 (1988), without latin descr. or type—Locality not stated.
scopa
f. erythrinus *R. Havlíček in Internoto,* 9(4): 125 (1988), without latin descr. or type—Locality not stated.
var. marchesii *W.R. Abraham in Kakt. And. Sukk.,* 40(7): 176 (1989)—Uruguay.
var. murielii *A. Hofacker in Internoto,* 10(2): 40 (1989), without latin descr. or type—Locality not stated.

×**NOTYLETTIA** hort. in Orchid Rev., 97(1145): centre page pull-out p. 9 (1989). *ORCHIDACEAE.* (Comparettia × Notylia).

NUTTALLANTHUS D.A. Sutton, Revis. tribe Antirrhineae: 455 (1988). *SCROPHULARIACEAE.*
canadensis *(L.) D.A. Sutton, Revis. tribe Antirrhineae:* 457 (1988): Antirrhinum canadense.
floridanus *(Chapman) D.A. Sutton, Revis. tribe Antirrhineae:* 461 (1988): Linaria floridana.
subandinus *(Diels) D.A. Sutton, Revis. tribe Antirrhineae:* 459 (1988): L. subandina.
texanus *(Scheele) D.A. Sutton, Revis. tribe Antirrhineae:* 460 (1988): L. texana.

NYMPHAEA (Nymphaeac.).
nouchali
var. caerulea *(Savigny) B. Verdcourt in Fl. Trop. E. Afr., Nymphaeac.:* 7 (1989): N. caerulea.
var. mutandaensis *B. Verdcourt in Fl. Trop. E. Afr., Nymphaeac.:* 10 (1989)—Uganda.
var. ovalifolia *(Conard) B. Verdcourt in Fl. Trop. E. Afr., Nymphaeac.:* 9 (1989): N. ovalifolia.
var. petersiana *(Klotzsch) B. Verdcourt in Fl. Trop. E. Afr., Nymphaeac.:* 11 (1989): N. petersiana.
var. zanzibariensis *(Casp.) B. Verdcourt in Fl. Trop. E. Afr., Nymphaeac.:* 10 (1989): N. zanzibariensis.
thermarum *E. Fischer in Feddes Repert.,* 99(9–10): 388 (1988)—Rwanda.

O

OBLIVIA J.L. Strother in Syst. Bot., 14(4): 541 (1989). *COMPOSITAE.*
mikanioides *(Britton) J.L. Strother in Syst. Bot.,* 14(4): 541 (1989): Salmea mikanioides.

OBOLINGA R.C. Barneby in Brittonia, 41(2): 170 (1989). *LEGUMINOSAE.*
zanonii *R.C. Barneby in Brittonia,* 41(2): 170 (1989)—Hispaniola (Dominican Republic, Haiti).

OCHTHOCOSMUS (Ixonanthac.).
micranthus *J.A. Steyermark in Ann. Missouri Bot. Gard.,* 75(1): 318 (1988)—Venezuela.

OCIMUM (Labiat.).
jamesii *O. Sebald in Stuttgarter Beitr. Naturk., A,* 405: 3 (1987), nom. nov.: O. tomentosum *Oliver.*

OCOTEA (Laurac.).
albescens *I. de Vattimo-Gil in Arq. Jard. Bot. Rio de Janeiro,* 28: 182 (1986–87 publ. 1988)—Brazil.
douradensis *I. de Vattimo-Gil in Rev. Brasil. Biol.,* 48(4): 951 (1988)—Brazil.
duplocolorata *I. de Vattimo-Gil in Arq. Jard. Bot. Rio de Janeiro,* 28: 183 (1986–87 publ. 1988)—Brazil.
flaventha *H. van der Werff in Ann. Missouri Bot. Gard.,* 76(2): 464 (1989)—Venezuela.
glabra *H. van der Werff in Ann. Missouri Bot. Gard.,* 76(2): 466 (1989)—Venezuela.
huberi *H. van der Werff in Ann. Missouri Bot. Gard.,* 76(2): 467 (1989)—Venezuela.
julianii *H. van der Werff in Ann. Missouri Bot. Gard.,* 76(2): 468 (1989), nom. nov.: Phoebe steyermarkiana *Allen.*
liesneri *H. van der Werff in Ann. Missouri Bot. Gard.,* 76(2): 468 (1989)—Venezuela.
megacarpa *H. van der Werff in Ann. Missouri Bot. Gard.,* 76(2): 469 (1989)—Venezuela.
roseopedunculata *H. van der Werff in Ann. Missouri Bot. Gard.,* 76(2): 470 (1989)—Venezuela.
silvae *I. de Vattimo-Gil in Rev. Brasil. Biol.,* 48(4): 952 (1988)—Brazil.
tomentosa *H. van der Werff in Ann. Missouri Bot. Gard.,* 76(2): 470 (1989)—Venezuela.
wilhelminae *I. de Vattimo-Gil in Rev. Brasil. Biol.,* 48(4): 954 (1988)—Surinam.

ODICARDIS (Scrophulariac.).
simensis *(Fresen.) K. Barringer in Taxon,* 38(1): 140 (1989): Veronica simensis.

ODONTITES (Scrophulariac.).
lanceolata
subsp. provincialis *M. Bolliger in Bot. Jahrb.,* 111(1): 19 (1989)—France.

ODONTOGLOSSUM (Orchidac.).
cristatellum
var. argus *(Rchb. f.) L. Bockemühl, Odontoglossum, Monogr. u. Ikonogr.*: 51 (1989): O. cristatum var. argus.
var. canari *(Rchb. f.) L. Bockemühl, Odontoglossum, Monogr. u. Ikonogr.*: 51 (1989): O. cristatum var. canari.
var. dayanum *(Rchb. f.) L. Bockemühl, Odontoglossum, Monogr. u. Ikonogr.*: 51 (1989): O. cristatum var. dayanum.
subsp. lehmannii *(Rchb. f.) L. Bockemühl, Odontoglossum, Monogr. u. Ikonogr.*: 53 (1989): O. lehmannii.
crocidipterum
subsp. dormanianum *(Rchb. f.) L. Bockemühl, Odontoglossum, Monogr. u. Ikonogr.*: 162 (1989): O. dormanianum.
gloriosum
var. flaviflorum *L. Bockemühl, Odontoglossum, Monogr. u. Ikonogr.*: 148 (1989)—Colombia.
lindleyanum
var. parviflorum *L. Bockemühl, Odontoglossum, Monogr. u. Ikonogr.*: 194 (1989)—Ecuador.
sceptrum
var. facetum *(Rchb. f.) L. Bockemühl, Odontoglossum, Monogr. u. Ikonogr.*: 116 (1989): O. facetum.
spectatissimum
var. cinctum *(Rchb. f.) L. Bockemühl, Odontoglossum, Monogr. u. Ikonogr.*: 68 (1989): O. triumphans var. cinctum.

OENANTHE (Umbellif.).
alatinervis *Y.Y. Qian in Guihaia*, 9(2): 117 (1989)—China.

OEONIA (Orchidac.).
brauniana
var. sarcanthoides *(Schltr.) J. Bosser in Bull. Mus. Nation. Hist. Nat., B, Adansonia*, Ser. 4, 11(2): 162 (1989): O. sarcanthoides.

OGLIFA (Compos.).
lojaconoi *S. Brullo in Doc. Phytosoc., N.S.*, 9: 8 (1985), in reprint, nom. nov.: Filago cossyrensis *Lojac.*

OLDENLANDIA (Rubiac.).
subgen. Phymatesta *B. Verdcourt in Fl. Zambes.*, 5(1): 121 (1989).
machingensis *B. Verdcourt in Fl. Zambes.*, 5(1): 140 (1989)—Malawi.
robinsonii *B. Verdcourt in Fl. Zambes.*, 5(1): 142 (1989)—Zambia.
verrucitesta *B. Verdcourt in Fl. Zambes.*, 5(1): 133 (1989)—Mozambique.

OLEA (Oleac.).
maroccana *W. Greuter & Burdet in Willdenowia*, 19(1): 37 (1989), nom. nov.: O. salicifolia *Barbero et al.*

OLEARIA (Compos.).
archeri *N.S. Lander in Muelleria*, 7(1): 117 (1989)—Tasmania.
astroloba *N.S. Lander & N.G. Walsh in Muelleria*, 7(1): 123 (1989)—Australia (Victoria).
gordonii *N.S. Lander in Nuytsia*, 7(1): 32 (1989)—Australia (Queensland).
humilis *N.S. Lander in Nuytsia*, 7(1): 26 (1989)—Australia (W. Australia).

OLYRA (Gramin.).
amapana *T.R. Soderstrom & F.O. Zuloaga in Smithsonian Contrib. Bot.*, 69: 5 (1989)—Brazil.
latispicula *T.R. Soderstrom & F.O. Zuloaga in Smithsonian Contrib. Bot.*, 69: 35 (1989)—Brazil.
retrorsa *T.R. Soderstrom & F.O. Zuloaga in Smithsonian Contrib. Bot.*, 69: 54 (1989)—Brazil.
tamanquareana *T.R. Soderstrom & F.O. Zuloaga in Smithsonian Contrib. Bot.*, 69: 58 (1989)—Brazil.

OMALOTHECA (Compos.).
supina
subsp. pusilla *(Haenke) S. Rivas-Martínez, F. Fernández González & D. Sánchez-Mata in Opusc. Bot. Pharm. Complutensis*, 2: 114 (1986), without basionym page: Gnaphalium pusillum.

ONCIDIUM (Orchidac.).
urophyllum
f. flavum *R.J. Midgett in Amer. Orchid Soc. Bull.*, 58(2): 124 (1989)—Leeward Is.

ONCOTHECA (Oncothecac.).
humboldtiana *(Guillaumin) P. Morat & J.M. Veillon in Fl. N. Caléd. & Dépend.*, 15: 96 (1988): Elaeodendron humboldtianum.

ONOBRYCHIS (Legumin.).
humilis
subsp. jahandiezii *(Širj.) W. Greuter & Burdet in Willdenowia*, 19(1): 33 (1989): O. jahandiezii.
subsp. matritensis *(Boiss. & Reuter) W. Greuter & Burdet in Willdenowia*, 19(1): 33 (1989): O. matritensis.

ONONIS (Legumin.).
alopecuroides
subsp. exalopecuroides *(G. López) W. Greuter & Burdet in Willdenowia*, 19(1): 33 (1989): O. exalopecuroides.
natrix
subsp. gibraltarica *(Boiss.) J.A. Devesa in Mem. Soc. Brot.*, 28: 34 (1988): O. gibraltarica.
spinosa
subsp. arvensis *(L.) W. Greuter & Burdet in Willdenowia*, 19(1): 33 (1989): O. arvensis.

ONONIS :-

subsp. australis *(Širj.) W. Greuter & Burdet in Willdenowia,* 19(1): 33 (1989): O. repens var. australis.

subsp. masquillierii *(Bertol.) W. Greuter & Burdet in Willdenowia,* 19(1): 34 (1989): O. masquillierii.

subsp. spinosiformis *(Simkovics) W. Greuter & Burdet in Willdenowia,* 19(1): 34 (1989): O. spinosiformis.

ONOSMA (Boraginac.).

nydeggeri *A. Huber-Morath in Bauhinia,* 9(1): 39 (1988)—Turkey.

stridii *H. Teppner in Phyton (Austria),* 28(2): 272 (1988)—Greece.

OPHRYS (Orchidac.).

×alibertiana *C. & A. Alibertis in Orchidophile,* 86: 83 (1989)—Crete. (O. spruneri × tenthredinifera).

×asterusica *C. & A. Alibertis in Orchidophile,* 86: 81 (1989)—Crete. (O. fleischmannii × omegaifera).

×biancoae *P. Medagli, S. d'Emerico & L. Ruggiero in Orchidophile,* 85: 29 (1989), contrary to Art. H.5.2 ICBN 1988—Italy. (O. garganica × sphegodes subsp. sphegodes).

bucephala *P. Gölz & H.R. Reinhard in Mitt. Arbeitskr. Heim. Orch. Baden-Württemberg,* 21(1): 13 (1989)—East Aegean Is.

×burneriana *C. & A. Alibertis in Orchidophile,* 86: 83 (1989)—Crete. (O. cretensis × spruneri).

candica

subsp. minoa *C. & A. Alibertis in Orchidophile,* 87: 111 (1989)—Crete.

×cesinensis *P. Medagli, S. d'Emerico & L. Ruggiero in Orchidophile,* 85: 30 (1989), contrary to Art. H.5.2 ICBN 1988—Italy. (O. fusca × lutea subsp. melena).

×clapensis *M. Balayer de Minden in Bull. Soc. Bot. France, Lett. Bot.,* 136(2): 158 (1989)—France. (O. bombyliflora × lutea).

×costei *H. Van Looken in Orchidophile,* 86: 84 (1989)—France. (O. aveyronensis × sphegodes).

cretensis *(Baumann & Künkele) H.F. Paulus in Mitt. Arbeitskr. Heim. Orch. Baden-Württemberg,* 20(4): 855 (1988): O. sphegodes subsp. cretensis.

×domus-maria *M.P. Grasso in Orchidee,* 40(3): 102 (1989)—Sardinia. (O. apifera × arachnitiformis).

exaltata

subsp. archipelagi *(Goelz & Rheinhard) C. Del Prete in C. Del Prete & G. Tosi, Orchidee Spontanee d'Italia:* 37 (1988): O. archipelagi.

subsp. tyrrhena *(Goelz & Reinh.) C. Del Prete in C. Del Prete & G. Tosi, Orchidee Spontanee d'Italia:* 36 (1988): O. tyrrhena.

OPHRYS :-

×franciniae *P. Bianco, P. Medagli, S. D'Emerico & L. Ruggiero in Webbia,* 42(2): 168 (1988)—Italy. (O. apulica × incubacea).

gortynia *(Baumann & Künkele) H.F. Paulus in Mitt. Arbeitskr. Heim. Orch. Baden-Württemberg,* 20(4): 855 (1988): O. sphegodes subsp. gortynia.

×kastelli

nothosubsp. antiskariensis *C. & A. Alibertis in Orchidophile,* 86: 79 (1989)—Crete. (O. bombyliflora × cretica subsp. cretica).

lesbis *P. Gölz & H.R. Reinhard in Mitt. Arbeitskr. Heim. Orch. Baden-Württemberg,* 21(1): 19 (1989)—East Aegean Is.

×lithinensis *C. & A. Alibertis in Orchidophile,* 86: 81 (1989)—Crete. (O. omegaifera × sitiaca).

×littoralis *J. & M. Baláž in Orchidee,* 40(3): 108 (1989)—Jugoslavia. (O. bertolonii × oestrifera subsp. oestrifera).

oestrifera

subsp. abchasica *H. Kümpel in Ber. Arbeitskr. Heim. Orch.,* 5(1–2): 25 (1988)—U.S.S.R. (Caucasus).

×parenzani *P. Medagli, S. d'Emerico & L. Ruggiero in Orchidophile,* 85: 29 (1989)—Italy. (O. fusca × tarentina).

×pauliana *C. & A. Alibertis in Orchidophile,* 86: 81 (1989)—Crete. (O. fleischmannii × sitiaca).

×plorae *C. & A. Alibertis in Orchidophile,* 86: 81 (1989), contrary to Art. H.5.2 ICBN (1988)—Crete. (O. cretica subsp. karpathensis × spruneri).

provincialis *(Baumann & Künkele) H.F. Paulus in Mitt. Arbeitskr. Heim. Orch. Baden-Württemberg,* 20(4): 855 (1988): O. sphegodes subsp. provincialis.

sitiaca *H.F. Paulus, C. Alibertis & A. Alibertis in Mitt. Arbeitskr. Heim. Orch. Baden-Württemberg,* 20(4): 839 (1988)—Crete.

×vallis-costae *H. Kümpel in Ber. Arbeitskr. Heim. Orch.,* 5(1–2): 27 (1988)—U.S.S.R. (Caucasus). (O. apifera × oestrifera subsp. abchasica).

vernixia

nothosubsp. innominata *(D. & B. Tyteca) P. Bouillie in Blancoana,* 7: 124 (1989): Basionym not stated.

OPUNTIA (Cactac.).

aureispina *(Brack & Heil) D.J. Pinkava & B.D. Parfitt in Sida,* 13(2): 128 (1988), without exact basionym page: O. macrocentra var. aureispina.

bakeri *J.E. Madsen in Fl. Ecuador,* 35(45): 47 (1989)—Ecuador.

chlorotica

var. gosseliniana *(Weber) D.J. Ferguson in Cact. Succ. J. (U.S.A.),* 60(4): 159 (1988): O. gosseliniana.

engelmannii

var. flavispina *(L. Benson) B.D. Parfitt & D.J. Pinkava in Madroño,* 35(4): 348 (1988 publ. 1989): O. phaeacantha var. flavispina.

OPUNTIA :–
var. flexospina *(Griffiths) B.D. Parfitt & D.J. Pinkava in Madroño,* 35(4): 348 (1988 publ. 1989): O. flexospina.
var. lindheimeri *(Engelm.) B.D. Parfitt & D.J. Pinkava in Madroño,* 35(4): 346 (1988 publ. 1989): O. lindheimeri.
var. linguiformis *(Griffiths) B.D. Parfitt & D.J. Pinkava in Madroño,* 35(4): 347 (1988 publ. 1989): O. linguiformis.

ORCHIS (Orchidac.).
boryi
var. alba *C. & A. Alibertis in Orchidophile,* 87: 112 (1989)—Crete.
×hermaniana *C. & A. Alibertis in Orchidophile,* 87: 108 (1989)—Crete. (O. anatolica × tridentata).
×mackaensis *C.A.J. Kreutz in Orchideeën,* 51(3): 72 (1989)—Turkey. (O. simia × stevenii).
×penzigiana
nothosubsp. sardoa *A. Scrugli & M.P. Grasso in Webbia,* 42(2): 192 (1988)—Sardinia. (O. mascula subsp. ichnusae × provincialis).
quadripunctata
var. albiflora *C. & A. Alibertis in Orchidophile,* 87: 112 (1989)—Crete.
×salkowskiana *C. & A. Alibertis in Orchidophile,* 87: 108 (1989), contrary to Art. H.5.2 ICBN (1988)—Crete. (O. anatolica subsp. sitiaca × collina).
×suckowii *H. Kümpel in Feddes Repert.,* 99(3–4): 94 (1988)—U.S.S.R. (Caucasus). (O. maxima × punctulata).
ustulata
var. aestivalis *H. Kümpel in Haussknechtia, Mitt. Thüring. Bot. Ges.,* 4: 23 (1988)—Germany.

×**ORCHISERAPIAS** (Orchidac.).
rousii *(Dupuy) H. Baumann & S. Künkele in Mitt. Arbeitskr. Heim. Orch. Baden-Württemberg,* 21(3): 840 (1989): Serapias × rousii.
timbalii *(K. Richt.) H. Baumann & S. Künkele in Mitt. Arbeitskr. Heim. Orch. Baden-Württemberg,* 21(3): 839 (1989): Serapias × timbalii.

OREOBOLOPSIS T. Koyama & E.R. Guaglianone in Darwiniana, 28(1–4): 79 (1987 publ. 1988). *CYPERACEAE.*
tepalifera *T. Koyama & E.R. Guaglianone in Darwiniana,* 28(1–4): 80 (1987 publ. 1988)—Bolivia.

OREOPANAX (Araliac.).
compactus *M.J. & J.F.M. Cannon in Bull. Brit. Mus. (Nat. Hist.), Bot.,* 19: 27 (1989)—Costa Rica, Panama.
spathulatus *M.J. & J.F.M. Cannon in Bull. Brit. Mus. (Nat. Hist.), Bot.,* 19: 29 (1989)—Panama.
striatus *M.J. & J.F.M. Cannon in Bull. Brit. Mus. (Nat. Hist.), Bot.,* 19: 27 (1989)—Panama, Costa Rica.

OREOPANAX :–
superoerstedianus *M.J. & J.F.M. Cannon in Bull. Brit. Mus. (Nat. Hist.), Bot.,* 19: 24 (1989)—Panama.

OREOTHAMNUS M.G. Baumann-Bodenheim, Syst. Fl. Neu-Caledonien, 5: 76 (1989), without latin descr. or type. *EPACRIDACEAE.*
amabilis *(Brongn. & Gris) M.G. Baumann-Bodenheim, Syst. Fl. Neu-Caledonien,* 5: 76 (1989), generic name not validly publ.: Dracophyllum amabile.
cosmelioides *(Panch. ex Oliver) M.G. Baumann-Bodenheim, Syst. Fl. Neu-Caledonien,* 5: 76 (1989), generic name not validly publ.: Dracophyllum cosmelioides.
ramosus *(Panch. ex Brongn. & Gris) M.G. Baumann-Bodenheim, Syst. Fl. Neu-Caledonien,* 4: 102 (1988 publ. Jan. 1989), without basionym ref.; *et Syst. Fl. Neu-Caledonien,* 5: 76, 97 (May 1989), with basionym ref., generic name not validly publ.: Dracophyllum ramosum.

ORITROPHIUM (Compos.).
marahuacense *J. Steyermark & B. Maguire in Acta Bot. Venezuel.,* 14(3): 28 (1984)—Venezuela.

ORIXA (Rutac.).
racemosa *Z.M. Tan in Bull. Bot. Res. North-East. Forest. Inst.,* 9(2): 44 (1989)—China.
subcoriacea *Z.M. Tan in Bull. Bot. Res. North-East. Forest. Inst.,* 9(2): 45 (1989)—China.

ORNITHOGALUM (Hyacinthac.).
dictaeum *T. Landström, Species of Ornithogalum L. subg. Ornithogalum (Hyacinthaceae) in Greece:* 5 (1989), name not considered as valid by author—Crete.
subsp. naxense *T. Landström, Species of Ornithogalum L. subg. Ornithogalum (Hyacinthaceae) in Greece:* 7 (1989), name not considered as valid by author—Greece.
hygrophilum *O.M. Hilliard & B.L. Burtt in Notes Roy. Bot. Gard. Edinburgh,* 45(2): 195 (1988 publ. 1989)—South Africa (Cape Province).
joschtiae *F. Speta in Phyton (Austria),* 29(1): 76 (1989)—Turkey.

OROBANCHE (Scrophulariac.).
lanuginosa *(C.A. Meyer) W. Greuter & Burdet in Willdenowia,* 19(1): 37 (1989): Phelypaea lanuginosa.
minor
var. maritima *(Pugsley) F.J. Rumsey & S.L. Jury in Watsonia,* 17(4): 444 (1989): O. maritima.
sinensis
var. cyanescens *(H. Smith) Z.Y. Zhang in Acta Phytotax. Sin.,* 26(5): 395 (1988): O. cyanescens.

OROSTACHYS (Crassulac.).
fimbriatus
var. grandiflorus *F.Z. Li & X.D. Chen in Bull. Bot. Res. North-East. Forest. Inst.*, 9(1): 74 (1989)—China.
var. shandongensis *F.Z. Li & X.D. Chen in Bull. Bot. Res. North-East. Forest. Inst.*, 9(1): 73 (1989)—China.

ORTGIESIA (Bromeliac.).
alegrensis *(Weber) L.B. Smith & W.J. Kress in Phytologia*, 66(1): 72 (1989): Aechmea alegrensis.
apocalyptica *(Reitz) L.B. Smith & W.J. Kress in Phytologia*, 66(1): 72 (1989): Aechmea apocalyptica.
bicolor *(L.B. Smith) L.B. Smith & W.J. Kress in Phytologia*, 66(1): 72 (1989): Aechmea bicolor.
blumenavii *(Reitz) L.B. Smith & W.J. Kress in Phytologia*, 66(1): 72 (1989): Aechmea blumenavii.
var. alba *(Reitz) L.B. Smith & W.J. Kress in Phytologia*, 66(1): 72 (1989): Aechmea blumenavii var. alba.
burle-marxii *(E. Pereira) L.B. Smith & W.J. Kress in Phytologia*, 66(1): 72 (1989): Aechmea burle-marxii.
calyculata *(E. Morren) L.B. Smith & W.J. Kress in Phytologia*, 66(1): 72 (1989): Hoplophytum calyculatum.
candida *(E. Morren ex Baker) L.B. Smith & W.J. Kress in Phytologia*, 66(1): 72 (1989): Aechmea candida.
caudata *(Lindman) L.B. Smith & W.J. Kress in Phytologia*, 66(1): 72 (1989): Aechmea caudata.
f. albiflora *(Weber & Roeth) L.B. Smith & W.J. Kress in Phytologia*, 66(1): 72 (1989): Aechmea caudata f. albiflora.
coelestis *(K. Koch) L.B. Smith & W.J. Kress in Phytologia*, 66(1): 72 (1989): Hoplophytum coeleste.
var. acutifolia *(E. Pereira) L.B. Smith & W.J. Kress in Phytologia*, 66(1): 72 (1989): Aechmea coelestis var. acutifolia.
var. albomarginata *(M.B. Foster) L.B. Smith & W.J. Kress in Phytologia*, 66(1): 72 (1989): Aechmea coelestis var. albomarginata.
cylindrata *(Lindman) L.B. Smith & W.J. Kress in Phytologia*, 66(1): 72 (1989): Aechmea cylindrata.
gamosepala *(Wittmack) L.B. Smith & W.J. Kress in Phytologia*, 66(1): 72 (1989): Aechmea gamosepala.
var. nivea *(Reitz) L.B. Smith & W.J. Kress in Phytologia*, 66(1): 72 (1989): Aechmea gamosepala var. nivea.
gracilis *(Lindman) L.B. Smith & W.J. Kress in Phytologia*, 66(1): 72 (1989): Aechmea gracilis.

ORTGIESIA :–
guaratubensis *(E. Pereira) L.B. Smith & W.J. Kress in Phytologia*, 66(1): 73 (1989): Aechmea guaratubensis.
kertesziae *(Reitz) L.B. Smith & W.J. Kress in Phytologia*, 66(1): 73 (1989): Aechmea kertesziae.
var. viridiaurata *(Reitz) L.B. Smith & W.J. Kress in Phytologia*, 66(1): 73 (1989): Aechmea kertesziae var. viridi-aurata.
lindenii *(E. Morren) L.B. Smith & W.J. Kress in Phytologia*, 66(1): 73 (1989): Hoplophytum lindenii.
var. makoyana *(Mez) L.B. Smith & W.J. Kress in Phytologia*, 66(1): 73 (1989): Aechmea lindenii var. makoyana.
lymanii *(Weber) L.B. Smith & W.J. Kress in Phytologia*, 66(1): 73 (1989): Aechmea lymanii.
organensis *(Wawra) L.B. Smith & W.J. Kress in Phytologia*, 66(1): 73 (1989): Aechmea organensis.
pimentivelosoi *(Reitz) L.B. Smith & W.J. Kress in Phytologia*, 66(1): 73 (1989): Aechmea pimenti-velosoi.
var. glabra *(Reitz) L.B. Smith & W.J. Kress in Phytologia*, 66(1): 73 (1989): Aechmea pimenti-velosoi var. glabra.
recurvata *(Klotzsch) L.B. Smith & W.J. Kress in Phytologia*, 66(1): 73 (1989): Macrochordium recurvatum.
var. benrathii *(Mez) L.B. Smith & W.J. Kress in Phytologia*, 66(1): 73 (1989): Aechmea benrathii.
var. ortgiesii *(Baker) L.B. Smith & W.J. Kress in Phytologia*, 66(1): 73 (1989): Aechmea ortgiesii.
seideliana *(Weber) L.B. Smith & W.J. Kress in Phytologia*, 66(1): 73 (1989): Aechmea seideliana.
winkleri *(Reitz) L.B. Smith & W.J. Kress in Phytologia*, 66(1): 73 (1989): Aechmea winkleri.

OSMANTHUS (Oleac.).
gracilinervis *Chia ex R.L. Lu in Acta Phytotax. Sin.*, 27(1): 71 (1989)—China.
marginatus
var. longissimus *(H.T. Chang) R.L. Lu in Acta Phytotax. Sin.*, 27(1): 72 (1989): O. longissimus.
var. pachyphyllus *(H.T. Chang) R.L. Lu in Acta Phytotax. Sin.*, 27(1): 73 (1989): O. pachyphyllus.

OURATEA (Ochnac.).
asisae *B. Maguire & J.A. Steyermark in Mem. New York Bot. Gard.*, 51: 62 (1989)—Venezuela.
ayacuchae *B. Maguire & J.A. Steyermark in Mem. New York Bot. Gard.*, 51: 62 (1989)—Venezuela.
brevicalyx *B. Maguire & J.A. Steyermark in Mem. New York Bot. Gard.*, 51: 62 (1989)—Venezuela.

OURATEA :-

brevipedicellata *B. Maguire & J.A. Steyermark in Mem. New York Bot. Gard.*, 51: 64 (1989)—Venezuela.

chaffanjonii *(van Tieghem) C. Sastre in Ernstia,* 56: 23 (1989): Cercouratea chaffanjonii.

croizatii *B. Maguire & J.A. Steyermark in Mem. New York Bot. Gard.*, 51: 65 (1989)—Venezuela.

culminicola *B. Maguire & J.A. Steyermark in Mem. New York Bot. Gard.*, 51: 65 (1989)—Venezuela.

deminuta *B. Maguire & J.A. Steyermark in Mem. New York Bot. Gard.*, 51: 66 (1989)—Venezuela.

discophora

subsp. pervenulosa *B. Maguire & J.A. Steyermark in Mem. New York Bot. Gard.*, 51: 67 (1989)—Venezuela.

evoluta *B. Maguire & J.A. Steyermark in Mem. New York Bot. Gard.*, 51: 67 (1989)—Venezuela.

fasciculata *B. Maguire & J.A. Steyermark in Mem. New York Bot. Gard.*, 51: 68 (1989)—Guyana.

gillyana

subsp. pachypoda *B. Maguire & J.A. Steyermark in Mem. New York Bot. Gard.*, 51: 71 (1989)—Guyana, Venezuela.

huberi *B. Maguire & J.A. Steyermark in Mem. New York Bot. Gard.*, 51: 74 (1989)—Venezuela.

inconformis *B. Maguire & J.A. Steyermark in Mem. New York Bot. Gard.*, 51: 75 (1989)—Venezuela.

longistyla *B. Maguire & J.A. Steyermark in Mem. New York Bot. Gard.*, 51: 78 (1989)—Venezuela.

macurucoensis *B. Maguire & J.A. Steyermark in Mem. New York Bot. Gard.*, 51: 80 (1989)—Venezuela.

marahuacensis *B. Maguire & J.A. Steyermark in Mem. New York Bot. Gard.*, 51: 80 (1989)—Venezuela.

medinae *B. Maguire & J.A. Steyermark in Mem. New York Bot. Gard.*, 51: 81 (1989)—Venezuela.

ornata *B. Maguire & J.A. Steyermark in Mem. New York Bot. Gard.*, 51: 82 (1989)—Venezuela.

papillata *B. Maguire & J.A. Steyermark in Mem. New York Bot. Gard.*, 51: 82 (1989)—Venezuela.

paruensis *B. Maguire & J.A. Steyermark in Mem. New York Bot. Gard.*, 51: 83 (1989)—Venezuela.

propinqua *B. Maguire & J.A. Steyermark in Mem. New York Bot. Gard.*, 51: 85 (1989)—Venezuela.

pseudotatei *B. Maguire & J.A. Steyermark in Mem. New York Bot. Gard.*, 51: 85 (1989)—Guyana.

pumila *B. Maguire & J.A. Steyermark in Mem. New York Bot. Gard.*, 51: 86 (1989)—Venezuela.

OURATEA :-

var. heterodoxa *B. Maguire & J.A. Steyermark in Mem. New York Bot. Gard.*, 51: 87 (1989)—Venezuela.

ramosissima *B. Maguire & J.A. Steyermark in Mem. New York Bot. Gard.*, 51: 87 (1989)—Venezuela.

rotundipetala *B. Maguire & J.A. Steyermark in Mem. New York Bot. Gard.*, 51: 90 (1989)—Venezuela.

sacleuxii *(van Tieghem) H.J. Beentje in Utafiti,* 1(2): 70 (1988): Cercanthemum sacleuxii.

sipapoensis *B. Maguire & J.A. Steyermark in Mem. New York Bot. Gard.*, 51: 92 (1989)—Venezuela.

subamplexicaulis *B. Maguire & J.A. Steyermark in Mem. New York Bot. Gard.*, 51: 96 (1989)—Venezuela.

wurdackiana *B. Maguire & J.A. Steyermark in Mem. New York Bot. Gard.*, 51: 100 (1989)—Venezuela.

OXALIS (Oxalidac.).

compacta

subsp. berteroana *(Barnéoud) A. Lourteig in Fl. Patagonica (Colecc. Cient. 8),* 5: 7 (1988): O. berteroana.

enneaphylla

subsp. ibarii *(Philippi) A. Lourteig in Fl. Patagonica (Colecc. Cient. 8),* 5: 9 (1988): O. ibarii.

fontana

var. rufa *(Small) T. Karlsson in Svensk Bot. Tidskr.*, 83(5): 313 (1989): O. rufa.

rigida *(Barnéoud) A. Lourteig in Fl. Patagonica (Colecc. Cient. 8),* 5: 24 (1988): O. laxa var. rigida.

OXYGONUM (Polygonac.).

altissimum *G. Germishuizen in Bothalia,* 19(2): 210 (1989)—Somali Republic.

dregeanum

subsp. canescens *(Sond.) G. Germishuizen in Bothalia,* 18(2): 175 (1988): O. canescens.

var. dissectum *G. Germishuizen in Bothalia,* 18(2): 176 (1988)—South Africa (Transvaal).

subsp. lanceolatum *G. Germishuizen in Bothalia,* 18(2): 178 (1988)—South Africa (Transvaal, Natal), Swaziland.

var. linearifolium *G. Germishuizen in Bothalia,* 18(2): 175 (1988)—South Africa (Transvaal, Natal).

var. lobophyllum *G. Germishuizen in Bothalia,* 18(2): 177 (1988)—South Africa (Transvaal), Swaziland.

var. pilosum *G. Germishuizen in Bothalia,* 18(2): 177 (1988)—South Africa (Transvaal).

subsp. streyi *G. Germishuizen in Bothalia,* 18(2): 173 (1988)—South Africa (Natal).

subsp. swazicum *(Burtt Davy) G. Germishuizen in Bothalia,* 18(2): 174 (1988): O. zeyheri var. swazicum.

OXYGYNE (Burmanniac.).
hyodoi *C. Abe & Y. Akasawa in J. Jap. Bot.*, 64(6): 161 (1989)—Japan.
shinzatoi *(Hatusima) C. Abe & Y. Akasawa in J. Jap. Bot.*, 64(6): 163 (1989): Saionia shinzatoi.

OXYLOBUS (Compos.).
preecei *B.L. Turner in Phytologia*, 65(5): 375 (1988)—Mexico.

OXYTROPIS (Legumin.).
sect. Arenicola *C.W. Chang in Acta Bot. Bor.-Occid. Sin.*, 9(1): 40 (1989).
sect. Falcicarpa *C.W. Chang in Acta Bot. Bor.-Occid. Sin.*, 9(1): 41 (1989).
sect. Sericopetala *C.W. Chang in Acta Bot. Bor.-Occid. Sin.*, 9(1): 40 (1989).
subsect. Uniflorae *B.A. Yurtsev in Bot. Zhurn.*, 73(12): 1653 (1988).
ajanensis
subsp. semiglobosa *(Jurtz.) N.S. Pavlova in Sosud. Rast. Sovet. Dal'nego Vostoka*, 4: 269 (1989): O. semiglobosa.
austrosachalinensis *Vass. ex N.S. Pavlova in Bot. Zhurn.*, 74(6): 904 (1989)—Sakhalin Is.
bajtulinii *Yu.A. Kotukhov in Bot. Zhurn.*, 74(3): 415 (1989)—U.S.S.R. (Middle Asia).
calcareorum *N.S. Pavlova in Bot. Zhurn.*, 74(6): 903 (1989)—Sakhalin Is.
coerulescens *(Jurtz. & Petrov.) A.P. Khokhryakov, Analiz Fl. Kolӯmskogo Nagor'ya:* 130 (1989): O. middendorfii subsp. coerulescens.
helenae *N.S. Pavlova in Bot. Zhurn.*, 74(6): 901 (1989)—Sakhalin Is.
janensis *(Jurtz.) A.P. Khokhryakov, Analiz Fl. Kolӯmskogo Nagor'ya:* 130 (1989): O. adamsiana subsp. janensis.
jarovoi *(Jurtz.) A.P. Khokhryakov, Analiz Fl. Kolӯmskogo Nagor'ya:* 130 (1989): O. middendorfii subsp. jarovoi.
pseudocoerulea *P.C. Li in Acta Bot. Yunnanica*, 11(3): 300 (1989)—China, Xizang.
substepposa *(Jurtz.) A.P. Khokhryakov, Analiz Fl. Kolӯmskogo Nagor'ya:* 67 (1989): O. vassilczenkoi subsp. substepposa.
uniflora *B.A. Yurtsev in Bot. Zhurn.*, 73(12): 1650 (1988)—U.S.S.R. (Soviet Far East).

P

PACHIRA (Bombacac.).
dolichocalyx *A. Robyns in Bull. Jard. Bot. Nation. Belg.*, 58(3–4): 535 (1988)—French Guiana.

PACHYPLEURUM (Umbellif.).
muliense *R.H. Shan & F.T. Pu in Acta Phytotax. Sin.*, 27(1): 62 (1989)—China.

PACHYSANDRA (Buxac.).
tricarpa *(Hayata) S.S. Ying, Coloured Illustr. Fl. Taiwian*, 3: 118 (1988): P. axillares var. tricarpa.

PAEPALANTHUS (Eriocaulac.).
holstii *J.A. Steyermark in Ann. Missouri Bot. Gard.*, 75(1): 311 (1988)—Venezuela.

PAGAMEA (Rubiac.).
aracaensis *B.M. Boom in Brittonia*, 41(2): 129 (1989)—Brazil.

PAIVAEA (Myrtac.).
phaea *(Berg) J.R. Mattos in Loefgrenia*, 94: 8 (1989): Abbevillea phaea.

PALICOUREA (Rubiac.).
albocaerulea *C.M. Taylor, Revis. Palicourea (Rubiac.) Mexico & C. Amer. (Syst. Bot. Monog., 26):* 15 (1989)—Costa Rica.
bellula *C.M. Taylor, Revis. Palicourea (Rubiac.) Mexico & C. Amer. (Syst. Bot. Monog., 26):* 24 (1989)—Costa Rica.
hammelii *C.M. Taylor, Revis. Palicourea (Rubiac.) Mexico & C. Amer. (Syst. Bot. Monog., 26):* 43 (1989)—Panama.
orosiana *C.M. Taylor, Revis. Palicourea (Rubiac.) Mexico & C. Amer. (Syst. Bot. Monog., 26):* 61 (1989)—Costa Rica, Panama.
pendula *C.M. Taylor, Revis. Palicourea (Rubiac.) Mexico & C. Amer. (Syst. Bot. Monog., 26):* 68 (1989)—Panama.
purpurea *C.M. Taylor, Revis. Palicourea (Rubiac.) Mexico & C. Amer. (Syst. Bot. Monog., 26):* 71 (1989)—Panama, Costa Rica.
skotakii *C.M. Taylor, Revis. Palicourea (Rubiac.) Mexico & C. Amer. (Syst. Bot. Monog., 26):* 76 (1989)—Costa Rica.
standleyana *C.M. Taylor, Revis. Palicourea (Rubiac.) Mexico & C. Amer. (Syst. Bot. Monog., 26):* 81 (1989), nom. nov.: Psychotria brenesii *Standley*.
steyermarkii *C.M. Taylor in Phytologia*, 66(6): 470 (1989), nom. nov.: P. garciae *Steyermark*.
tilaranensis *C.M. Taylor, Revis. Palicourea (Rubiac.) Mexico & C. Amer. (Syst. Bot. Monog., 26):* 84 (1989)—Costa Rica.

PANDANUS (Pandanac.).
sect. Cuneoidea *H. St. John, Pandanus of Cook Is. (Rev. Gen. Pandanus, 47):* 2 (1989).
sect. Vadosa *H. St. John, Pandanus of Samoa (Revis. Gen. Pandanus, 54):* 22 (1989).
absonus *H. St. John, Rev. Gen. Pandanus, 50, Austral Is., French Polynesia:* 3 (1989)—Tubuai Is.
aitutakiensis *H. St. John, Pandanus of Cook Is. (Rev. Gen. Pandanus, 47):* 4 (1989)—Cook Is.
akiakiensis *H. St. John, Pandanus of Tuamotu Archip. (Revis. Gen. Pandanus, 53):* 3 (1989)—Tuamotu Is.

PANDANUS :-

alifer *H. St. John, Pandanus Bismarck & Admiralty Is. (Rev. Gen. Pandanus, 51):* 29 (1989)—New Ireland.

amicalis *H. St. John, Pandanus of Tonga, Polynesia (Revis. Gen. Pandanus, 56):* 3 (1989)—Tonga.

amplexus *H. St. John, Rev. Gen. Pandanus, 50, Austral Is., French Polynesia:* 5 (1989)—Tubuai Is.

angulosus *H. St. John, Pandanus of Tonga, Polynesia (Revis. Gen. Pandanus, 56):* 4 (1989)—Tonga.

anisos *H. St. John, Pandanus of Pacific Equatorial (Line) Is. (Revis. Gen. Pandanus, 58):* 2 (1989)—Christmas Is. Pacific Ocean.

arapepe *H. St. John, Pandanus of Cook Is. (Rev. Gen. Pandanus, 47):* 2 (1989)—Cook Is.

arrectialatus *H. St. John, Pandanus of New Hebrides (Revis. Gen. Pandanus, 60):* 12 (1989)—Vanuatu.

asauensis *H. St. John, Pandanus of Samoa (Revis. Gen. Pandanus, 54):* 22 (1989)—Samoa.

ater *H. St. John, Pandanus in Leeward Soc. Is., Meetia, & Henderson (Rev. Gen. Pandanus, 49):* 3 (1989)—Society Is.

bassus *H. St. John, Rev. Gen. Pandanus, 50, Austral Is., French Polynesia:* 7 (1989)—Tubuai Is.

bathys *H. St. John, Pandanus of Samoa (Revis. Gen. Pandanus, 54):* 3 (1989)—Samoa.

bicurvatus *H. St. John, Pandanus of Tuamotu Archip. (Revis. Gen. Pandanus, 53):* 4 (1989)—Tuamotu Is.

bismarckensis *H. St. John, Pandanus Bismarck & Admiralty Is. (Rev. Gen. Pandanus, 51):* 30 (1989)—St. Matthias Is.

boraboraensis *H. St. John, Pandanus in Leeward Soc. Is., Meetia, & Henderson (Rev. Gen. Pandanus, 49):* 5 (1989)—Society Is.

cacuminatus *H. St. John, Pandanus in Leeward Soc. Is., Meetia, & Henderson (Rev. Gen. Pandanus, 49):* 6 (1989), nom. nov.: P. tahitensis var. exiguus *J.W. Moore.*

christophersenii *H. St. John, Pandanus of Samoa (Revis. Gen. Pandanus, 54):* 4 (1989)—Samoa.

collatus *H. St. John, Rev. Gen. Pandanus, 50, Austral Is., French Polynesia:* 10 (1989)—Tubuai Is.

complanatus *H. St. John, Pandanus in Leeward Soc. Is., Meetia, & Henderson (Rev. Gen. Pandanus, 49):* 7 (1989)—Society Is.

cooperi *(Martelli) H. St. John, Pandanus of Pacific Equatorial (Line) Is. (Revis. Gen. Pandanus, 58):* 3 (1989): P. pulposus var. cooperi.

crassus *H. St. John, Pandanus of Tuamotu Archip. (Revis. Gen. Pandanus, 53):* 6 (1989)—Tuamotu Is.

PANDANUS :-

cuneiformis *H. St. John, Pandanus of New Hebrides (Revis. Gen. Pandanus, 60):* 13 (1989)—Vanuatu.

cymatilis *H. St. John, Pandanus of Tuamotu Archip. (Revis. Gen. Pandanus, 53):* 7 (1989)—Tuamotu Is.

divaricatus *H. St. John, Pandanus in Leeward Soc. Is., Meetia, & Henderson (Rev. Gen. Pandanus, 49):* 9 (1989)—Society Is.

dotyi *H. St. John, Pandanus of Tuamotu Archip. (Revis. Gen. Pandanus, 53):* 8 (1989)—Tuamotu Is.

drakei *H. St. John, Pandanus in Leeward Soc. Is., Meetia, & Henderson (Rev. Gen. Pandanus, 49):* 10 (1989)—Society Is.

edwinii *H. St. John, Pandanus of Samoa (Revis. Gen. Pandanus, 54):* 6 (1989)—Samoa.

efateensis *H. St. John, Pandanus of New Hebrides (Revis. Gen. Pandanus, 60):* 15 (1989)—Vanuatu.

elevatus *H. St. John, Pandanus of Tuamotu Archip. (Revis. Gen. Pandanus, 53):* 10 (1989)—Tuamotu Is.

evexus *H. St. John, Pandanus of Cook Is. (Rev. Gen. Pandanus, 47):* 7 (1989)—Cook Is.

exilis *H. St. John, Pandanus of Samoa (Revis. Gen. Pandanus, 54):* 7 (1989), nom. nov.: P. upoluensis var. angulosus *Martelli.*

var. juddii *H. St. John, Pandanus of Samoa (Revis. Gen. Pandanus, 54):* 8 (1989)—Samoa.

fahina *H. St. John, Pandanus of Tonga, Polynesia (Revis. Gen. Pandanus, 56):* 6 (1989), nom. nov.: P. odoratissimus var. setchellii *Martelli.*

faramaa *H. St. John, Pandanus in Leeward Soc. Is., Meetia, & Henderson (Rev. Gen. Pandanus, 49):* 12 (1989)—Society Is.

fatuhivaensis *H. St. John, Pandanus in Marquesas Is. (Revis. Gen. Pandanus, 57):* 1 (1989)—Marquesas Is.

filiciatilis *H. St. John, Pandanus in Marquesas Is. (Revis. Gen. Pandanus, 57):* 2 (1989)—Marquesas Is.

flintinsularis *H. St. John, Remaining Polynesian spp. (Revis. Gen. Pandanus, 59):* 1 (1989)—Cook Is.

freetownensis *K.L. Huynh in Bot. Helvetica,* 98(2): 187 (1988)—Sierra Leone.

futunaensis *H. St. John, Remaining Polynesian spp. (Revis. Gen. Pandanus, 59):* 3 (1989)—Futuna Is.

gasicus *K.L. Huynh in Bot. Helvetica,* 99(1): 21 (1989)—Mozambique.

gazelleensis *H. St. John, Pandanus Bismarck & Admiralty Is. (Rev. Gen. Pandanus, 51):* 31 (1989)—New Britain.

glaphyros *H. St. John, Pandanus Bismarck & Admiralty Is. (Rev. Gen. Pandanus, 51):* 33 (1989)—New Britain.

glomerosus *H. St. John, Pandanus on Niue Is., Polynesia (Rev. Gen. Pandanus, 52):* 1 (1989), nom. nov.: P. tahitensis var. niueana *B.C. Stone.*

PANDANUS :-

grantii *H. St. John, Pandanus in Leeward Soc. Is., Meetia, & Henderson (Rev. Gen. Pandanus, 49):* 13 (1989)—Society Is.

granulosus *H. St. John, Pandanus Bismarck & Admiralty Is. (Rev. Gen. Pandanus, 51):* 1 (1989)—New Ireland.

haapaiensis *H. St. John, Pandanus of Tonga, Polynesia (Revis. Gen. Pandanus, 56):* 6 (1989)—Tonga.

hendersonensis *H. St. John, Pandanus in Leeward Soc. Is., Meetia, & Henderson (Rev. Gen. Pandanus, 49):* 15 (1989)—Henderson Is.

hivaoaensis *H. St. John, Pandanus in Marquesas Is. (Revis. Gen. Pandanus, 57):* 3 (1989)—Marquesas Is.

horneinsularum *H. St. John, Remaining Polynesian spp. (Revis. Gen. Pandanus, 59):* 5 (1989)—Futuna Is.

houmaensis *H. St. John, Pandanus of Tonga, Polynesia (Revis. Gen. Pandanus, 56):* 8 (1989), nom. nov.: P. odoratissimus var. parksii *Martelli.*

humericus *H. St. John, Pandanus on Niue Is., Polynesia (Rev. Gen. Pandanus, 52):* 7 (1989)—Cook Is.

humilior *H. St. John, Pandanus Bismarck & Admiralty Is. (Rev. Gen. Pandanus, 51):* 34 (1989)—New Ireland.

inarmatus *H. St. John, Pandanus in Leeward Soc. Is., Meetia, & Henderson (Rev. Gen. Pandanus, 49):* 17 (1989)—Society Is.

inflexus *H. St. John, Rev. Gen. Pandanus, 50, Austral Is., French Polynesia:* 11 (1989)—Tubuai Is.

interior *H. St. John, Pandanus of Cook Is. (Rev. Gen. Pandanus, 47):* 8 (1989)—Cook Is.

intralaevis *H. St. John, Pandanus in Leeward Soc. Is., Meetia, & Henderson (Rev. Gen. Pandanus, 49):* 18 (1989)—Society Is.

isis *H. St. John, Pandanus Bismarck & Admiralty Is. (Rev. Gen. Pandanus, 51):* 17 (1989)—New Ireland.

jonesii *(F. Brown) H. St. John, Pandanus in Marquesas Is. (Revis. Gen. Pandanus, 57):* 5 (1989): P. tectorius var. jonesii.

kaviengensis *H. St. John, Pandanus Bismarck & Admiralty Is. (Rev. Gen. Pandanus, 51):* 35 (1989)—New Ireland.

laculatus *H. St. John, Rev. Gen. Pandanus, 50, Austral Is., French Polynesia:* 13 (1989)—Tubuai Is.

latior *H. St. John, Pandanus of Cook Is. (Rev. Gen. Pandanus, 47):* 10 (1989)—Cook Is.

licinus *H. St. John, Rev. Gen. Pandanus, 50, Austral Is., French Polynesia:* 14 (1989)—Tubuai Is.

longissimus *H. St. John, Pandanus Bismarck & Admiralty Is. (Rev. Gen. Pandanus, 51):* 36 (1989)—New Britain.

makateaensis *H. St. John, Pandanus of Tuamotu Archip. (Revis. Gen. Pandanus, 53):* 12 (1989)—Tuamotu Is.

PANDANUS :-

mariaensis *H. St. John, Rev. Gen. Pandanus, 50, Austral Is., French Polynesia:* 15 (1989)—Tubuai Is.

marquesasensis *H. St. John, Pandanus in Marquesas Is. (Revis. Gen. Pandanus, 57):* 5 (1989)—Marquesas Is.

mauricei *H. St. John, Pandanus of New Hebrides (Revis. Gen. Pandanus, 60):* 1 (1989)—Vanuatu.

meetiaensis *H. St. John, Pandanus in Leeward Soc. Is., Meetia, & Henderson (Rev. Gen. Pandanus, 49):* 19 (1989)—Society Is.

metius *H. St. John, Pandanus in Leeward Soc. Is., Meetia, & Henderson (Rev. Gen. Pandanus, 49):* 20 (1989)—Society Is.

minimus *H. St. John, Pandanus of New Hebrides (Revis. Gen. Pandanus, 60):* 16 (1989)—Vanuatu.

montaguei *H. St. John, Pandanus in Leeward Soc. Is., Meetia, & Henderson (Rev. Gen. Pandanus, 49):* 21 (1989)—Henderson Is.

mooreaensis *H. St. John, Pandanus in Leeward Soc. Is., Meetia, & Henderson (Rev. Gen. Pandanus, 49):* 23 (1989)—Society Is.

motuensis *H. St. John, Rev. Gen. Pandanus, 50, Austral Is., French Polynesia:* 17 (1989)—Tubuai Is.

multidentatus *H. St. John, Pandanus of New Hebrides (Revis. Gen. Pandanus, 60):* 3 (1989)—Vanuatu.

mussauensis *H. St. John, Pandanus Bismarck & Admiralty Is. (Rev. Gen. Pandanus, 51):* 10 (1989)—St. Matthias Is.

neomecklenburgensis *H. St. John, Pandanus Bismarck & Admiralty Is. (Rev. Gen. Pandanus, 51):* 38 (1989)—New Ireland.

ngunaensis *H. St. John, Pandanus of New Hebrides (Revis. Gen. Pandanus, 60):* 17 (1989)—Vanuatu.

niueensis *H. St. John, Pandanus on Niue Is., Polynesia (Rev. Gen. Pandanus, 52):* 3 (1989)—Niue.

nogarete *H. St. John, Pandanus of New Hebrides (Revis. Gen. Pandanus, 60):* 10 (1989)—Vanuatu.

notialis *H. St. John, Rev. Gen. Pandanus, 50, Austral Is., French Polynesia:* 18 (1989)—Tubuai Is.

novibritannicus *H. St. John, Pandanus Bismarck & Admiralty Is. (Rev. Gen. Pandanus, 51):* 39 (1989)—New Britain.

oblaticonvexus *H. St. John, Pandanus of Tonga, Polynesia (Revis. Gen. Pandanus, 56):* 9 (1989), nom. nov.: P. odoratissimus f. major *Martelli.*

oblongicapitellatus *K.L. Huynh in Bot. Helvetica,* 98(2): 177 (1988)—Guinea.

onesuaensis *H. St. John, Pandanus of New Hebrides (Revis. Gen. Pandanus, 60):* 18 (1989)—Vanuatu.

orarius *H. St. John, Pandanus in Leeward Soc. Is., Meetia, & Henderson (Rev. Gen. Pandanus, 49):* 24 (1989)—Society Is.

PANDANUS :-
otemanuensis *H. St. John, Pandanus in Leeward Soc. Is., Meetia, & Henderson (Rev. Gen. Pandanus, 49):* 25 (1989)—Society Is.
pachys *H. St. John, Pandanus in Leeward Soc. Is., Meetia, & Henderson (Rev. Gen. Pandanus, 49):* 26 (1989)—Society Is.
pakari *H. St. John, Pandanus of Cook Is. (Rev. Gen. Pandanus, 47):* 11 (1989)—Cook Is.
palmyraensis *H. St. John, Pandanus of Pacific Equatorial (Line) Is. (Revis. Gen. Pandanus, 58):* 9 (1989), nom. nov.: P. fischerianus f. bryanii *Stone.*
paogo *H. St. John, Pandanus on Niue Is., Polynesia (Rev. Gen. Pandanus, 52):* 4 (1989)—Niue, Samoa.
parkinsonii
var. kukuwai *H. St. John, Pandanus Bismarck & Admiralty Is. (Rev. Gen. Pandanus, 51):* 12 (1989)—New Britain.
parou *H. St. John, Pandanus of New Hebrides (Revis. Gen. Pandanus, 60):* 19 (1989)—Vanuatu.
paucicarpellatus *H. St. John, Pandanus of Samoa (Revis. Gen. Pandanus, 54):* 9 (1989)—Samoa.
peekelii *H. St. John, Pandanus Bismarck & Admiralty Is. (Rev. Gen. Pandanus, 51):* 3 (1989)—New Ireland.
planus *H. St. John, Rev. Gen. Pandanus, 50, Austral Is., French Polynesia:* 20 (1989)—Tubuai Is.
problematicus *K.L. Huynh in Bot. Helvetica,* 98(2): 190 (1988)—Sierra Leone.
prolixus *H. St. John, Pandanus of Tuamotu Archip. (Revis. Gen. Pandanus, 53):* 13 (1989)—Tuamotu Is.
pseudolinnaei *(Martelli) H. St. John, Pandanus of Tonga, Polynesia (Revis. Gen. Pandanus, 56):* 11 (1989): P. odoratissimus var. pseudo-Linnaei.
pukapukaensis *H. St. John, Remaining Polynesian spp. (Revis. Gen. Pandanus, 59):* 2 (1989)—Cook Is.
puniceus *H. St. John, Pandanus of Samoa (Revis. Gen. Pandanus, 54):* 12 (1989)—Samoa.
pusillus *H. St. John, Rev. Gen. Pandanus, 50, Austral Is., French Polynesia:* 22 (1989)—Tubuai Is.
pweleensis *H. St. John, Pandanus of New Hebrides (Revis. Gen. Pandanus, 60):* 21 (1989)—Vanuatu.
pyramidos *H. St. John, Pandanus of New Hebrides (Revis. Gen. Pandanus, 60):* 22 (1989)—Vanuatu.
quinarius *H. St. John, Pandanus of New Hebrides (Revis. Gen. Pandanus, 60):* 23 (1989)—Vanuatu.
rabaulensis *H. St. John, Pandanus Bismarck & Admiralty Is. (Rev. Gen. Pandanus, 51):* 42 (1989)—New Britain.
radiatus *H. St. John, Rev. Gen. Pandanus, 50, Austral Is., French Polynesia:* 24 (1989)—Tubuai Is.

PANDANUS :-
radifer *H. St. John, Pandanus of New Hebrides (Revis. Gen. Pandanus, 60):* 24 (1989)—Vanuatu.
raiateaensis *H. St. John, Pandanus in Leeward Soc. Is., Meetia, & Henderson (Rev. Gen. Pandanus, 49):* 27 (1989)—Society Is.
raroiaensis *H. St. John, Pandanus of Tuamotu Archip. (Revis. Gen. Pandanus, 53):* 15 (1989)—Tuamotu Is.
rimataraensis *H. St. John, Rev. Gen. Pandanus, 50, Austral Is., French Polynesia:* 27 (1989)—Tubuai Is.
rurutuensis *H. St. John, Rev. Gen. Pandanus, 50, Austral Is., French Polynesia,* 28 (1989)—Tubuai Is.
saltuarius *H. St. John, Rev. Gen. Pandanus, 50, Austral Is., French Polynesia:* 30 (1989)—Tubuai Is.
samoanus *(Martelli) H. St. John, Pandanus of Samoa (Revis. Gen. Pandanus, 54):* 14 (1989): P. calostigma f. samoana.
savaiiensis *(Martelli) H. St. John, Pandanus of Samoa (Revis. Gen. Pandanus, 54):* 14 (1989): P. tectorius var. savaiiensis.
sierraleonensis *K.L. Huynh in Bot. Helvetica,* 98(2): 174 (1988)—Sierra Leone.
subaequalis *H. St. John, Remaining Polynesian spp. (Revis. Gen. Pandanus, 59):* 6 (1989)—Wallis Is.
sykesii *H. St. John, Pandanus of Cook Is. (Rev. Gen. Pandanus, 47):* 13 (1989)—Cook Is.
taepa *(F. Brown) H. St. John, Pandanus in Marquesas Is. (Revis. Gen. Pandanus, 57):* 10 (1989): P. tectorius var. taepa.
tahaaensis *H. St. John, Pandanus in Leeward Soc. Is., Meetia, & Henderson (Rev. Gen. Pandanus, 49):* 28 (1989)—Society Is.
takaroaensis *H. St. John, Pandanus of Tuamotu Archip. (Revis. Gen. Pandanus, 53):* 17 (1989)—Tuamotu Is.
tapeinos *H. St. John, Pandanus in Leeward Soc. Is., Meetia, & Henderson (Rev. Gen. Pandanus, 49):* 31 (1989)—Society Is.
tikeiensis *H. St. John, Pandanus of Tuamotu Archip. (Revis. Gen. Pandanus, 53):* 19 (1989)—Tuamotu Is.
tima *H. St. John, Pandanus of Tuamotu Archip. (Revis. Gen. Pandanus, 53):* 20 (1989)—Tuamotu Is.
timoeensis *H. St. John, Pandanus of Tuamotu Archip. (Revis. Gen. Pandanus, 53):* 21 (1989)—Tuamotu Is.
tongaensis *H. St. John, Pandanus of Tonga, Polynesia (Revis. Gen. Pandanus, 56):* 12 (1989), nom. nov.: P. odoratissimus var. pyriformis *Martelli in Univ. Calif. Publ. Bot., 12: 358 (1930).*
tongatapuensis *H. St. John, Pandanus of Tonga, Polynesia (Revis. Gen. Pandanus, 56):* 14 (1989)—Tonga.
trapaneus *H. St. John, Pandanus of Tuamotu Archip. (Revis. Gen. Pandanus, 53):* 23 (1989)—Tuamotu Is.
triangularis *H. St. John ex K.L. Huynh in Bot. Helvetica,* 98(2): 182 (1988)—Sierra Leone.

PANDANUS :-
tupaiensis *H. St. John, Pandanus in Leeward Soc. Is., Meetia, & Henderson (Rev. Gen. Pandanus, 49):* 33 (1989)—Society Is.
uea *H. St. John, Pandanus in Leeward Soc. Is., Meetia, & Henderson (Rev. Gen. Pandanus, 49):* 34 (1989)—Society Is.
vahitahiensis *H. St. John, Pandanus of Tuamotu Archip. (Revis. Gen. Pandanus, 53):* 25 (1989)—Tuamotu Is.
vavauensis *H. St. John, Pandanus of Tonga, Polynesia (Revis. Gen. Pandanus, 56):* 15 (1989)—Tonga.
virginalis *H. St. John, Rev. Gen. Pandanus, 50, Austral Is., French Polynesia:* 33 (1989)—Tubuai Is.
viri *H. St. John, Pandanus of Tuamotu Archip. (Revis. Gen. Pandanus, 53):* 26 (1989)—Tuamotu Is.
wilderi *H. St. John, Pandanus of Cook Is. (Rev. Gen. Pandanus, 47):* 15 (1989)—Cook Is.
yunckeri *H. St. John, Pandanus on Niue Is., Polynesia (Rev. Gen. Pandanus, 52):* 5 (1989)—Niue.

PANICUM (Gramin.).
sect. Valida *F. Zuloaga & O. Morrone in Syst. Bot.,* 14(2): 228 (1989).
bullockii *S.A. Renvoize in Kew Bull.,* 44(3): 543 (1989)—Zambia.
curviflorum
var. suishaense *(Hayata) J.F. Veldkamp in Blumea,* 34(1): 81 (1989): Panicum suishaense.
glandulopaniculatum *S.A. Renvoize in Kew Bull.,* 44(3): 544 (1989)—Malawi, Zambia, Zimbabwe, Mozambique, South Africa (Natal).
kasumense *S.A. Renvoize in Kew Bull.,* 44(3): 546 (1989)—Zimbabwe.
longiloreum *M.M. Rahman in Kew Bull.,* 44(3): 485 (1989)—Thailand, Vietnam.
nymphoides *S.A. Renvoize in Kew Bull.,* 44(3): 545 (1989)—Malawi.
perangustatum *S.A. Renvoize in Kew Bull.,* 44(3): 545 (1989)—Zambia.
phippsii *S.A. Renvoize in Kew Bull.,* 44(3): 546 (1989)—Zambia.
pseudoracemosum *S.A. Renvoize in Kew Bull.,* 44(3): 544 (1989)—Zambia, Tanzania, Burundi, Zaïre.
smithii *M.M. Rahman in Blumea,* 34(1): 173 (1989)—Vietnam.
zambesiense *S.A. Renvoize in Kew Bull.,* 44(3): 544 (1989)—Zimbabwe, Zambia.

PAPAVER (Papaverac.).
arachnoideum *J.W. Kadereit in Notes Roy. Bot. Gard. Edinburgh,* 45(2): 250 (1988 publ. 1989)—Turkey.
commutatum
subsp. euxinum *J.W. Kadereit in Notes Roy. Bot. Gard. Edinburgh,* 45(2): 257 (1988 publ. 1989)—Turkey.

PAPAVER :-
dubium
subsp. erosum *(Litv.) J.W. Kadereit in Notes Roy. Bot. Gard. Edinburgh,* 45(2): 246 (1988 publ. 1989): P. laevigatum var. erosum.
subsp. glabrum *(Royle) J.W. Kadereit in Notes Roy. Bot. Gard. Edinburgh,* 45(2): 247 (1988 publ. 1989): P. glabrum.
subsp. laevigatum *(M. Bieb.) J.W. Kadereit in Notes Roy. Bot. Gard. Edinburgh,* 45(2): 244 (1988 publ. 1989): P. laevigatum.
nudicaule
var. glabricarpum *P.Y. Fu in Bull. Bot. Res. North-East. Forest. Inst.,* 9(2): 17 (1989)—China.
var. seticarpum *P.Y. Fu in Bull. Bot. Res. North-East. Forest. Inst.,* 9(2): 18 (1989)—China.
purpureomarginatum *J.W. Kadereit in Notes Roy. Bot. Gard. Edinburgh,* 45(2): 235 (1988 publ. 1989)—Turkey, Karpathos, East Aegean Is., Crete, Greece, Cyprus.
rechingeri *J.W. Kadereit in Notes Roy. Bot. Gard. Edinburgh,* 45(2): 249 (1988 publ. 1989)—Iraq, Iran.
victoris *M. Škornik & T. Wraber in Biol. Vestn.,* 36(3): 82 (1988)—Jugoslavia.

PAPHINIA (Orchidac.).
seegeri *G. Gerlach in Orchidee,* 40(1): 20 (1989)—Colombia.
var. semialba *G. Gerlach in Orchidee,* 40(1): 21 (1989), as 'semi-alba'—Colombia.

PAPHIOPEDILUM (Orchidac.).
×expansum *J. Atwood in Orchid Rev.,* 97(1148): 183 (1989)—Philippine Is. (P. hennisianum × philippinense).
mohrianum *G.J. Braem in Orchid Dig.,* 53(2): 73 (1989)—Lesser Sunda Is.
richardianum *J.H. Asher & J.H. Beaman in Orchid Dig.,* 52(2): 61 (1988)—Celebes.

PARABOEA (Gesneriac.).
barbatipes *K.Y. Pan in Acta Phytotax. Sin.,* 26(6): 441 (1988)—China.

PARAMAMMEA J.F. Leroy in Compt. R. Hebd. Séances Acad. Sci., D (Paris), 284(16): 1524 (1977). *GUTTIFERAE.*
megaphylla *J.F. Leroy in Compt. R. Hebd. Séances Acad. Sci., D (Paris),* 284(16): 1524 (1977)—Madagascar.

PARAPRENANTHES Chang ex C. Shih in Acta Phytotax. Sin., 26(6): 418 (1988). *COMPOSITAE.*
auriculiformis *C. Shih in Acta Phytotax. Sin.,* 26(6): 421 (1988)—China.
heptantha *C. Shih & D.J. Liou in Acta Phytotax. Sin.,* 26(6): 423 (1988)—China.
longiloba *Ling & C. Shih in Acta Phytotax. Sin.,* 26(6): 421 (1988)—China.
multiformis *C. Shih in Acta Phytotax. Sin.,* 26(6): 420 (1988)—China.

PARAPRENANTHES :-
pilipes *(Migo) C. Shih in Acta Phytotax. Sin.*, 26(6): 424 (1988): Mycelis sororia var. pilipes.
prenanthoides *(Hemsl.) C. Shih in Acta Phytotax. Sin.*, 26(6): 423 (1988): Crepis prenanthoides.
sagittiformis *C. Shih in Acta Phytotax. Sin.*, 26(6): 420 (1988)—China.
sororia *(Miq.) C. Shih in Acta Phytotax. Sin.*, 26(6): 422 (1988): Lactuca sororia.
sylvicola *C. Shih in Acta Phytotax. Sin.*, 26(6): 419 (1988)—China.
thirionii *(Lévl.) C. Shih in Acta Phytotax. Sin.*, 26(6): 424 (1988): Lactuca thirionii.
yunnanensis *(Franch.) C. Shih in Acta Phytotax. Sin.*, 26(6): 421 (1988): Lactuca yunnanensis.

PARIETARIA (Urticac.).
cardiostegia *W. Greuter in Fl. Australia,* 3: 190, 92 (1989)—Australia (New South Wales, W. Australia, N. Terr., S. Australia), Tasmania.

PARNASSIA (Parnassiac.).
tenuis *(Wahlb.) A.P. Khokhryakov, Analiz Fl. Kolÿmskogo Nagor'ya:* 48 (1989): P. palustris var. tenuis.

PARODIA (Cactac.).
krahnii *W. Weskamp in Kakt. And. Sukk.*, 40(3): 60 (1989)—Bolivia.
tillii *W. Weskamp in Succulenta,* 67(10): 210 (1988)—Bolivia.

PARODIOLYRA T.R. Soderstrom & F.O. Zuloaga in Smithsonian Contrib. Bot., 69: 64 (1989). *GRAMINEAE.*
lateralis *(Presl ex Nees von Esenbeck) T.R. Soderstrom & F.O. Zuloaga in Smithsonian Contrib. Bot.,* 69: 66 (1989): Panicum laterale.
luetzelburgii *(Pilger) T.R. Soderstrom & F.O. Zuloaga in Smithsonian Contrib. Bot.,* 69: 70 (1989): Olyra luetzelburgii.
ramosissima *(Trinius) T.R. Soderstrom & F.O. Zuloaga in Smithsonian Contrib. Bot.,* 69: 73 (1989): Olyra ramosissima.

PARRYA (Crucif.).
australis *N.V. Pavlov in Vestn. Akad. Nauk Kaz. SSR,* 1949(1): 28 (1949)—U.S.S.R. (Middle Asia).

PARVATIA (Lardizabalac.).
elliptica *(Hemsl.) M.P. Nayar & T.K. Paul in Fasc. Fl. India,* 19: 36 (1988): Stauntonia elliptica.

PASANIA (Fagac.).
brevicaudata
var. arisanensis *(Hayata) S.S. Ying, Coloured Illustr. Fl. Taiwan,* 3: 248 (1988): Quercus arisanensis.

PASANIA :-
kuarunensis *(Tomiya) S.S. Ying, Coloured Illustr. Fl. Taiwan,* 3: 252 (1988): Synaedrys kuarunensis.
ternaticupula
f. subreticulata *(Hayata) S.S. Ying, Coloured Illustr. Fl. Taiwan,* 3: 252 (1988): Quercus subreticulata.

PASPALUM (Gramin.).
cultratum *(Trin.) A.G. Burman in Acta Bot. Venezuel.,* 14(4): 90 (1987), without basionym ref.: Basionym not stated.
dispar
var. marahuacense *H. Rodríguez R. in Ernstia,* 52: 1 (1989)—Venezuela.
multinervium *A.G. Burman in Acta Bot. Venezuel.,* 14(4): 90 (1987), nom. nov.: Thrasya reticulata *Swallen.*
peregrinum *A.G. Burman & Filgueiras in Acta Bot. Venezuel.,* 14(4): 91 (1987), nom. nov.: Thrasya villosa *Hitchc.*
venezuelanum *(Chase) A.G. Burman in Acta Bot. Venezuel.,* 14(4): 90 (1987): Thrasya venezuelana.

PASSIFLORA (Passiflorac.).
sect. Colombiana *L.K. Escobar in Fl. Colombia,* 10: 26 (1988).
ser. Colombianae *L.K. Escobar in Fl. Colombia,* 10: 41 (1988).
sect. Fimbriatistipula *L.K. Escobar in Fl. Colombia,* 10: 61 (1988).
ser. Leptomischae *L.K. Escobar in Fl. Colombia,* 10: 27 (1988).
subgen. Manicata *(Harms) L.K. Escobar in Fl. Colombia,* 10: 92 (1988): P. ser. Manicatae.
sect. Parritana *(Harms) L.K. Escobar in Fl. Colombia,* 10: 58 (1988): P. ser. Parritanae.
subgen. Porphyropathanthus *L.K. Escobar in Ann. Missouri Bot. Gard.,* 76(3): 884 (1989).
ser. Quindiensae *L.K. Escobar in Fl. Colombia,* 10: 37 (1988).
amazonica *L.K. Escobar in Ann. Missouri Bot. Gard.,* 76(3): 877 (1989)—Peru.
caatingae *L.K. Escobar in Ann. Missouri Bot. Gard.,* 76(3): 880 (1989)—Brazil.
chlorina *L.K. Escobar in Phytologia,* 67(2): 132 (1989)—Brazil.
citrina *J.M. MacDougal in Ann. Missouri Bot. Gard.,* 76(1): 354 (1989)—Honduras, Guatemala.
eglandulosa *J.M. MacDougal in Ann. Missouri Bot. Gard.,* 75(4): 1658 (1988 publ. 1989)—Guatemala, Salvador, Honduras.
fernandezii *L.K. Escobar in Phytologia,* 66(1): 80 (1989)—Bolivia.
gilbertiana *J.M. MacDougal in Ann. Missouri Bot. Gard.,* 76(2): 610 (1989)—Costa Rica.
jardinensis *L.K. Escobar in Mutisia,* 71: 4 (1988)—Colombia.
linearistipula *L.K. Escobar in Fl. Colombia,* 10: 39 (1988)—Colombia.
mayarum *J.M. MacDougal in Ann. Missouri Bot. Gard.,* 76(2): 615 (1989)—Mexico, Guatemala, Belize.

PASSIFLORA :-
nubicola *J.M. MacDougal in Ann. Missouri Bot. Gard.*, 76(2): 608 (1989)—Costa Rica.
pacifica *L.K. Escobar in Mutisia,* 71: 1 (1988)—Colombia.
pascoensis *L.K. Escobar in Ann. Missouri Bot. Gard.*, 76(3): 880 (1989)—Peru.
pusilla *J.M. MacDougal in Ann. Missouri Bot. Gard.*, 75(1): 392 (1988)—Nicaragua, Costa Rica.
sierrae *L.K. Escobar in Ann. Missouri Bot. Gard.*, 76(3): 884 (1989)—Colombia.
solomonii *L.K. Escobar in Ann. Missouri Bot. Gard.*, 76(3): 884 (1989)—Bolivia.
tenerifensis *L.K. Escobar in Mutisia,* 71: 6 (1988)—Colombia.
uribei *L.K. Escobar in Fl. Colombia,* 10: 64 (1988), nom. nov.: P. albicans *L. Uribe.*
viridescens *L.K. Escobar in Phytologia,* 66(1): 81 (1989)—Peru.

PAVONIA (Malvac.).
alia *P.A. Fryxell, Malvac. of Mexico (Syst. Bot. Monog., 25):* 315 (1988)—Mexico.
baumliana *P.A. Fryxell, Malvac. of Mexico (Syst. Bot. Monog., 25):* 317 (1988)—Mexico.
durangensis *P.A. Fryxell, Malvac. of Mexico (Syst. Bot. Monog., 25):* 321 (1988)—Mexico.
×gledhillii *M. Cheek in Kew Bull.*, 44(1): 146 (1989)—Cultivation. (P. makoyana × multiflora).
paludicola *Nicolson ex P.A. Fryxell in R.A. Howard, Fl. Lesser Antilles (Dicot. 2)*, 5: 241 (1989), nom. nov.: Malache scabra *Vogel.*
windischii *A. Krapovickas in Prog. Resum. Soc. Bot. Brasil, XXXVI Congr. Nac. Bot., Curitiba:* 361 (1985), without latin descr. or type—Brazil.

PECTEILIS (Orchidac.).
cochinchinensis *(Gagnep.) L.V. Aver'yanov, Prelim. list Vietnam. Orchids,* 2: 119 (1988): Parahabenaria cochinchinensis.

PECTIS (Compos.).
amplifolia *D.J. Keil in Phytologia,* 67(4): 346 (1989)—Mexico.
elongata
var. fasciculiflora *(DC.) D.J. Keil in Phytologia,* 67(4): 352 (1989): P. fasciculiflora.
exilis *D.J. Keil in Phytologia,* 67(4): 349 (1989)—Mexico.
holochaeta *(S.F. Blake) D.J. Keil in Phytologia,* 67(4): 352 (1989): P. propetes var. holochaeta.
var. cana *D.J. Keil in Phytologia,* 67(4): 354 (1989)—Mexico.

PEDICULARIS (Scrophulariac.).
alaica *A.D. Li in Opred. Rast. Sred. Azii,* 9: 362 (1987)—U.S.S.R. (Middle Asia).
ruoergaiensis *H.P. Yang in Acta Phytotax. Sin.*, 27(3): 222 (1989)—China.

PEDICULARIS :-
sibthorpii
var. eriantha *(Boiss. & Buhse) T.N. Popova in Fl. Armenii,* 8: 348 (1987), without basionym ref.: P. comosa var. eriantha.
var. ochrorrhyncha *(Galushko & T.N. Pop.) T.N. Popova in Fl. Armenii,* 8: 348 (1987): P. ochrorrhyncha.
yanyuanensis *H.P. Yang in Acta Phytotax. Sin.*, 27(3): 224 (1989)—China.

PEDILOCHILUS (Orchidac.).
hermonii *P.J. Cribb & B. Lewis in Orchid Rev.*, 97(1150): 251 (1989)—Vanuatu.

PEDIOMELUM (Legumin.).
digitatum
var. parvifolium *(Shinners) K.N. Gandhi & L.E. Brown in Sida,* 13(3): 373 (1989): Psoralea digitata var. parvifolia.
latestipulatum
var. appressum *(Ockendon) K.N. Gandhi & L.E. Brown in Sida,* 13(3): 373 (1989): Psoralea latestipulata var. appressa.

PELLIONIA (Urticac.).
radicans
var. chikushiensis *(Yamamoto) S.S. Ying, Coloured Illustr. Fl. Taiwan,* 3: 451 (1988): P. chikushiensis.

PENAEA (Penaeac.).
dahlgrenii *J.P. Rourke in S. Afr. J. Bot.*, 55(4): 400 (1989)—South Africa (Cape Province).

PENSTEMON (Scrophulariac.).
idahoensis *N.D. Atwood & S.L. Welsh in Great Basin Nat.*, 48(4): 496 (1988)—U.S.A. (Idaho).

PENTAS (Rubiac.).
zanzibarica
var. haroniensis *B. Verdcourt in Fl. Zambes.*, 5(1): 76 (1989)—Zimbabwe.
subsp. milangiana *(Verdc.) B. Verdcourt in Fl. Zambes.*, 5(1): 76 (1989): P. zanzibarica var. milangiana.

PENTOSSAEA W.S. Judd in Ann. Missouri Bot. Gard., 76(2): 486 (1989). *MELASTOMATACEAE.*
angustifolia *(DC.) W.S. Judd in Ann. Missouri Bot. Gard.*, 76(2): 490 (1989): Leandra angustifolia.
brachystachya *(DC.) W.S. Judd in Ann. Missouri Bot. Gard.*, 76(2): 490 (1989): Clidemia brachystachya.
confertiflora *(DC.) W.S. Judd in Ann. Missouri Bot. Gard.*, 76(2): 490 (1989): Clidemia confertiflora.
congestiflora *(Naudin) W.S. Judd in Ann. Missouri Bot. Gard.*, 76(2): 490 (1989): Clidemia congestiflora.
coriacea *(Naudin) W.S. Judd in Ann. Missouri Bot. Gard.*, 76(2): 490 (1989), without basionym ref.: Basionym not stated.

PENTOSSAEA :–
heteronervis *(Naudin) W.S. Judd in Ann. Missouri Bot. Gard.*, 76(2): 490 (1989): Sagraea heteronervis.
marginata *(Desr.) W.S. Judd in Ann. Missouri Bot. Gard.*, 76(2): 490 (1989): Melastoma marginata.

PEPEROMIA (Piperac.).
adscendens *(Endl.) K. Schumann in Notizbl. Bot. Gart. Berlin*, 1: 47 (1895), illegit. later homonym of C. DC. (1866): Piper adscendens.

PEPINIA (Bromeliac.).
carnososepala *(Rauh & Gross) H.E. Luther in Phytologia*, 67(4): 312 (1989), as 'carnosasepala': Pitcairnia carnososepala.
hammelii *(Luther) H.E. Luther in J. Bromeliad Soc.*, 39(4): 173 (1989): Pitcairnia hammelii.

PERISTROPHE (Acanthac.).
decorticans *K. Balkwill in S. Afr. J. Bot.*, 55(2): 254 (1989)—South Africa (Transvaal), Botswana.

PERISTYLUS (Orchidac.).
aliformis *(C. Schweinf.) J. Renz & S. Vodonaivalu in Blumea*, 34(1): 88 (1989 - received at K on 30 June): Habenaria aliformis.
aliformis *(C. Schweinf.) P.J. Kores in Allertonia*, 5(1): 14 (August 1989), isonym see J. Renz & S. Vodonaivalu *in Blumea*, 34(1): 88 (1989): Habenaria aliformis.
cymbochilus *J. Renz in Davis & Hedge Festschrift (ed. Kit Tan):* 185 (1989)—New Guinea (Papua New Guinea).
hatusimanus *T. Hashimoto in Ann. Tsukuba Bot. Gard.*, 7: 161 (1988)—Japan.
maculifer *(C. Schweinf.) J. Renz & S. Vodonaivalu in Blumea*, 34(1): 89 (1989 - received at K on 30 June), as 'maculiferus': Habenaria maculifera.
maculifer *(C. Schweinf.) P.J. Kores in Allertonia*, 5(1): 13 (August 1989), isonym see J. Renz & S. Vodonaivalu *in Blumea*, 34(1): 89 (1989): Habenaria maculifera.
pachyneuroides *J. Renz in Davis & Hedge Festschrift (ed. Kit. Tan):* 187 (1989)—New Guinea (Papua New Guinea).
scrotiformis *(L.O. Williams) J. Renz & S. Vodonaivalu in Blumea*, 34(1): 89 (1989): Habenaria scrotiformis.
stenodon *(Reichenb. f.) P.J. Kores in Allertonia*, 5(1): 13 (August 1989): Habenaria stenodon.
stenodon *(Reichb. f.) Renz & Vodonaivalu ex B. Lewis & P. Cribb, Orchids of Vanuatu:* 41 (5 October 1989), as 'stenodontus', isonym see P.J. Kores *in Allertonia*, 5(1): 13 (August 1989): Habenaria stenodon.
tradescantifolius *(Reichenb. f.) P.J. Kores in Allertonia*, 5(1): 12 (1989): Habenaria tradescantifolia.
wheatleyi *P.J. Cribb & B. Lewis in Orchid Rev.*, 97(1150): 251 (1989)—Vanuatu.

PERITYLE (Compos.).
scopulorum *(M.E. Jones) A.M. Powell & B.L. Turner in Phytologia*, 66(6): 505 (1989): Laphamia scopulorum.
vandevenderi *B.L. Turner in Phytologia*, 66(6): 504 (1989)—Mexico.

PERSEA (Laurac.).
areolatocostae *(Allen) H. van der Werff in Ann. Missouri Bot. Gard.*, 76(2): 471 (1989): Phoebe areolatocostae.
croatii *H. van der Werff in Ann. Missouri Bot. Gard.*, 76(2): 471 (1989)—Venezuela.
croizatii *H. van der Werff in Ann. Missouri Bot. Gard.*, 76(2): 472 (1989)—Venezuela.
fluviatilis *H. van der Werff in Ann. Missouri Bot. Gard.*, 76(2): 474 (1989)—Venezuela.
glaucescens *(Wt.) N.P. Singh, Fl. East. Karnataka*, 2: 548 (1988): Machilus glaucescens.
jariensis *I. de Vattimo-Gil in Rev. Brasil. Biol.*, 48(4): 956 (1988)—Brazil.
julianae *H. van der Werff in Ann. Missouri Bot. Gard.*, 76(3): 939 (1989)—Surinam.

PERSICARIA (Polygonac.).
sect. Amblygonon *(Meissn.) N.N. Tsvelev in Novosti Sist. Vyssh. Rast.*, 25: 185 (1988); *et in Sosud. Rast. Sovet. Dal'nego Vostoka*, 4: 67 (1989): Polygonum sect. Amblygonon.
affinis *(D. Don) L.P. Ronse Decraene in Bot. J. Linn. Soc.*, 98(4): 368 (1988): Polygonum affine.
angustifolia *(Pall.) L.P. Ronse Decraene in Bot. J. Linn. Soc.*, 98(4): 367 (1988): Polygonum angustifolium.
bistorta
subsp. carnea *(C. Koch) W. Greuter & Burdet in Willdenowia*, 19(1): 41 (1989): Polygonum carneum.
campanulata *(Hook. f.) L.P. Ronse Decraene in Bot. J. Linn. Soc.*, 98(4): 367 (1988): Polygonum campanulatum.
hookeri *(Meissn.) L.P. Ronse Decraene in Bot. J. Linn. Soc.*, 98(4): 367 (1988): Polygonum hookeri.
hypanica *(Klok.) N.N. Tsvelev in Novosti Sist. Vyssh. Rast.*, 25: 186 (1988): Polygonum hypanicum.
lanata *(Roxb.) N.N. Tsvelev in Novosti Sist. Vyssh. Rast.*, 25: 186 (1988): Polygonum lanatum.
lapathifolia
var. tomentosa *(Schrank) N.N. Tsvelev in Sosud. Rast. Sovet. Dal'nego Vostoka*, 4: 70 (1989): Polygonum tomentosum.
obtusifolia *(Täckh. & Boulos) W. Greuter & Burdet in Willdenowia*, 19(1): 41 (1989): Polygonum obtusifolium.
saporoviensis *(Klok.) N.N. Tsvelev in Novosti Sist. Vyssh. Rast.*, 25: 186 (1988): Polygonum saporoviense.

PERSICARIA :–
scabra
var. amurica *N.N. Tsvelev in Sosud. Rast. Sovet. Dal'nego Vostoka,* 4: 71 (1989)—U.S.S.R. (Soviet Far East).
var. incana *(F.W. Schmidt) N.N. Tsvelev in Sosud. Rast. Sovet. Dal'nego Vostoka,* 4: 71 (1989): Polygonum incanum.
tripterocarpa *(A. Gray) L.P. Ronse Decraene in Bot. J. Linn. Soc.,* 98(4): 367 (1988): Polygonum tripterocarpum.
vacciniifolia *(Wall. ex Meissn.) L.P. Ronse Decraene in Bot. J. Linn. Soc.,* 98(4): 368 (1988): Polygonum vacciniifolium.
vivipara *(L.) L.P. Ronse Decraene in Bot. J. Linn. Soc.,* 98(4): 368 (1988): Polygonum viviparum.
wallichii *W. Greuter & Burdet in Willdenowia,* 19(1): 41 (1989), nom. nov.: Polygonum polystachyum *Meissner.*
weyrichii *(F. Schmidt Petrop.) L.P. Ronse Decraene in Bot. J. Linn. Soc.,* 98(4): 367 (1988): Polygonum weyrichii.

PERTYA (Compos.).
cordifolia
var. desmocephala *(Diels) Z. Wei & Y.B. Chang in Bull. Bot. Res. North-East. Forest. Inst.,* 9(2): 37 (1989): P. desmocephala.
pubescens *Y. Ling in Contrib. Inst. Bot. Nation. Acad. Peiping,* 6(1): 32 (1948 publ. 1949)—China.
tsoongiana *Ling in Contrib. Bot. Surv. Northwest China,* 1(2): 40 (1939); *cf. Y. Ling in Contrib. Inst. Bot. Nation. Acad. Peiping,* 6(1): 29 (1948 publ. 1949)— China.

PERYMENIUM (Compos.).
oaxacanum *B.L. Turner in Phytologia,* 65(5): 371 (1988)—Mexico.

PEUCEDANUM (Umbellif.).
gabrielae *R. Frey in Candollea,* 44(1): 275 (26 July 1989); *et in Mitt. Bot. Mus. Univ. Zürich,* 383: 28 (9 August 1989), not effectively published as thesis in 1988—Greece.
officinale
subsp. longifolium *(Waldstein & Kitaibel) R. Frey in Candollea,* 44(1): 272 (26 July 1989); *et in Mitt. Bot. Mus. Univ. Zürich,* 383: 24 (9 August 1989), not effectively published as thesis in 1988: P. longifolium.
subsp. paniculatum *(Loiseleur) R. Frey in Candollea,* 44(1): 273 (26 July 1989); *et in Mitt. Bot. Mus. Univ. Zürich,* 383: 25 (9 August 1989), not effectively published as thesis in 1988: P. paniculatum.
sonderi *(Hiroe) B.L. Burtt in Notes Roy. Bot. Gard. Edinburgh,* 45(2): 200 (1988 publ. 1989): Seseli sonderi.
songpanense *R.H. Shan & F.T. Pu in Acta Phytotax. Sin.,* 27(1): 65 (1989)—China.

PHELLODENDRON (Rutac.).
amurense
var. molle *(Nakai) S.H. Li & S.Z. Liou in Fl. Liaoningica, chief ed. S.X. Li,* 1: 1051 (1988): P. molle.

PHILODENDRON (Arac.).
subsect. Bulaoanum *S.J. Mayo in Bot. J. Linn. Soc.,* 100(2): 168 (1989), nom. nov.: P. "Gruppe" § Doratophyllum *Engler.*
subsect. Canniphyllum *(Schott) S.J. Mayo in Bot. J. Linn. Soc.,* 100(2): 168 (1989): P. grex Canniphyllum.
subsect. Eucardium *(Engler) S.J. Mayo in Bot. J. Linn. Soc.,* 100(2): 168 (1989): P. "Gruppe" § Eucardium.
subsect. Oligocarpidium *(Engler) S.J. Mayo in Bot. J. Linn. Soc.,* 100(2): 168 (1989): P. "Gruppe" § Oligocarpidium.
subgen. Pteromischum *(Schott) S.J. Mayo in Bot. J. Linn. Soc.,* 100(2): 168 (1989): P. grex Pteromischum.
dardanianum *S.J. Mayo in Bot. J. Linn. Soc.,* 100(2): 164 (1989), without latin descr. or type—Brazil.
triplum *G.S. Bunting in Ann. Missouri Bot. Gard.,* 76(3): 917 (1989)—Venezuela.

PHLEUM (Gramin.).
hirsutum
subsp. dactylinum *A.G. Helm in Ber. Bot.-Zool. Ges. Liechtenstein-S.-W.,* 17: 11 (1988)—Switzerland.
paniculatum
subsp. annuum *(M. Bieb.) J. Holub in Folia Geobot. Phytotax.,* 23(4): 413 (1988): P. annuum.

PHLOMOIDES (Labiat.).
agraria *(Bge.) T.A. Adÿlov, R. Kam. & A. Makhmedov in Opred. Rast. Sred. Azii,* 9: 106 (1987): Phlomis agraria.
alaica *(Knorr.) T.A. Adÿlov, R. Kam. & A. Makhmedov in Opred. Rast. Sred. Azii,* 9: 107 (1987): Phlomis alaica.
alberti *(Rgl.) T.A. Adÿlov, R. Kam. & A. Makhmedov in Opred. Rast. Sred. Azii,* 9: 101 (1987): Eremostachys alberti.
alpina *(Pall.) T.A. Adÿlov, R. Kam. & A. Makhmedov in Opred. Rast. Sred. Azii,* 9: 104 (1987): Phlomis alpina.
ambigua *(M. Pop. ex Pazij & Vved.) T.A. Adÿlov, R. Kam. & A. Makhmedov in Opred. Rast. Sred. Azii,* 9: 95 (1987): Eremostachys ambigua.
angreni *(M. Pop.) T.A. Adÿlov, R. Kam. & A. Makhmedov in Opred. Rast. Sred. Azii,* 9: 101 (1987): Eremostachys angreni.
arctifolia *(M. Pop.) T.A. Adÿlov, R. Kam. & A. Makhmedov in Opred. Rast. Sred. Azii,* 9: 102 (1987): Eremostachys arctiifolia.
baburii *(Adÿl.) T.A. Adÿlov in Opred. Rast. Sred. Azii,* 9: 100 (1987): Eremostachys baburii.

PHLOMOIDES :-
baissunensis *(M. Pop.) T.A. Adÿlov, R. Kam. & A. Makhmedov in Opred. Rast. Sred. Azii,* 9: 96 (1987): Eremostachys baissunensis.
baldschuanica *(Rgl.) T.A. Adÿlov, R. Kam. & A. Makhmedov in Opred. Rast. Sred. Azii,* 9: 95 (1987): Eremostachys baldschuanica.
beckeri *(Rgl.) T.A. Adÿlov, R. Kam. & A. Makhmedov in Opred. Rast. Sred. Azii,* 9: 92 (1987): Eremostachys beckeri.
boissieriana *(Rgl.) T.A. Adÿlov, R. Kam. & A. Makhmedov in Opred. Rast. Sred. Azii,* 9: 91 (1987): Eremostachys boissieriana.
botschantzevii *(Adÿl.) T.A. Adÿlov in Opred. Rast. Sred. Azii,* 9: 94 (1987): Eremostachys botschantzevii.
brachystegia *(Bge.) T.A. Adÿlov, R. Kam. & A. Makhmedov in Opred. Rast. Sred. Azii,* 9: 103 (1987): Phlomis brachystegia.
canescens *(Rgl.) T.A. Adÿlov, R. Kam. & A. Makhmedov in Opred. Rast. Sred. Azii,* 9: 104 (1987): Phlomis canescens.
cephalariifolia *(M. Pop.) T.A. Adÿlov, R. Kam. & A. Makhmedov in Opred. Rast. Sred. Azii,* 9: 96 (1987): Eremostachys cephalariifolia.
czuiliensis *(Golosk.) T.A. Adÿlov, R. Kam. & A. Makhmedov in Opred. Rast. Sred. Azii,* 9: 96 (1987): Eremostachys czuiliensis.
dszumrutensis *(Afan.) T.A. Adÿlov, R. Kam. & A. Makhmedov in Opred. Rast. Sred. Azii,* 9: 105 (1987): Phlomis dszumrutensis.
ebracteolata *(M. Pop.) T.A. Adÿlov, R. Kam. & A. Makhmedov in Opred. Rast. Sred. Azii,* 9: 100 (1987): Eremostachys ebracteolata.
eriocalyx *(Rgl.) T.A. Adÿlov, R. Kam. & A. Makhmedov in Opred. Rast. Sred. Azii,* 9: 99 (1987): Eremostachys eriocalyx.
ferganensis *(M. Pop.) T.A. Adÿlov, R. Kam. & A. Makhmedov in Opred. Rast. Sred. Azii,* 9: 104 (1987): Phlomis ferganensis.
fetisowii *(Rgl.) T.A. Adÿlov, R. Kam. & A. Makhmedov in Opred. Rast. Sred. Azii,* 9: 102 (1987): Eremostachys fetisowii.
fulgens *(Bge.) T.A. Adÿlov, R. Kam. & A. Makhmedov in Opred. Rast. Sred. Azii,* 9: 94 (1987): Eremostachys fulgens.
gymnocalyx *(Schrenk) T.A. Adÿlov, R. Kam. & A. Makhmedov in Opred. Rast. Sred. Azii,* 9: 99 (1987): Eremostachys gymnocalyx.
gypsacea *(M. Pop.) T.A. Adÿlov, R. Kam. & A. Makhmedov in Opred. Rast. Sred. Azii,* 9: 100 (1987): Eremostachys gypsacea.
hissarica *(Rgl.) T.A. Adÿlov, R. Kam. & A. Makhmedov in Opred. Rast. Sred. Azii,* 9: 101 (1987): Eremostachys hissarica.
iliensis *(Rgl.) T.A. Adÿlov, R. Kam. & A. Makhmedov in Opred. Rast. Sred. Azii,* 9: 98 (1987): Eremostachys iliensis.
impressa *(Pazij & Vved.) T.A. Adÿlov, R. Kam. & A. Makhmedov in Opred. Rast. Sred. Azii,* 9: 100 (1987): Eremostachys impressa.
integior *(Pazij & Vved.) T.A. Adÿlov, R. Kam. & A. Makhmedov in Opred. Rast. Sred. Azii,* 9: 98 (1987): Eremostachys integior.

PHLOMOIDES :-
kaufmanniana *(Rgl.) T.A. Adÿlov, R. Kam. & A. Makhmedov in Opred. Rast. Sred. Azii,* 9: 92 (1987): Eremostachys kaufmanniana.
kirghisorum *T.A. Adÿlov, R. Kam. & A. Makhmedov in Opred. Rast. Sred. Azii,* 9: 99 (1987), nom. nov.: Eremostachys ferganensis *Ubuk.*
knorringiana *(M. Pop.) T.A. Adÿlov, R. Kam. & A. Makhmedov in Opred. Rast. Sred. Azii,* 9: 107 (1987): Phlomis knorringiana.
korovinii *(M. Pop.) T.A. Adÿlov, R. Kam. & A. Makhmedov in Opred. Rast. Sred. Azii,* 9: 102 (1987): Eremostachys korovinii.
labiosa *(Bge.) T.A. Adÿlov, R. Kam. & A. Makhmedov in Opred. Rast. Sred. Azii,* 9: 92 (1987): Eremostachys labiosa.
labiosiformis *(M. Pop.) T.A. Adÿlov, R. Kam. & A. Makhmedov in Opred. Rast. Sred. Azii,* 9: 97 (1987): Eremostachys laciniata var. labiosaeformis.
labiosissima *(Pazij & Vved.) T.A. Adÿlov, R. Kam. & A. Makhmedov in Opred. Rast. Sred. Azii,* 9: 94 (1987): Eremostachys labiosissima.
lehmanniana *(Bge.) T.A. Adÿlov, R. Kam. & A. Makhmedov in Opred. Rast. Sred. Azii,* 9: 101 (1987): Eremostachys lehmanniana.
leiocalyx *(Pazij & Vved.) T.A. Adÿlov, R. Kam. & A. Makhmedov in Opred. Rast. Sred. Azii,* 9: 100 (1987): Eremostachys leiocalyx.
lyrata *(M. Pop.) T.A. Adÿlov in Opred. Rast. Sred. Azii,* 9: 97 (1987): Eremostachys sarawschanica f. lyrata.
mihaelis *T.A. Adÿlov, R. Kam. & A. Makhmedov in Opred. Rast. Sred. Azii,* 9: 99 (1987), nom. nov.: Eremostachys canescens *M. Pop.*
napuligera *(Franch.) T.A. Adÿlov, R. Kam. & A. Makhmedov in Opred. Rast. Sred. Azii,* 9: 93 (1987): Eremostachys napuligera.
nuda *(Rgl.) T.A. Adÿlov, R. Kam. & A. Makhmedov in Opred. Rast. Sred. Azii,* 9: 97 (1987): Eremostachys nuda.
oreophila *(Kar. & Kir.) T.A. Adÿlov, R. Kam. & A. Makhmedov in Opred. Rast. Sred. Azii,* 9: 105 (1987): Phlomis oreophila.
ostrowskiana *(Rgl.) T.A. Adÿlov, R. Kam. & A. Makhmedov in Opred. Rast. Sred. Azii,* 9: 103 (1987): Phlomis ostrowskiana.
pauciflora *(Kuntze) T.A. Adÿlov, R. Kam. & A. Makhmedov in Opred. Rast. Sred. Azii,* 9: 91 (1987): Eremostachys pauciflora.
pectinata *(M. Pop.) T.A. Adÿlov, R. Kam. & A. Makhmedov in Opred. Rast. Sred. Azii,* 9: 98 (1987): Eremostachys pectinata.
popovii *(Gontsch.) T.A. Adÿlov, R. Kam. & A. Makhmedov in Opred. Rast. Sred. Azii,* 9: 94 (1987): Eremostachys popovii.
pratensis *(Kar. & Kir.) T.A. Adÿlov, R. Kam. & A. Makhmedov in Opred. Rast. Sred. Azii,* 9: 105 (1987): Phlomis pratensis.
puberula *(Kryl. & Serg.) T.A. Adÿlov, R. Kam. & A. Makhmedov in Opred. Rast. Sred. Azii,* 9: 106 (1987): Phlomis puberula.

PHLOMOIDES :-
pulchra *(M. Pop.) T.A. Adÿlov, R. Kam. & A. Makhmedov in Opred. Rast. Sred. Azii,* 9: 94 (1987): Eremostachys pulchra.
regeliana *(Aitch. & Hemsl.) T.A. Adÿlov, R. Kam. & A. Makhmedov in Opred. Rast. Sred. Azii,* 9: 91 (1987): Eremostachys regeliana.
sarawschanica *(Rgl.) T.A. Adÿlov, R. Kam. & A. Makhmedov in Opred. Rast. Sred. Azii,* 9: 96 (1987): Eremostachys sarawschanica.
schugnanica *(M. Pop.) T.A. Adÿlov, R. Kam. & A. Makhmedov in Opred. Rast. Sred. Azii,* 9: 97 (1987): Eremostachys speciosa var. schugnanica.
septentrionalis *(M. Pop.) T.A. Adÿlov, R. Kam. & A. Makhmedov in Opred. Rast. Sred. Azii,* 9: 98 (1987): Eremostachys nuda var. septentrionalis.
speciosa *(Rupr.) T.A. Adÿlov, R. Kam. & A. Makhmedov in Opred. Rast. Sred. Azii,* 9: 95 (1987): Eremostachys speciosa.
subspicata *(M. Pop.) T.A. Adÿlov, R. Kam. & A. Makhmedov in Opred. Rast. Sred. Azii,* 9: 92 (1987): Eremostachys subspicata.
tadschikistanica *(B. Fedtsch.) T.A. Adÿlov, R. Kam. & A. Makhmedov in Opred. Rast. Sred. Azii,* 9: 92 (1987): Eremostachys tadschikistanica.
tianschanica *(M. Pop.) T.A. Adÿlov, R. Kam. & A. Makhmedov in Opred. Rast. Sred. Azii,* 9: 102 (1987): Eremostachys tianschanica.
tschimganica *(Vved.) T.A. Adÿlov, R. Kam. & A. Makhmedov in Opred. Rast. Sred. Azii,* 9: 106 (1987): Phlomis tschimganica.
tytthaster *(Vved.) T.A. Adÿlov, R. Kam. & A. Makhmedov in Opred. Rast. Sred. Azii,* 9: 104 (1987): Phlomis tytthaster.
uniflora *(Rgl.) T.A. Adÿlov, R. Kam. & A. Makhmedov in Opred. Rast. Sred. Azii,* 9: 100 (1987): Eremostachys uniflora.
urodonta *(M. Pop.) T.A. Adÿlov, R. Kam. & A. Makhmedov in Opred. Rast. Sred. Azii,* 9: 106 (1987): Phlomis urodonta.
vavilovii *(M. Pop.) T.A. Adÿlov, R. Kam. & A. Makhmedov in Opred. Rast. Sred. Azii,* 9: 103 (1987): Phlomis vavilovii.
zenaidae *(M. Pop.) T.A. Adÿlov, R. Kam. & A. Makhmedov in Opred. Rast. Sred. Azii,* 9: 101 (1987): Eremostachys zenaidae.

PHOEBE (Laurac.).
glaucifolia *S.K. Lee & F.N. Wei in Guihaia,* 8(4): 298 (1988)—China, Xizang.
scortechinii *(Gamb.) K.M. Kochummen in Tree Fl. Malaya,* 4: 177 (1989), with incorrect basionym ref.: Machilus scortechinii.

PHOLIDOTA (Orchidac.).
sect. Acanthoglossum *(Blume) E.F. de Vogel in Orchid Monogr.,* 3: 9 (1988): gen. Acanthoglossum.
sect. Advena *E.F. de Vogel in Orchid Monogr.,* 3: 18 (1988).

PHOLIDOTA :-
sect. Chinenses *(Rchb. f.) E.F. de Vogel in Orchid Monogr.,* 3: 41 (1988): Coelogyne sect. Chinenses.
sect. Repentes *E.F. de Vogel in Orchid Monogr.,* 3: 67 (1988).
camelostalix
var. vaginata *(Carr) E.F. de Vogel in Orchid Monogr.,* 3: 30 (1988): P. vaginata.
carnea
var. parviflora *(Hooker f.) E.F. de Vogel in Orchid Monogr.,* 3: 55 (1988): P. parviflora.
var. pumila *(Ridley) E.F. de Vogel in Orchid Monogr.,* 3: 56 (1988): P. parviflora var. pumila.
longilabra *E.F. de Vogel in Orchid Monogr.,* 3: 34 (1988), nom. nov.: P. rupestris *J.J. Smith.*
mediocris *E.F. de Vogel in Orchid Monogr.,* 3: 12 (1988)—Borneo (Sarawak, Kalimantan).

PHORADENDRON (Viscac.).
leucarpum *(Raf.) J.L. Reveal & M.C. Johnston in Taxon,* 38(1): 107 (1989): Viscum leucarpum.

PHREATIA (Orchidac.).
bigibbula *P.J. Kores in Allertonia,* 5(1): 173 (1989)—Fiji.
flavovirens *P.J. Kores in Allertonia,* 5(1): 178 (1989)—Fiji.
gillespiei *P.J. Kores in Allertonia,* 5(1): 172 (1989)—Fiji.
pentagona *P.J. Kores in Allertonia,* 5(1): 176 (1989)—Fiji.

PHRYMA (Phrymac.).
asiatica *(Hara) N.S. Probatova in Sosud. Rast. Sovet. Dal'nego Vostoka,* 4: 365 (1989): P. leptostachya var. asiatica.

PHYLLANTHUS (Euphorbiac.).
gentryi *G.L. Webster in Ann. Missouri Bot. Gard.,* 75(3): 1096 (1988)—Panama.

PHYLLOCEPHALUM (Compos.).
ritchiei *(Hook. f.) Almeida in J. Bombay Nat. Hist. Soc.,* 85(3): 522 (1988 publ. 1989): Centratherum ritchiei.

PHYLLOPODIUM (Scrophulariac.).
anomalum *O.M. Hilliard in Notes Roy. Bot. Gard. Edinburgh,* 45(3): 484 (1988 publ. 1989)—South Africa (Cape Province).
caespitosum *O.M. Hilliard in Notes Roy. Bot. Gard. Edinburgh,* 45(3): 484 (1988 publ. 1989)—South Africa (Cape Province).
collinum *(Hiern) O.M. Hilliard in Notes Roy. Bot. Gard. Edinburgh,* 45(3): 484 (1988 publ. 1989): Polycarena collina.
dolomiticum *O.M. Hilliard in Notes Roy. Bot. Gard. Edinburgh,* 45(3): 484 (1988 publ. 1989)—South Africa (Cape Province).
elegans *(Choisy) O.M. Hilliard in Notes Roy. Bot. Gard. Edinburgh,* 45(3): 484 (1988 publ. 1989): Selago elegans.

PHYLLOPODIUM :-
lupuliforme *(Thellung) O.M. Hilliard in Notes Roy. Bot. Gard. Edinburgh,* 45(3): 485 (1988 publ. 1989): Polycarena lupuliformis.
maxii *(Hiern) O.M. Hilliard in Notes Roy. Bot. Gard. Edinburgh,* 45(3): 485 (1988 publ. 1989): Polycarena maxii.
micranthum *(Schltr.) O.M. Hilliard in Notes Roy. Bot. Gard. Edinburgh,* 45(3): 485 (1988 publ. 1989): Polycarena rariflora var. micrantha.
mimetes *O.M. Hilliard in Notes Roy. Bot. Gard. Edinburgh,* 45(3): 485 (1988 publ. 1989)—South Africa (Cape Province).
namaense *(Thellung) O.M. Hilliard in Notes Roy. Bot. Gard. Edinburgh,* 45(3): 485 (1988 publ. 1989): Polycarena namaensis.
pubiflorum *O.M. Hilliard in Notes Roy. Bot. Gard. Edinburgh,* 45(3): 485 (1988 publ. 1989)—South Africa (Cape Province).
rustii *(Rolfe) O.M. Hilliard in Notes Roy. Bot. Gard. Edinburgh,* 45(3): 486 (1988 publ. 1989): Selago rustii.
tweedense *O.M. Hilliard in Notes Roy. Bot. Gard. Edinburgh,* 45(3): 486 (1988 publ. 1989)—South Africa (Cape Province).
viscidissimum *O.M. Hilliard in Notes Roy. Bot. Gard. Edinburgh,* 45(3): 486 (1988 publ. 1989)—South Africa (Cape Province).

PHYLLOSTACHYS (Gramin.).
carnea *G.H. Ye & Z.P. Wang in Acta Phytotax. Sin.,* 27(3): 228 (1989)—China.
nigrivagina *T.H. Wen in J. Bamboo Res.,* 8(1): 15 (1989)—China.
viridis
f. viridisulcata *P.X. Chang in J. Bamboo Res.,* 8(4): 40 (1989)—China.

PHYMASPERMUM (Compos.).
argenteum *F. Brusse in Bothalia,* 19(1): 29 (1989)—South Africa (Transvaal).

PHYTEUMA (Campanulac.).
spicatum
subsp. nigrum *(F.W. Schmidt) E.J. Weeda in Gorteria,* 15(1): 24 (1989): P. nigrum.

PICEA (Pinac.).
subgen. Casicta *(Mayr) P.A. Schmidt in Haussknechtia, Mitt. Thüring. Bot. Ges.,* 4: 38 (1988), isonym; *see E. Murray in Kalmia,* 14: 11 (1984): P. sect. Casicta.
sect. Pungentes *(Bobr.) P.A. Schmidt in Haussknechtia, Mitt. Thüring. Bot. Ges.,* 4: 38 (1988): P. ser. Pungentes.
abies
var. alpestris *(Bruegger) P.A. Schmidt in Haussknechtia, Mitt. Thüring. Bot. Ges.,* 4: 38 (1988): P. alpestris.
engelmannii
subsp. mexicana *(Martinez) P.A. Schmidt in Haussknechtia, Mitt. Thüring. Bot. Ges.,* 4: 38 (1988): P. mexicana.

PICEA :-
×fennica
nothosubsp. uralensis *(Tepl.) P.A. Schmidt in Haussknechtia, Mitt. Thüring. Bot. Ges.,* 4: 38 (1988): P. vulgaris var. uralensis.
glauca
subsp. albertiana *(S. Brown) P.A. Schmidt in Haussknechtia, Mitt. Thüring. Bot. Ges.,* 4: 38 (1988): P. albertiana.
jezoensis
subsp. hondoensis *(Mayr) P.A. Schmidt in Haussknechtia, Mitt. Thüring. Bot. Ges.,* 4: 38 (1988): P. hondoensis.
martinezii *T.F. Patterson in Sida,* 13(2): 131 (1988)—Mexico.

PICRASMA (Simaroubac.).
quassioides
var. dasycarpa *(Kitag.) S.Z. Liou in Fl. Liaoningica, chief ed. S.X. Li,* 1: 1054 (1988): P. ailanthoides var. dasycarpa.

PILEA (Urticac.).
funkikensis
var. rotundinucula *(Hayata) S.S. Ying, Coloured Illustr. Fl. Taiwan,* 3: 431 (1988): P. rotundinucula.
var. somai *(Hayata) S.S. Ying, Coloured Illustr. Fl. Taiwan,* 3: 432 (1988): P. somai.
gesnerioides *I.A. Grudzinskaya in Novosti Sist. Vyssh. Rast.,* 25: 80 (1988)—Cuba.
pedroi *I.A. Grudzinskaya in Novosti Sist. Vyssh. Rast.,* 25: 82 (1988)—Cuba.
plataniflora
var. taitoensis *(Hayata) S.S. Ying, Coloured Illustr. Fl. Taiwan,* 3: 430 (1988): P. taitoensis.

PILOSOCEREUS (Cactac.).
subgen. Floribunda *(Ritter) P.J. Braun in Bradleya,* 6: 90 (1988): gen. Floribunda.
subgen. Lagenopsis *(F. Buxbaum) P.J. Braun in Bradleya,* 6: 89 (1988): Coleocephalocereus subgen. Lagenopsis.
diersianus *(Esteves) P.J. Braun in Bradleya,* 6: 88 (1988): Pseudopilocereus diersianus.
parvus *(Diers & Esteves) P.J. Braun in Bradleya,* 6: 88 (1988): Pseudopilocereus parvus.
pleurocarpus *(Ritter) P.J. Braun in Bradleya,* 6: 88 (1988): Cipocereus pleurocarpus.
pusilliflorus *(Ritter) P.J. Braun in Bradleya,* 6: 90 (1988): Floribunda pusilliflora.
vilaboensis *(Diers & Esteves) P.J. Braun in Bradleya,* 6: 88 (1988): Pseudopilocereus vilaboensis.

PIMELEA (Thymelaeac.).
pelinos *B.L. Rye in Nuytsia,* 7(1): 59 (1989)—Australia (W. Australia).

PIMPINELLA (Umbellif.).
ahmarensis *D. Abebe in Kew Bull.,* 44(2): 346 (1989)—Ethiopia.
heywoodii *D. Abebe in Kew Bull.,* 44(2): 341 (1989)—Ethiopia.

PIMPINELLA :–
petrosa *D. Abebe in Kew Bull.*, 44(2): 343 (1989)—Ethiopia.
rhomboidea
var. tenuiloba *R.H. Shan & F.T. Pu in Acta Phytotax. Sin.*, 27(1): 63 (1989)—China.
schimperi *D. Abebe in Kew Bull.*, 44(2): 343 (1989)—Ethiopia.

PINANGA (Palmae).
bicolana *E.S. Fernando in Principes*, 32(4): 172 (1988)—Philippine Is.

PINGUICULA (Lentibulariac.).
crassifolia *S. Zamudio in Acta Bot. Mex.*, 3: 21 (1988)—Mexico.
gracilis *S. Zamudio in Acta Bot. Mex.*, 3: 25 (1988)—Mexico.
laueana *F. Speta & F. Fuchs in Phyton (Austria)*, 29(1): 94 (1989)—Mexico.
lutea
f. alba *G.W. Folkerts & J.D. Freeman in Castanea*, 54(1): 40 (1989)—U.S.A. (Florida).
potosiensis *F. Speta & F. Fuchs in Phyton (Austria)*, 29(1): 100 (1989)—Mexico.
rectifolia *F. Speta & F. Fuchs in Phyton (Austria)*, 29(1): 97 (1989)—Mexico.

PINUS (Pinac.).
subsect. Coxinoides *P. Landry in Phytologia*, 65(6): 479 (1989).
subsect. Gojae *P. Landry in Phytologia*, 65(6): 470 (1989).
subsect. Nat-clehianae *P. Landry in Phytologia*, 65(6): 471 (1989).
subsect. Nucifragae *P. Landry in Phytologia*, 65(6): 477 (1989).
subsect. Resinosae *W. Klaus in Pl. Syst. Evol.*, 162(1–4): 159 (1989).
subsect. Rzedowskiae *W. Klaus in Pl. Syst. Evol.*, 162(1–4): 142 (1989).
densiflora
f. nigricorticalis *Q.L. Wang in Fl. Liaoningica, chief ed. S.X. Li*, 1: 1361, 152 (1988)—China.
nigra
var. yaltirikiana *C.Ü. Alptekin in Istanbul Univ. Orman Fak. Dergisi, A*, 36(2): 147 (1986 publ. 1987)—Turkey.
×novaemexicana *P. Landry in Phytologia*, 65(6): 479 (1989), without type—Locality not stated. (P. flexilis × strobiformis).
×parapumila *Ishii ex P. Landry in Phytologia*, 65(6): 479 (1989), without type—Locality not stated. (P. cembra × pumila).
tabulaeformis
f. umbraculifera *(Liou & Wang) Q.L. Wang in Fl. Liaoningica, chief ed. S.X. Li*, 1: 156 (1988): P. tabulaeformis var. umbraculifera.

PIPER (Piperac.).
arboreum
subsp. tuberculatum *(Jacq.) M.C. Tebbs in Bull. Brit. Mus. (Nat. Hist.), Bot.*, 19: 156 (1989): P. tuberculatum.

PIPER :–
brevipedicellatum *A.J. Bornstein in J. Arnold Arbor.*, 70(1): 23 (1989)—Mexico.
cihuatlanense *A.J. Bornstein in J. Arnold Arbor.*, 70(1): 25 (1989)—Mexico.
colotlipanense *A.J. Bornstein in J. Arnold Arbor.*, 70(1): 27 (1989)—Mexico.
macvaughii *A.J. Bornstein in J. Arnold Arbor.*, 70(1): 30 (1989), as 'mcvaughii'—Mexico.
novogalicianum *A.J. Bornstein in J. Arnold Arbor.*, 70(1): 40 (1989)—Mexico.
truman-yunckeri *R. Callejas in Brittonia*, 41(3): 325 (1989)—Ecuador.

PIPTATHERUM (Gramin.).
holciforme
subsp. abyssinicum *H. Freitag in Davis & Hedge Festschrift (ed. Kit Tan)*: 117 (1989)—Ethiopia, Saudi Arabia, Yemen.

PIPTOCHAETIUM (Gramin.).
featherstonei *(Hitchcock) O. Tovar in Opusc. Bot. Pharm. Complutensis*, 4: 104 (1988): Stipa featherstonei.

PIRIQUETA (Turnerac.).
alba *J.R. Mattos in Loefgrenia*, 92: 1 (1988)—Brazil.
mortonii *S.D. Koch & P.A. Fryxell in Brittonia*, 41(2): 109 (1989)—Mexico.

PITCAIRNIA (Bromeliac.).
caduciflora *W. Rauh & E. Gross in Abh. Akad. Wiss. Lit. Mainz, Math.-Naturwiss. Klasse, Trop. Subtrop. Pflanzenwelt*, 65: 22 (1988)—Ecuador.
compostelae *R. McVaugh, Fl. Novo-Galiciana*, 15: 27 (1989)—Mexico.
piepenbringii *Rauh & Gross in Bromelie*, 1988(1): 1 (1988)—Brazil.
utcubambensis *W. Rauh in Abh. Akad. Wiss. Lit. Mainz, Math.-Naturwiss. Klasse, Trop. Subtrop. Pflanzenwelt*, 65: 33 (1988)—Peru.
violascens
var. bogneri *W. Rauh in Abh. Akad. Wiss. Lit. Mainz, Math.-Naturwiss. Klasse, Trop. Subtrop. Pflanzenwelt*, 65: 36 (1988)—Ecuador.

PITHECELLOBIUM (Legumin.).
paucipinnatum *(Schery) A.H. Gentry & Dodson in Ann. Missouri Bot. Gard.*, 75(4): 1438 (1988 publ. 1989): Albizia paucipinnata.

PITTOSPORUM (Pittosporac.).
daphniphylloides
var. adaphniphylloides *(Hu & Wang) W.T. Wang in Bull. Bot. Res. North-East. Forest. Inst.*, 9(1): 4 (1989): P. adaphniphylloides.

PLANTAGO (Plantaginac.).
afra
subsp. zwierleinii *(Nicotra) S. Brullo in Doc. Phytosoc., N.S.*, 9: 9 (1985), in reprint: P. psillium var. zwierleinii.

PLANTAGO :–
arachnoidea
var. lorata *J.Z. Liu in Acta Phytotax. Sin.*, 27(4): 300 (1989)—Xinjiang.
argentea
subsp. liburnica *V. Ravnik in Biol. Vestn.*, 36(3): 56 (1988)—Jugoslavia, Italy.
densiflora *J.Z. Liu in Acta Phytotax. Sin.*, 27(4): 298 (1989)—China.
pilosa *A.J. Cavanilles, Icon.*, 3: 25 (1795)—Spain.
subulata
subsp. humilis *(Guss.) W. Greuter & Burdet in Willdenowia,* 19(1): 39 (1989): P. humilis.

PLATANTHERA (Orchidac.).
hyperborea
f. alba *M.H.S. Light in Lindleyana,* 4(3): 158 (1989)—Canada (Quebec).

PLATANUS (Platanac.).
subgen. Castaneophyllum *J.F. Leroy in Compt. R. Séances Acad. Sci., Sér. 3 (Paris),* 295(3): 251, 254 (1982).

PLATYAECHMEA (Baker) L.B. Smith & W.J. Kress in Phytologia, 66(1): 73 (1989), without basionym ref.: Basionym not stated (Bromeliac.).
anomala *(L.B. Smith) L.B. Smith & W.J. Kress in Phytologia,* 66(1): 73 (1989), generic name not validly publ.: Aechmea anomala.
caesia *(E. Morren ex Baker) L.B. Smith & W.J. Kress in Phytologia,* 66(1): 73 (1989), generic name not validly publ.: Aechmea caesia.
chantinii *(Carrière) L.B. Smith & W.J. Kress in Phytologia,* 66(1): 73 (1989), generic name not validly publ.: Billbergia chantinii.
contracta *(Martius ex Schultes f.) L.B. Smith & W.J. Kress in Phytologia,* 66(1): 74 (1989), generic name not validly publ.: Billbergia contracta.
dealbata *(E. Morren ex Baker) L.B. Smith & W.J. Kress in Phytologia,* 66(1): 74 (1989), generic name not validly publ.: Aechmea dealbata.
dichlamidea *(Baker) L.B. Smith & W.J. Kress in Phytologia,* 66(1): 74 (1989), generic name not validly publ.: Aechmea dichlamidea.
var. pariaensis *(Pittendrigh) L.B. Smith & W.J. Kress in Phytologia,* 66(1): 74 (1989), generic name not validly publ.: Aechmea dichlamidea var. pariaensis.
var. trinitensis *(L.B. Smith) L.B. Smith & W.J. Kress in Phytologia,* 66(1): 74 (1989), generic name not validly publ.: Aechmea dichlamidea var. trinitensis.
distichantha *(Lemaire) L.B. Smith & W.J. Kress in Phytologia,* 66(1): 74 (1989), generic name not validly publ.: Aechmea distichantha.
f. albiflora *(L.B. Smith) L.B. Smith & W.J. Kress in Phytologia,* 66(1): 74 (1989), generic name not validly publ.: Aechmea distichantha f. albiflora.

PLATYAECHMEA :–
var. glaziovii *(Baker) L.B. Smith & W.J. Kress in Phytologia,* 66(1): 74 (1989), generic name not validly publ.: Aechmea glaziovii.
var. schlumbergeri *(E. Morren ex Mez) L.B. Smith & W.J. Kress in Phytologia,* 66(1): 74 (1989), generic name not validly publ.: Aechmea distichantha var. schlumbergeri.
var. vernicosa *(E. Pereira) L.B. Smith & W.J. Kress in Phytologia,* 66(1): 74 (1989), generic name not validly publ.: Aechmea distichantha var. vernicosa.
fasciata *(Lindley) L.B. Smith & W.J. Kress in Phytologia,* 66(1): 74 (1989), generic name not validly publ.: Billbergia fasciata.
var. flavivittata *(Reitz) L.B. Smith & W.J. Kress in Phytologia,* 66(1): 74 (1989), generic name not validly publ.: Aechmea fasciata var. flavi-vittata.
var. pruinosa *(Reitz) L.B. Smith & W.J. Kress in Phytologia,* 66(1): 74 (1989), generic name not validly publ.: Aechmea fasciata var. pruinosa.
var. purpurea *(Guillon) L.B. Smith & W.J. Kress in Phytologia,* 66(1): 74 (1989), generic name not validly publ.: Billbergia rhodocyanea var. purpurea.
flavorosea *(E. Pereira) L.B. Smith & W.J. Kress in Phytologia,* 66(1): 74 (1989), generic name not validly publ.: Aechmea flavo-rosea.
moorei *(Luther) L.B. Smith & W.J. Kress in Phytologia,* 66(1): 74 (1989), generic name not validly publ.: Aechmea moorei.
retusa *(L.B. Smith) L.B. Smith & W.J. Kress in Phytologia,* 66(1): 74 (1989), generic name not validly publ.: Aechmea retusa.
romeroi *(L.B. Smith) L.B. Smith & W.J. Kress in Phytologia,* 66(1): 74 (1989), generic name not validly publ.: Aechmea romeroi.
serrata *(L.) L.B. Smith & W.J. Kress in Phytologia,* 66(1): 75 (1989), generic name not validly publ.: Tillandsia serrata.
smithiorum *(Mez) L.B. Smith & W.J. Kress in Phytologia,* 66(1): 75 (1989), generic name not validly publ.: Aechmea smithiorum.
tessmannii *(Harms) L.B. Smith & W.J. Kress in Phytologia,* 66(1): 75 (1989), generic name not validly publ.: Aechmea tessmannii.
tillandsioides *(Martius ex Schultes f.) L.B. Smith & W.J. Kress in Phytologia,* 66(1): 75 (1989), generic name not validly publ.: Billbergia tillandsioides.
var. kienastii *(E. Morren ex Mez) L.B. Smith & W.J. Kress in Phytologia,* 66(1): 75 (1989), generic name not validly publ.: Aechmea kienastii.
wittmackiana *(Regel) L.B. Smith & W.J. Kress in Phytologia,* 66(1): 75 (1989), generic name not validly publ.: Quesnelia wittmackiana.
zebrina *(L.B. Smith) L.B. Smith & W.J. Kress in Phytologia,* 66(1): 75 (1989), generic name not validly publ.: Aechmea zebrina.

PLEUROTHALLIS (Orchidac.).
atrohiata *D.D. Dod in Moscosoa,* 5: 226 (1989)—Hispaniola (Haiti).

PLINIA (Myrtac.).
povedae *P.E. Sánchez-Vindas in Brenesia,* 25–26: 313 (1986 publ. 1988)—Costa Rica.

PLUCHEA (Compos.).
sect. Amplectifolium *G.L. Nesom in Phytologia,* 67(2): 162 (1989).
sect. Pterocaulis *G.L. Nesom in Phytologia,* 67(2): 164 (1989).
mexicana *(Godfrey) G.L. Nesom in Phytologia,* 67(2): 159 (1989): P. rosea var. mexicana.
yucatanensis *G.L. Nesom in Phytologia,* 67(2): 160 (1989)—Mexico, Belize, U.S.A. (Mississippi).

PLUMERIA (Apocynac.).
magna *T.A. Zanoni & M.M. Mejía P. in Moscosoa,* 5: 89 (1989)—Hispaniola (Dominican Republic).

POA (Gramin.).
angustifolia
var. angustiglumis *(Roshev.) V.N. Voroshilov in Byull. Glavn. Bot. Sada (Moscow),* 150: 56 (1988): P. angustiglumis.
eminens
var. alexeji *(Sofeikova & Worosch.) V.N. Voroshilov in Byull. Glavn. Bot. Sada (Moscow),* 150: 58 (1988): P. alexeji.
glauca
subsp. extremiorientalis *(Ohwi) V.N. Voroshilov in Byull. Glavn. Bot. Sada (Moscow),* 150: 57 (1988): P. extremiorientalis.
hedbergii *S.M. Phillips in Kew Bull.,* 44(1): 134 (1989)—Ethiopia.
macrocalyx
var. koriakiensis *V.N. Voroshilov in Byull. Glavn. Bot. Sada (Moscow),* 150: 56 (1988)—U.S.S.R. (Soviet Far East).
remota
subsp. raduliformis *(Probat.) V.N. Voroshilov in Byull. Glavn. Bot. Sada (Moscow),* 150: 56 (1988): P. raduliformis.

POCHOTA (Bombacac.).
amazonica *(Robyns) J.A. Steyermark & W.D. Stevens in Ann. Missouri Bot. Gard.,* 75(1): 397 (1988): Bombacopsis amazonica.
coriacea *(Martius & Zucc.) J.A. Steyermark & W.D. Stevens in Ann. Missouri Bot. Gard.,* 75(1): 397 (1988): Bombax coriaceum.
subsp. orinocensis *(Robyns) J.A. Steyermark & W.D. Stevens in Ann. Missouri Bot. Gard.,* 75(1): 397 (1988): Rhodognaphalopsis coriacea subsp. orinocensis.
cowanii *(Robyns) J.A. Steyermark & W.D. Stevens in Ann. Missouri Bot. Gard.,* 75(1): 397 (1988): Bombacopsis cowanii.
ewelii *J.A. Steyermark in Ann. Missouri Bot. Gard.,* 75(2): 1077 (1988)—Brazil.

POCHOTA :–
flaviflora *(Pulle) J.A. Steyermark & W.D. Stevens in Ann. Missouri Bot. Gard.,* 75(1): 397 (1988): Bombax flaviflorum.
fuscolepidota *J.A. Steyermark in Ann. Missouri Bot. Gard.,* 75(3): 1077 (1988)—Venezuela.
gracilis *(Robyns) J.A. Steyermark & W.D. Stevens in Ann. Missouri Bot. Gard.,* 75(1): 397 (1988): Rhodognaphalopsis gracilis.
subsp. bolivarensis *J.A. Steyermark in Ann. Missouri Bot. Gard.,* 75(3): 1078 (1988)—Venezuela.
humilis *(Spruce ex Decne.) J.A. Steyermark & W.D. Stevens in Ann. Missouri Bot. Gard.,* 75(1): 397 (1988): Pachira humilis.
liesneri *J.A. Steyermark in Ann. Missouri Bot. Gard.,* 75(3): 1079 (1988)—Venezuela.
maguirei *(Robyns) J.A. Steyermark & W.D. Stevens in Ann. Missouri Bot. Gard.,* 75(1): 397 (1988): Rhodognaphalopsis maguirei.
mawarinumae *J.A. Steyermark in Ann. Missouri Bot. Gard.,* 75(3): 1079 (1988)—Venezuela.
minor *(Sims) J.A. Steyermark & W.D. Stevens in Ann. Missouri Bot. Gard.,* 75(1): 397 (1988): Carolinea minor.
nitida *(Kunth) J.A. Steyermark & W.D. Stevens in Ann. Missouri Bot. Gard.,* 75(1): 397 (1988): Pachira nitida.
obovata *(Robyns) J.A. Steyermark & W.D. Stevens in Ann. Missouri Bot. Gard.,* 75(1): 397 (1988): Bombacopsis obovata.
orinocensis *(Robyns) J.A. Steyermark & W.D. Stevens in Ann. Missouri Bot. Gard.,* 75(1): 398 (1988): Bombacopsis orinocensis.
pseudamazonica *(Robyns) J.A. Steyermark & W.D. Stevens in Ann. Missouri Bot. Gard.,* 75(1): 398 (1988): Bombacopsis pseudamazonica.
pseudofaroensis *(Robyns) J.A. Steyermark & W.D. Stevens in Ann. Missouri Bot. Gard.,* 75(1): 398 (1988): Rhodognaphalopsis pseudofaroensis.
redmondii *J.A. Steyermark in Ann. Missouri Bot. Gard.,* 75(3): 1081 (1988)—Venezuela.
robynsii *J.A. Steyermark & W.D. Stevens in Ann. Missouri Bot. Gard.,* 75(4): 1586 (1988 publ. 1989), nom. nov.: Bombacopsis coriacea *Robyns.*
sordida *(R. Schultes) J.A. Steyermark & W.D. Stevens in Ann. Missouri Bot. Gard.,* 75(1): 398 (1988): Bombax sordidum.
tepuiensis *J.A. Steyermark in Ann. Missouri Bot. Gard.,* 75(3): 1081 (1988)—Venezuela.
trinitensis *(Urban) J.A. Steyermark & W.D. Stevens in Ann. Missouri Bot. Gard.,* 75(1): 398 (1988): Pachira trinitensis.
wurdackii *(Robyns) J.A. Steyermark & W.D. Stevens in Ann. Missouri Bot. Gard.,* 75(1): 398 (1988): Bombacopsis wurdackii.

PODAECHMEA (Mez) L.B. Smith & W.J. Kress in Phytologia, 66(1): 70 (1989), without basionym ref.: Basionym not stated (Bromeliac.).

PODAECHMEA :–
ferruginea *(L.B. Smith) L.B. Smith & W.J. Kress in Phytologia,* 66(1): 70 (1989), generic name not validly publ.: Aechmea ferruginea.
galeottii *(Baker) L.B. Smith & W.J. Kress in Phytologia,* 66(1): 70 (1989), generic name not validly publ.: Aechmea galeottii.
lueddemanniana *(K. Koch) L.B. Smith & W.J. Kress in Phytologia,* 66(1): 70 (1989), generic name not validly publ.: Pironneava lueddemanniana.
macvaughii *(L.B. Smith) L.B. Smith & W.J. Kress in Phytologia,* 66(1): 70 (1989), generic name not validly publ.: Aechmea macvaughii.
mexicana *(Baker) L.B. Smith & W.J. Kress in Phytologia,* 66(1): 70 (1989), generic name not validly publ.: Aechmea mexicana.

PODANDROGYNE (Capparac.).
colombiana *H.H. Iltis & T.S. Cochrane in Rev. Acad. Colomb. Cienc. Exact. Fis. Nat.,* 65: 266 (1989)—Colombia.
densiflora *(Benth.) H.H. Iltis & T.S. Cochrane in Rev. Acad. Colomb. Cienc. Exact. Fis. Nat.,* 65: 268 (1989): Gynandropsis densiflora.
f. pallens *(Tr. & Pl.) H.H. Iltis & T.S. Cochrane in Rev. Acad. Colomb. Cienc. Exact. Fis. Nat.,* 65: 268 (1989): Cleome densiflora var. β pallens.
trichopus *(Benth.) H.H. Iltis & T.S. Cochrane in Rev. Acad. Colomb. Cienc. Exact. Fis. Nat.,* 65: 270 (1989): Gynandropsis trichopus.

PODOCARPUS (Podocarpac.).
angustifolius
var. aristulatus *(Parl.) J. Staszkiewicz in Fragm. Flor. Geobot.,* 33(1–2): 77 (1988), without basionym ref.: P. aristulatus.
subsp. buchii *(Urb.) J. Staszkiewicz in Fragm. Flor. Geobot.,* 33(1–2): 77 (1988): P. buchii.
var. leonii *(Car.) J. Staszkiewicz in Fragm. Flor. Geobot.,* 33(1–2): 77 (1988), without basionym ref.: P. leonii.
f. victorinianus *(Car.) J. Staszkiewicz in Fragm. Flor. Geobot.,* 33(1–2): 77 (1988), without basionym ref.: P. victorinianus.
aracensis *D.J. De Laubenfels & J. Silba in Phytologia,* 65(5): 330 (1988)—Brazil.
caespitosus *(Brongn. & Gris) M.G. Baumann-Bodenheim, Syst. Fl. Neu-Caledonien,* 5: 76 (1989): P. alpinus var. caespitosus.
chinensis
var. wardii *D.J. De Laubenfels & J. Silba in Phytologia,* 65(5): 331 (1988)—Burma.
macrophyllus
var. piliramulus *Z.X. Chen & Z.Q. Li in Bull. Bot. Res. North-East. Forest. Inst.,* 9(3): 69 (1989)—China.

PODOTHECA (Compos.).
pritzelii *P.S. Short in Muelleria,* 7(1): 49 (1989)—Australia (W. Australia).

PODOTHECA :–
uniseta *P.S. Short in Muelleria,* 7(1): 51 (1989)—Australia (W. Australia).
wilsonii *P.S. Short in Muelleria,* 7(1): 46 (1989)—Australia (W. Australia).

POECILANTHE (Legumin.).
itapuana *G.P. Lewis in Kew Bull.,* 44(1): 167 (1989)—Brazil.

POIKILACANTHUS (Acanthac.).
capitatus *(Leonard) T.P. Ramamoorthy in Syst. Bot.,* 14(2): 151 (1989): Jacobinia capitata.

POLEMANNIOPSIS B.L. Burtt in Notes Roy. Bot. Gard. Edinburgh, 45(3): 498 (1988 publ. 1989). *UMBELLIFERAE.*
marlothii *(H. Wolff) B.L. Burtt in Notes Roy. Bot. Gard. Edinburgh,* 45(3): 499 (1988 publ. 1989): Polemannia (?) marlothii.

POLEMONIUM (Polemoniac.).
chinense
var. hirticaulum *G.H. Liu & Y.Q. Ma in Acta Sci. Nat. Univ. Intramongolicae,* 20(3): 392 (1989)—China.
pulcherrimum
subsp. lindleyi *(Wherry) V. Grant in Bot. Gaz.,* 150(2): 163 (1989): P. lindleyi.
sumushanense *G.H. Liu & Y.Q. Ma in Acta Sci. Nat. Univ. Intramongolicae,* 20(3): 392 (1989)—China.

POLHILLIA (Legumin.).
brevicalyx *(C.H. Stirton) B.E. van Wyk & A.L. Schutte in Kew Bull.,* 44(3): 420 (1989): Argyrolobium brevicalyx.
involucratum *(Thunb.) B.E. van Wyk & A.L. Schutte in Kew Bull.,* 44(3): 420 (1989): Psoralea involucrata.

POLIANTHES (Agavac.).
geminiflora
var. clivicola *R. McVaugh, Fl. Novo-Galiciana,* 250 (1989)—Mexico.
var. graminifolia *(Rose) R. McVaugh, Fl. Novo-Galiciana,* 15: 251 (1989): P. graminifolia.

POLYALTHIA (Annonac.).
ovalifolia *S.H. Rogstad in J. Arnold Arbor.,* 70(2): 214 (1989)—Borneo (Sarawak, Brunei).

POLYCARENA (Scrophulariac.).
aemulans *O.M. Hilliard in Notes Roy. Bot. Gard. Edinburgh,* 45(3): 486 (1988 publ. 1989)—South Africa (Cape Province).
batteniana *O.M. Hilliard in Notes Roy. Bot. Gard. Edinburgh,* 45(3): 486 (1988 publ. 1989)—South Africa (Cape Province).
comptonii *O.M. Hilliard in Notes Roy. Bot. Gard. Edinburgh,* 45(3): 487 (1988 publ. 1989)—South Africa (Cape Province).
exigua *O.M. Hilliard in Notes Roy. Bot. Gard. Edinburgh,* 45(3): 487 (1988 publ. 1989)—South Africa (Cape Province).

POLYCARENA :–
formosa *O.M. Hilliard in Notes Roy. Bot. Gard. Edinburgh,* 45(3): 487 (1988 publ. 1989)—South Africa (Cape Province).
gracilis *O.M. Hilliard in Notes Roy. Bot. Gard. Edinburgh,* 45(3): 487 (1988 publ. 1989)—South Africa (Cape Province).
lilacina *O.M. Hilliard in Notes Roy. Bot. Gard. Edinburgh,* 45(3): 488 (1988 publ. 1989)—South Africa (Cape Province).
var. difficilis *O.M. Hilliard in Notes Roy. Bot. Gard. Edinburgh,* 45(3): 488 (1988 publ. 1989)—South Africa (Cape Province).
nardouwensis *O.M. Hilliard in Notes Roy. Bot. Gard. Edinburgh,* 45(3): 488 (1988 publ. 1989)—South Africa (Cape Province).
subtilis *O.M. Hilliard in Notes Roy. Bot. Gard. Edinburgh,* 45(3): 488 (1988 publ. 1989)—South Africa (Cape Province).

POLYGALA (Polygalac.).
vulgaris
f. angustifolia *Tosco ex G. Ariello, G.G. Bellia & M.T. Della Beffa in Rea,* 11: 10 (1981–87 publ. 1988)—Italy.
f. latifolia *Tosco ex G. Ariello, G.G. Bellia & M.T. Della Beffa in Rea,* 11: 10 (1981–87 publ. 1988)—Italy.

POLYGONATUM (Convallariac.).
tsinlingense *Y.W. Tsui in Contrib. Inst. Bot. Nation. Acad. Peiping,* 6(3): 135 (1949)—China.

POLYGONUM (Polygonac.).
amplexicaulis *(D. Don) L.P. Ronse Decraene in Bot. J. Linn. Soc.,* 98(4): 368 (1988): Polygonum amplexicaule.
botuliforme *V. Mozaffarian in Iranian J. Bot.,* 4(1): 62 (1988)—Iran.
fusco-ochreatum
f. stans *(Kitag.) C.F. Fang in Fl. Liaoningica, chief ed. S.X. Li,* 1: 319 (1988), without basionym ref.: P. stans.
humifusum
f. yamatutae *(Kitag.) C.F. Fang in Fl. Liaoningica, chief ed. S.X. Li,* 1: 320 (1988): P. yamatutae.
papillosum *P. Hartvig in Willdenowia,* 19(1): 75 (1989)—Greece.
pronum *C.F. Fang in Fl. Liaoningica, chief ed. S.X. Li,* 1: 335 (1988), nom. nov.: P. procumbens *Chang & Li.*
psammophilum *(Bordz. ex Tzvel.) N.N. Tsvelev in Novosti Sist. Vyssh. Rast.,* 25: 188 (1988): P. novoascanicum subsp. psammophilum.
roseoviride
var. manshuricola *(Kitag.) C.F. Fang in Fl. Liaoningica, chief ed. S.X. Li,* 1: 334 (1988): Persicaria manshuricola.
samsunicum *Ş. Yildirimli & E. Leblebici in Willdenowia,* 19(1): 87 (1989)—Turkey.

POLYGONUM :–
sivasicum *Kit Tan & B. Yildiz in Notes Roy. Bot. Gard. Edinburgh,* 45(3): 439 (1988 publ. 1989)—Turkey.
volchovense *N.N. Tsvelev in Novosti Sist. Vyssh. Rast.,* 25: 187 (1988)—U.S.S.R. (European Russia).

POLYOSMA (Escalloniac.).
robusta *(Ridl.) L.G. Saw in Tree Fl. Malaya,* 4: 468 (1989): P. laete-virens var. robusta.

POLYSCIAS (Araliac.).
schmidii *P.P. Lowry in Bull. Mus. Nation. Hist. Nat., B, Adansonia,* Ser. 4, 11(2): 140 (1989)—Vanuatu.

POMADERRIS (Rhamnac.).
gilmourii *N.G. Walsh in Muelleria,* 7(1): 81 (1989)—Australia (New South Wales).
var. cana *N.G. Walsh in Muelleria,* 7(1): 82 (1989)—Australia (New South Wales).
humilis *N.G. Walsh in Muelleria,* 7(1): 84 (1989)—Australia (Victoria).

POPULUS (Salicac.).
nigra
var. grandifolia *N. Janjić in Glasn. Zem. Muz. Bosne Herceg. Prir. Nauke,* 23: 270 (1984), as 'comb. nov.'—Jugoslavia.
f. grandifolia *N. Janjić in Glasn. Zem. Muz. Bosne Herceg. Prir. Nauke,* 23: 271 (1984)—Jugoslavia.
var. narentana *N. Janjić in Glasn. Zem. Muz. Bosne Herceg. Prir. Nauke,* 23: 272 (1984), as 'comb. nov.'—Jugoslavia.
f. parvifolia *N. Janjić in Glasn. Zem. Muz. Bosne Herceg. Prir. Nauke,* 23: 272 (1984), as 'comb. nov.'—Jugoslavia.
f. pseudocaudina *N. Janjić in Glasn. Zem. Muz. Bosne Herceg. Prir. Nauke,* 23: 271 (1984)—Jugoslavia.
f. torulosa *N. Janjić in Glasn. Zem. Muz. Bosne Herceg. Prir. Nauke,* 23: 271 (1984), as 'nom. nov.'—Jugoslavia.
tomentosa
f. yixianensis *H.M. Jiang & J.X. Huang in Bull. Bot. Res. North-East. Forest. Inst.,* 9(3): 75 (1989)—China.
tremula
f. flagelliformis *N. Karhu in Mem. Soc. Fauna Flora Fenn.,* 65(1): 7 (1989)—Finland.
f. sallensis *L. Hämet-Ahti in Mem. Soc. Fauna Flora Fenn.,* 65(1): 7 (1989)—Finland.
×wettsteinii *L. Hämet-Ahti in Mem. Soc. Fauna Flora Fenn.,* 65(1): 8 (1989)—Cultivation. (P. tremula × tremuloides).

POROPHYLLUM (Compos.).
pygmaeum *D.J. Keil & J.D. Morefield in Syst. Bot.,* 14(4): 583 (1989)—U.S.A. (Nevada).

PORROGLOSSUM (Orchidac.).
dactylum *C.A. Luer, Ic. Pleurothallid. V (Monogr. Syst. Bot. Missouri Bot. Gard. 26):* 108 (1988)—Ecuador.
hystrix *C.A. Luer, Ic. Pleurothallid. V (Monogr. Syst. Bot. Missouri Bot. Gard. 26):* 110 (1988)—Ecuador.

PORTULACA (Portulacac.).
insignis *J.A. Steyermark in Ann. Missouri Bot. Gard.*, 75(3): 1058 (1988)—Venezuela.
pygmaea *J.A. Steyermark in Ann. Missouri Bot. Gard.*, 75(3): 1058 (1988)—Venezuela.

POTAMOGETON (Potamogetonac.).
xinganensis *Y.Q. Ma in Acta Sci. Nat. Univ. Intramongolicae,* 20(2): 281 (1989)—China.

POTENTILLA (Rosac.).
agrimonioides
subsp. hispanica *(Zimm.) J. Soják in Čas. Nár. Muz. (Prague),* 154(2): 74 (1985 publ. 1988): P. hispanica.
subsp. oreodoxa *(Soják) J. Soják in Čas. Nár. Muz. (Prague),* 154(2): 74 (1985 publ. 1988): P. oreodoxa.
bifurca
var. misera *J. Soják in Candollea,* 43(2): 451 (1988)—Xizang.
brassii
var. simplex *C. Kalkman in Blumea,* 34(1): 151 (1989)—New Guinea (W. Irian).
var. strigosa *C. Kalkman in Blumea,* 34(1): 151 (1989)—New Guinea (W. Irian).
crantzii
subsp. gelida *(C.A. Mey.) J. Soják in Čas. Nár. Muz. (Prague),* 154(2): 74 (1985 publ. 1988): P. gelida.
doubjonneana
subsp. ossetica *J. Soják in Candollea,* 43(2): 452 (1988)—U.S.S.R. (Caucasus).
gelida
subsp. borissii *(Ovcz. & Koczk.) J. Soják in Candollea,* 43(2): 453 (1988): P. borissii.
geranioides
subsp. syriaca *(Boiss.) J. Soják in Čas. Nár. Muz. (Prague),* 154(2): 74 (1985 publ. 1988): P. geranioides β. syriaca.
glabriuscula *(Yü & Li) J. Soják in Candollea,* 43(2): 453 (1988): Sibbaldia glabriuscula.
gorokana *C. Kalkman in Blumea,* 34(1): 155 (1989)—New Guinea (Papua New Guinea).
griffithii
subsp. beauvaisii *(Cardot) J. Soják in Candollea,* 43(2): 453 (1988): P. beauvaisii.
var. decurrens *J. Soják in Candollea,* 43(2): 452 (1988)—Assam.
var. intercedens *J. Soják in Candollea,* 43(2): 452 (1988)—Bhutan.
var. metallica *J. Soják in Candollea,* 43(2): 452 (1988)—Nepal.
var. reticulata *(Franchet) J. Soják in Candollea,* 43(2): 453 (1988): P. leschenaultiana var. reticulata.

POTENTILLA :–
subsp. yunnanensis *J. Soják in Candollea,* 43(2): 452 (1988)—China.
var. yunnanensis *J. Soják in Candollea,* 43(2): 452 (1988)—China.
indivisa *C. Kalkman in Blumea,* 34(1): 155 (1989)—New Guinea (W. Irian).
irianensis *C. Kalkman in Blumea,* 34(1): 156 (1989)—New Guinea (W. Irian).
mangenii *C. Kalkman in Blumea,* 34(1): 158 (1989)—New Guinea (W. Irian).
monanthes
var. alata *J. Soják in Candollea,* 43(2): 453 (1988)—Nepal.
var. violacea *J. Soják in Candollea,* 43(2): 452 (1988)—Bhutan.
multifida
subsp. bimundorum *(Soják) J. Soják in Čas. Nár. Muz. (Prague),* 154(2): 74 (1985 publ. 1988): P. bimundorum.
pulchella
subsp. gracilicaulis *(A.E. Porsild) B.A. Yurtsev in Byull. Mosk. Obshch. Ispȳt. Prir., Biol.,* 94(3): 86 (1989): P. pulchella var. gracilicaulis.
turczaninowiana
subsp. kuramensis *(Wolf) J. Soják in Candollea,* 43(2): 453 (1988): P. kuramensis.
subsp. nephogena *J. Soják in Candollea,* 43(2): 452 (1988)—U.S.S.R. (Middle Asia).

POTHUAVA (Bromeliac.).
aciculosa *(Mez & Sodiro) L.B. Smith & W.J. Kress in Phytologia,* 66(1): 75 (1989): Aechmea aciculosa.
allenii *(L.B. Smith) L.B. Smith & W.J. Kress in Phytologia,* 66(1): 75 (1989): Aechmea allenii.
alopecurus *(Mez) L.B. Smith & W.J. Kress in Phytologia,* 66(1): 75 (1989): Aechmea alopecurus.
bocainensis *(E. Pereira & Leme) L.B. Smith & W.J. Kress in Phytologia,* 66(1): 75 (1989): Aechmea bocainensis.
drakeana *(André) L.B. Smith & W.J. Kress in Phytologia,* 66(1): 75 (1989): Aechmea drakeana.
fraseri *(Baker) L.B. Smith & W.J. Kress in Phytologia,* 66(1): 75 (1989): Aechmea fraseri.
guaraparaiensis *(E. Pereira & Leme) L.B. Smith & W.J. Kress in Phytologia,* 66(1): 75 (1989): Aechmea guaraparaiensis.
involucrata *(André) L.B. Smith & W.J. Kress in Phytologia,* 66(1): 75 (1989): Aechmea involucrata.
kleinii *(Reitz) L.B. Smith & W.J. Kress in Phytologia,* 66(1): 75 (1989): Aechmea kleinii.
mariae-reginae *(H. Wendland) L.B. Smith & W.J. Kress in Phytologia,* 66(1): 76 (1989): Aechmea mariae-reginae.

POTHUAVA :–

nudicaulis

var. aequalis *(L.B. Smith & Reitz) L.B. Smith & W.J. Kress in Phytologia,* 66(1): 76 (1989): Aechmea nudicaulis var. aequalis.

var. aureorosea *(Antoine) L.B. Smith & W.J. Kress in Phytologia,* 66(1): 76 (1989): Hoplophytum aureo-roseum.

var. cuspidata *(Baker) L.B. Smith & W.J. Kress in Phytologia,* 66(1): 76 (1989): Aechmea nudicaulis var. cuspidata.

var. flavomarginata *(E. Pereira) L.B. Smith & W.J. Kress in Phytologia,* 66(1): 76 (1989): Aechmea nudicaulis var. flavomarginata.

var. simulans *(E. Pereira) L.B. Smith & W.J. Kress in Phytologia,* 66(1): 76 (1989): Aechmea nudicaulis var. simulans.

var. tabuleirensis *(Reitz) L.B. Smith & W.J. Kress in Phytologia,* 66(1): 76 (1989): Aechmea nudicaulis var. tabuleirensis.

ornata *(Baker) L.B. Smith & W.J. Kress in Phytologia,* 66(1): 76 (1989): Aechmea ornata.

var. hoehneana *(L.B. Smith) L.B. Smith & W.J. Kress in Phytologia,* 66(1): 76 (1989): Aechmea ornata var. hoehneana.

var. nationalis *(Reitz) L.B. Smith & W.J. Kress in Phytologia,* 66(1): 76 (1989): Aechmea ornata var. nationalis.

pectinata *(Baker) L.B. Smith & W.J. Kress in Phytologia,* 66(1): 76 (1989): Aechmea pectinata.

pineliana *(Brongniart ex Planchon) L.B. Smith & W.J. Kress in Phytologia,* 66(1): 76 (1989): Echinostachys pineliana.

var. minuta *(M.B. Foster) L.B. Smith & W.J. Kress in Phytologia,* 66(1): 76 (1989): Aechmea pineliana var. minuta.

roberto-anselmii *(E. Pereira & Leme) L.B. Smith & W.J. Kress in Phytologia,* 66(1): 76 (1989): Aechmea roberto-anselmii.

squarrosa *(Baker) L.B. Smith & W.J. Kress in Phytologia,* 66(1): 76 (1989): Aechmea squarrosa.

subpetiolata *(L.B. Smith) L.B. Smith & W.J. Kress in Phytologia,* 66(1): 76 (1989): Aechmea subpetiolata.

tonduzii *(Mez & Pittier ex Mez) L.B. Smith & W.J. Kress in Phytologia,* 66(1): 76 (1989): Aechmea tonduzii.

triticina *(Mez) L.B. Smith & W.J. Kress in Phytologia,* 66(1): 77 (1989): Aechmea triticina.

var. capensis *(L.B. Smith) L.B. Smith & W.J. Kress in Phytologia,* 66(1): 77 (1989): Aechmea triticina var. capensis.

vanhouteana *(Van Houte) L.B. Smith & W.J. Kress in Phytologia,* 66(1): 77 (1989): Echinostachys vanhouteana.

weberbaueri *(Harms) L.B. Smith & W.J. Kress in Phytologia,* 66(1): 77 (1989): Aechmea weberbaueri.

POUZOLZIA (Urticac.).

matsudae *(Yamamoto) S.S. Ying, Coloured Illustr. Fl. Taiwan,* 3: 479 (1988), as 'matsudai': Memorialis matsudai.

neurocarpa *(Yamamoto) S.S. Ying, Coloured Illustr. Fl. Taiwan,* 3: 481 (1988): Memorialis neurocarpus.

pentandra

var. akoensis *(Yamamoto) S.S. Ying, Coloured Illustr. Fl. Taiwan,* 3: 477 (1988): Memorialis pentandra var. akoensis.

var. hypericifolia *(Blume) S.S. Ying, Coloured Illustr. Fl. Taiwan,* 3: 476 (1988): P. hypericifolia.

PRANGOS (Umbellif.).

herderi

subsp. xinjiangensis *X.Y. Chen & Q.X. Liu in Bull. Bot. Res. North-East. Forest. Inst.,* 9(3): 99 (1989)—Xinjiang.

PRASOPHYLLUM (Orchidac.).

calcicola *R. Bates in J. Adelaide Bot. Gard.,* 11(2): 179 (1989)—Australia (S. Australia, W. Australia).

fecundum *R. Bates in J. Adelaide Bot. Gard.,* 11(2): 189 (1989)—Australia (S. Australia).

occultans *R. Bates in J. Adelaide Bot. Gard.,* 11(2): 192 (1989)—Australia (S. Australia).

ringens *(Reichb. f.) R. Bates in J. Adelaide Bot. Gard.,* 11(2): 182 (1989): Prasophyllum gracile var. ringens.

PRAXELIS (Compos.).

sanctopaulensis *(B.L. Robinson) R.M. King & H. Robinson in Phytologia,* 65(6): 488 (1989): Eupatorium sanctopaulense.

PRESTOEA (Palmae).

brachyclada *(Burret) R.G. Bernal, G. Galeano-Garcés & A. Henderson in Taxon,* 38(1): 102 (1989): Euterpe brachyclada.

dasystachys *(Burret) R.G. Bernal, G. Galeano-Garcés & A. Henderson in Taxon,* 38(1): 102 (1989): Euterpe dasystachys.

integrifolia *G. de Nevers & A.J. Henderson in Ann. Missouri Bot. Gard.,* 75(1): 208 (1988)—Panama.

latisecta *(Burret) R.G. Bernal, G. Galeano-Garcés & A. Henderson in Taxon,* 38(1): 102 (1989): Euterpe latisecta.

semispicata *G. de Nevers & A.J. Henderson in Ann. Missouri Bot. Gard.,* 75(1): 213 (1988)—Panama.

simplicifrons *(Burret) A.J. Henderson & G. de Nevers in Ann. Missouri Bot. Gard.,* 75(1): 210 (1988): Euterpe simplicifrons.

PRIMULA (Primulac.).

acaulis

subsp. atlantica *(Maire & Wilczek) W. Greuter & Burdet in Willdenowia,* 19(1): 42 (1989): P. vulgaris var. atlantica.

PRIMULA :–
subsp. balearica *(Willk.) W. Greuter & Burdet in Willdenowia,* 19(1): 42 (1989): P. vulgaris var. balearica.
subsp. rubra *(Sm.) W. Greuter & Burdet in Willdenowia,* 19(1): 42 (1989): P. acaulis var. rubra.

PRISMATOMERIS (Rubiac.).
connata *Y.Z. Ruan in Acta Phytotax. Sin.,* 26(6): 447 (1988)—China.
subsp. hainanensis *Y.Z. Ruan in Acta Phytotax. Sin.,* 26(6): 448 (1988)—Hainan.
tetrandra
subsp. multiflora *(Ridley) Y.Z. Ruan in Acta Phytotax. Sin.,* 26(6): 447 (1988): P. multiflora.

PRISTIGLOTTIS (Orchidac.).
degeneri *(L.O. Williams) P.J. Kores in Allertonia,* 5(1): 29 (1989): Odontochilus degeneri.
longiflora *(Reichenb. f.) P.J. Kores in Allertonia,* 5(1): 28 (1989): Anecochilus longiflorus.

PRITCHARDIA (Palmae).
waialealeana *R.W. Read in Principes,* 32(4): 135 (1988)—Hawaiian Is.

PROCKIA (Tiliac.).
pentamera *A.H. Gentry in Ann. Missouri Bot. Gard.,* 75(4): 1430 (1988 publ. 1989)—Peru, Ecuador.

PROSTANTHERA (Labiat.).
albiflora *B.J. Conn in Nuytsia,* 6(3): 401 (1988)—Australia (W. Australia).
althoferi *B.J. Conn in Nuytsia,* 6(3): 374 (1988)—Australia (W. Australia, S. Australia, N. Terr.).
subsp. longifolia *B.J. Conn in Nuytsia,* 6(3): 376 (1988)—Australia (S. Australia, N. Terr.).
ammophila *B.J. Conn in Nuytsia,* 6(3): 380 (1988)—Australia (S. Australia).
centralis *B.J. Conn in Nuytsia,* 6(3): 382 (1988)—Australia (N. Terr., W. Australia).
nanophylla *B.J. Conn in Nuytsia,* 6(3): 394 (1988)—Australia (W. Australia).
petrophila *B.J. Conn in Nuytsia,* 6(3): 391 (1988)—Australia (W. Australia).
sericea *(J.M. Black) B.J. Conn in Nuytsia,* 6(3): 364 (1988): P. baxteri var. sericea.
splendens *B.J. Conn in Nuytsia,* 6(3): 388 (1988)—Australia (W. Australia).
verticillaris *B.J. Conn in Nuytsia,* 6(3): 355 (1988)—Australia (W. Australia).

PROTASPARAGUS (Asparagac.).
humilis *(Engl.) B. Mathew in Kew Bull.,* 44(1): 181 (1989): Asparagus humilis.
longicladus *(N.E. Br.) B. Mathew in Kew Bull.,* 44(1): 181 (1989): Asparagus longicladus.

PROTIUM (Burserac.).
nitidifolium *(Cuatrec.) D.C. Daly in Brittonia,* 41(1): 25 (1989): Paraprotium nitidifolium.
pilosum *(Cuatrec.) D.C. Daly in Brittonia,* 41(1): 24 (1989): Tetragastris pilosa.
vestitum *(Cuatrec.) D.C. Daly in Brittonia,* 41(1): 24 (1989): Paraprotium vestitum.

PRUNUS (Rosac.).
cercocarpifolia *J.A. Villarreal Q. in Sida,* 13(3): 273 (1989)—Mexico.
fortunensis *G. McPherson in Ann. Missouri Bot. Gard.,* 75(1): 373 (1988)—Panama.
gentryi
f. flavipulpa *J.E. Laferrière in Phytologia,* 65(6): 482 (1989)—Mexico.
salicina
var. cordata *Y. He & J.Y. Zhang in Bull. Bot. Res. North-East. Forest. Inst.,* 9(3): 71 (1989)—China.
triloba
f. petzoldii *(Koch) Q.L. Wang in Fl. Liaoningica, chief ed. S.X. Li,* 1: 829 (1988): P. petzoldii.

PSEPHELLUS (Compos.).
ser. Absinthifolii *A.A. Alieva in A.I. Galushko (ed.), Fl. Severn. Kavkaza i Voprosÿ ee istorii:* 44 (1983).
ser. Andini *A.A. Alieva in A.I. Galushko (ed.), Fl. Severn. Kavkaza i Voprosÿ ee istorii:* 33 (1983).
ser. Boissierani *A.A. Alieva in A.I. Galushko (ed.), Fl. Severn. Kavkaza i Voprosÿ ee istorii:* 28 (1983).
ser. Daghestanici *A.A. Alieva in A.I. Galushko (ed.), Fl. Severn. Kavkaza i Voprosÿ ee istorii:* 38 (1983).
ser. Dealbati *A.A. Alieva in A.I. Galushko (ed.), Fl. Severn. Kavkaza i Voprosÿ ee istorii:* 40 (1983).
ser. Galushkoani *A.A. Alieva in A.I. Galushko (ed.), Fl. Severn. Kavkaza i Voprosÿ ee istorii:* 31 (1983).
ser. Paucilobi *A.A. Alieva in A.I. Galushko (ed.), Fl. Severn. Kavkaza i Voprosÿ ee istorii:* 47 (1983).
ser. Prokhanoviani *A.A. Alieva in A.I. Galushko (ed.), Fl. Severn. Kavkaza i Voprosÿ ee istorii:* 35 (1983).
ser. Salviaefolii *A.A. Alieva in A.I. Galushko (ed.), Fl. Severn. Kavkaza i Voprosÿ ee istorii:* 48 (1983).
ser. Schistosi *A.A. Alieva in A.I. Galushko (ed.), Fl. Severn. Kavkaza i Voprosÿ ee istorii:* 24 (1983).
circassicus *(Albov) A.I. Galushko, Fl. Severn. Kavkaza,* 3: 229 (1980): P. hypoleucus var. circassicus.
hymenolepis
var. subintegra *(Trautv.) A.A. Alieva in A.I. Galushko (ed.), Fl. Severn. Kavkaza i Voprosÿ ee istorii:* 28 (1983): Centaurea hymenolepis var. subintegra.

PSEPHELLUS :-
schistosus *(Sosn.) A.A. Alieva in A.I. Galushko (ed.), Fl. Severn. Kavkaza i Voprosy̆ ee istorii:* 24 (1983): Centaurea schistosa.

PSEUDERANTHEMUM (Acanthac.).
lapathifolium *(Vahl) B. Hansen in Nordic J. Bot.*, 9(2): 214 (1989): Justicia lapathifolia.
latifolium *(Vahl) B. Hansen in Nordic J. Bot.*, 9(2): 213 (1989): Justicia latifolia.

PSEUDOJACOBAEA (Hook. f.) R. Mathur in R.R. Rao et al., Fl. Ind. Enumerat. - Asterac.: 59 (1988): Senecio sect. Pseudojacobaea (Compos.).
lavandulaefolius *(DC.) R. Mathur in R.R. Rao et al., Fl. Ind. Enumerat. - Asterac.:* 59 (1988): Senecio lavandulaefolius.

PSEUDOLYSIMACHION (Scrophulariac.).
spicatum
subsp. transcaucasicum *(Bordz.) É.Ts. Gabriélyan in Fl. Armenii,* 8: 259 (1987): Veronica spicata subsp. transcaucasica.

PSEUDOROEGNERIA (Gramin.).
libanotica *(Hackel) D.R. Dewey in Gene Manipulat. in Pl. Improv. (ed. J.P. Gustafson):* 272 (1984): Agropyron libanoticum.

PSEUDOSASA (Gramin.).
altiligulata *T.H. Wen in J. Bamboo Res.*, 8(1): 18 (1989)—China.

PSEUDOTRIMEZIA (Iridac.).
cathartica *(Kl.) P. Ravenna in Onira, Bot. Leafl.*, 1(2): 23 (1988): Lansbergia cathartica.
concava *P. Ravenna in Onira, Bot. Leafl.*, 1(7): 49 (1988)—Brazil.
subsp. clinostyla *P. Ravenna in Onira, Bot. Leafl.*, 1(7): 50 (1988)—Brazil.
elegans *P. Ravenna in Onira, Bot. Leafl.*, 1(7): 48 (1988)—Brazil.
planifolia *P. Ravenna in Onira, Bot. Leafl.*, 1(7): 51 (1988)—Brazil.
setacea *P. Ravenna in Onira, Bot. Leafl.*, 1(2): 22 (1988)—Brazil.
sublateralis *P. Ravenna in Onira, Bot. Leafl.*, 1(2): 22 (1988)—Brazil.
subtilis *P. Ravenna in Onira, Bot. Leafl.*, 1(7): 51 (1988), nom. nov.: Lansbergia setacea *Klatt.*
xyridea *P. Ravenna in Onira, Bot. Leafl.*, 1(7): 50 (1988)—Brazil.

PSEUDOXYTENANTHERA T.R. Soderstrom & R.P. Ellis in Smithsonian Contrib. Bot., 72: 52 (1988). *GRAMINEAE.*
monadelpha *(Thwaites) T.R. Soderstrom & R.P. Ellis in Smithsonian Contrib. Bot.*, 72: 52 (1988): Dendrocalamus monadelphus.

PSIDIUM (Myrtac.).
acutangulum
var. oblongata *J.R. Mattos in Loefgrenia,* 94: 12 (1989)—Brazil.
canum *J.R. Mattos in Loefgrenia,* 94: 11 (1989)—Brazil.
hagelundianum *J.R. Mattos in Loefgrenia,* 94: 11 (1989)—Brazil.
subrostrifolium *J.R. Mattos in Loefgrenia,* 90: 3 (1986), as 'sub-rostrifolium'—Brazil.
turbinatum *J.R. Mattos in Loefgrenia,* 94: 12 (1989)—Brazil.

PSILOSTROPHE (Compos.).
gnaphalodes
var. mexicana *(R.C. Brown) B.L. Turner in Phytologia,* 65(3): 234 (1988): P. mexicana.
tagetina
var. cerifera *(A. Nels.) B.L. Turner in Phytologia,* 65(3): 234 (1988): P. cerifera var. cerifera.

PSYCHOTRIA (Rubiac.).
amboniana
subsp. mosambicensis *(Petit) B. Verdcourt in Fl. Zambes.*, 5(1): 24 (1989): P. albidocalyx var. mosambicensis.
anartiothrix *J.A. Steyermark in Ann. Missouri Bot. Gard.*, 75(1): 340 (1988)—Venezuela.
aquatilis
var. divaricatus *S.H. Sohmer in Bishop Mus. Bull. Bot.*, 1: 32 (1988)—New Guinea (Papua New Guinea).
archboldii *S.H. Sohmer in Bishop Mus. Bull. Bot.*, 1: 34 (1988)—Louisiade Archipelago, New Guinea (Papua New Guinea), New Britain.
var. multinervia *S.H. Sohmer in Bishop Mus. Bull. Bot.*, 1: 37 (1988)—New Guinea (Papua New Guinea).
asekiensis *S.H. Sohmer in Bishop Mus. Bull. Bot.*, 1: 37 (1988)—New Guinea (Papua New Guinea).
beaufortiensis *Valeton ex S.H. Sohmer in Bishop Mus. Bull. Bot.*, 1: 44 (1988)—New Guinea (W. Irian).
boloboensis
var. balimensis *(Merrill & Perry) S.H. Sohmer in Bishop Mus. Bull. Bot.*, 1: 49 (1988): P. balimensis.
breweri *J. Steyermark in Acta Bot. Venezuel.*, 14(3): 25 (1984)—Venezuela.
butibumensis *S.H. Sohmer in Bishop Mus. Bull. Bot.*, 1: 54 (1988)—New Guinea (Papua New Guinea).
capensis
var. puberula *(Petit) B. Verdcourt in Fl. Zambes.*, 5(1): 14 (1989): P. riparia var. puberula.
subsp. riparia *(K. Schum. & K. Krause) B. Verdcourt in Fl. Zambes.*, 5(1): 13 (1989): Grumilea riparia.

PSYCHOTRIA :-

chonantha *(Gilli) S.H. Sohmer in Bishop Mus. Bull. Bot.,* 1: 57 (1988): Caelospermum chonanthum.

chrysanthoides *S.H. Sohmer in Bishop Mus. Bull. Bot.,* 1: 63 (1988)—New Guinea (Papua New Guinea, W. Irian).

conglobatioides *S.H. Sohmer in Bishop Mus. Bull. Bot.,* 1: 67 (1988)—New Guinea (W. Irian).

crassipedunculata *S.H. Sohmer in Bishop Mus. Bull. Bot.,* 1: 69 (1988)—New Britain, New Guinea (Papua New Guinea).

crassiramula *S.H. Sohmer in Bishop Mus. Bull. Bot.,* 1: 71 (1988)—New Guinea (Papua New Guinea).

croftiana *S.H. Sohmer in Bishop Mus. Bull. Bot.,* 1: 73 (1988)—New Guinea (Papua New Guinea).

damasiana *S.H. Sohmer in Bishop Mus. Bull. Bot.,* 1: 77 (1988)—New Britain, New Guinea (Papua New Guinea), Louisiade Archipelago.

diplococca

var. tauriensis *S.H. Sohmer in Bishop Mus. Bull. Bot.,* 1: 88 (1988)—New Guinea (Papua New Guinea).

dipteropodioides *S.H. Sohmer in Bishop Mus. Bull. Bot.,* 1: 92 (1988)—New Guinea (Papua New Guinea).

diversinodula *(Verdc.) B. Verdcourt in Fl. Zambes.,* 5(1): 29 (1989): P. kirkii var. diversinodula.

edaphothrix *J.A. Steyermark in Ann. Missouri Bot. Gard.,* 75(1): 342 (1988)—Venezuela.

flaviramula *S.H. Sohmer in Bishop Mus. Bull. Bot.,* 1: 103 (1988)—New Guinea (W. Irian).

foremanii *S.H. Sohmer in Bishop Mus. Bull. Bot.,* 1: 105 (1988)—New Guinea (Papua New Guinea).

frodinii *S.H. Sohmer in Bishop Mus. Bull. Bot.,* 1: 107 (1988)—New Guinea (Papua New Guinea).

galorei *S.H. Sohmer in Bishop Mus. Bull. Bot.,* 1: 109 (1988)—New Britain, Louisiade Archipelago.

giluwensis *S.H. Sohmer in Bishop Mus. Bull. Bot.,* 1: 112 (1988)—New Guinea (Papua New Guinea).

goodenoughiensis *S.H. Sohmer in Bishop Mus. Bull. Bot.,* 1: 116 (1988)—Louisiade Archipelago.

haumugaensis *S.H. Sohmer in Bishop Mus. Bull. Bot.,* 1: 118 (1988)—New Guinea (Papua New Guinea).

hellwigiensis *Valeton ex S.H. Sohmer in Bishop Mus. Bull. Bot.,* 1: 313 (1988)—New Guinea (W. Irian).

hentyi *S.H. Sohmer in Bishop Mus. Bull. Bot.,* 1: 122 (1988)—New Guinea (Papua New Guinea, W. Irian).

hollandiae

var. pioraensis *S.H. Sohmer in Bishop Mus. Bull. Bot.,* 1: 129 (1988)—New Guinea (Papua New Guinea, W. Irian).

PSYCHOTRIA :-

var. whitei *(Moore) S.H. Sohmer in Bishop Mus. Bull. Bot.,* 1: 132 (1988): P. whitei.

johnsii *S.H. Sohmer in Bishop Mus. Bull. Bot.,* 1: 135 (1988)—New Guinea (Papua New Guinea).

kairoana *S.H. Sohmer in Bishop Mus. Bull. Bot.,* 1: 137 (1988)—New Guinea (Papua New Guinea).

karemaensis *S.H. Sohmer in Bishop Mus. Bull. Bot.,* 1: 143 (1988)—New Guinea (Papua New Guinea).

katikii *S.H. Sohmer in Bishop Mus. Bull. Bot.,* 1: 145 (1988)—New Guinea (Papua New Guinea, W. Irian).

leleana *S.H. Sohmer in Bishop Mus. Bull. Bot.,* 1: 152 (1988)—New Britain, Admiralty Is.

leleanoides *S.H. Sohmer in Bishop Mus. Bull. Bot.,* 1: 154 (1988)—New Britain, British Solomon Is., Bougainville Is.

leptothyrsa

var. friabilis *S.H. Sohmer in Bishop Mus. Bull. Bot.,* 1: 163 (1988)—New Guinea (Papua New Guinea).

var. multifurca *(Valeton) S.H. Sohmer in Bishop Mus. Bull. Bot.,* 1: 165 (1988): P. multifurca.

longipaniculata *S.H. Sohmer in Bishop Mus. Bull. Bot.,* 1: 171 (1988)—New Guinea (Papua New Guinea).

magnasepala *S.H. Sohmer in Bishop Mus. Bull. Bot.,* 1: 179 (1988)—New Guinea (Papua New Guinea).

marafungaensis *S.H. Sohmer in Bishop Mus. Bull. Bot.,* 1: 181 (1988)—New Guinea (Papua New Guinea).

merrilliana *S.H. Sohmer in Bishop Mus. Bull. Bot.,* 1: 188 (1988), nom. nov.: P. trichostoma var. macrophylla *Merrill & Perry.*

misolensis *Valeton ex S.H. Sohmer in Bishop Mus. Bull. Bot.,* 1: 318 (1988)—New Guinea (W. Irian).

monopedicellata *S.H. Sohmer in Bishop Mus. Bull. Bot.,* 1: 200 (1988)—New Britain.

morobensis *S.H. Sohmer in Bishop Mus. Bull. Bot.,* 1: 203 (1988)—New Guinea (Papua New Guinea).

multipedunculata *S.H. Sohmer in Bishop Mus. Bull. Bot.,* 1: 209 (1988)—New Britain, New Ireland, Louisiade Archipelago.

mur-murensis *S.H. Sohmer in Bishop Mus. Bull. Bot.,* 1: 211 (1988)—New Guinea (Papua New Guinea).

mwinilungae *B. Verdcourt in Fl. Zambes.,* 5(1): 23 (1989)—Zambia.

nanifrutex *S.H. Sohmer in Bishop Mus. Bull. Bot.,* 1: 216 (1988)—New Guinea (Papua New Guinea).

nigra

var. peninsularis *(Hook. f.) D.B. Deb & M.G. Gangopadhyay in J. Econ. Taxon. Bot., Addit. Ser.* 7: 51 (1989): P. thwaitesii var. peninsularis.

novohiberiensis *S.H. Sohmer in Bishop Mus. Bull. Bot.,* 1: 219 (1988)—New Ireland.

PSYCHOTRIA :-
pavairiensis *S.H. Sohmer in Bishop Mus. Bull. Bot.*, 1: 223 (1988)—Bougainville Is.
pectinata *J.A. Steyermark in Ann. Missouri Bot. Gard.*, 75(1): 344 (1988)—Venezuela.
peduncularis
var. angustibracteata *B. Verdcourt in Fl. Zambes.*, 5(1): 22 (1989)—Zimbabwe, Mozambique.
var. rufonyassana *B. Verdcourt in Fl. Zambes.*, 5(1): 22 (1989)—Zambia, Malawi, Tanzania.
phaeochlamysioides *S.H. Sohmer in Bishop Mus. Bull. Bot.*, 1: 231 (1988)—Louisiade Archipelago, New Guinea (Papua New Guinea).
praecox *Valeton ex S.H. Sohmer in Bishop Mus. Bull. Bot.*, 1: 326 (1988)—New Guinea (W. Irian).
pulleniana *S.H. Sohmer in Bishop Mus. Bull. Bot.*, 1: 233 (1988)—New Guinea (W. Irian, Papua New Guinea).
pumila
var. subumbellata *(Petit) B. Verdcourt in Fl. Zambes.*, 5(1): 27 (1989): P. albidocalyx var. subumbellata.
purariensis *S.H. Sohmer in Bishop Mus. Bull. Bot.*, 1: 235 (1988)—New Guinea (Papua New Guinea).
ramadecumbens *S.H. Sohmer in Bishop Mus. Bull. Bot.*, 1: 238 (1988)—New Guinea (Papua New Guinea).
ramuensis *S.H. Sohmer in Bishop Mus. Bull. Bot.*, 1: 240 (1988)—New Guinea (Papua New Guinea).
var. pubescentia *S.H. Sohmer in Bishop Mus. Bull. Bot.*, 1: 243 (1988)—New Guinea (Papua New Guinea).
randiana
var. misimensis *(Merrill & Perry) S.H. Sohmer in Bishop Mus. Bull. Bot.*, 1: 252 (1988): P. misimensis.
var. tafaensis *(Merrill & Perry) S.H. Sohmer in Bishop Mus. Bull. Bot.*, 1: 250 (1988): P. tafaensis.
reflexapedunculata *S.H. Sohmer in Bishop Mus. Bull. Bot.*, 1: 252 (1988)—Louisiade Archipelago.
rosseliensis *S.H. Sohmer in Bishop Mus. Bull. Bot.*, 1: 256 (1988)—Louisiade Archipelago.
saidoriensis *S.H. Sohmer in Bishop Mus. Bull. Bot.*, 1: 260 (1988)—New Guinea (Papua New Guinea), New Britain.
sarmiensis *S.H. Sohmer in Bishop Mus. Bull. Bot.*, 1: 260 (1988)—New Guinea (W. Irian).
steinii *J.A. Steyermark in Ann. Missouri Bot. Gard.*, 75(1): 345 (1988)—Venezuela.
stevensiana *S.H. Sohmer in Bishop Mus. Bull. Bot.*, 1: 271 (1988)—New Britain.
streimannii *S.H. Sohmer in Bishop Mus. Bull. Bot.*, 1: 273 (1988)—New Guinea (Papua New Guinea).
talasensis *S.H. Sohmer in Bishop Mus. Bull. Bot.*, 1: 275 (1988)—New Britain, New Guinea (Papua New Guinea).

PSYCHOTRIA :-
testacea *S.H. Sohmer in Bishop Mus. Bull. Bot.*, 1: 280 (1988)—New Guinea (W. Irian, Papua New Guinea).
thesceloantha *J.A. Steyermark in Ann. Missouri Bot. Gard.*, 75(1): 347 (1988)—Venezuela.
tripedunculata *S.H. Sohmer in Bishop Mus. Bull. Bot.*, 1: 286 (1988)—New Guinea (W. Irian).
vaccinioides
var. kostermansii *S.H. Sohmer in Bishop Mus. Bull. Bot.*, 1: 290 (1988)—New Guinea (W. Irian).
vaccinioidifolia *S.H. Sohmer in Bishop Mus. Bull. Bot.*, 1: 290 (1988)—New Guinea (W. Irian).
valetoniana *S.H. Sohmer in Bishop Mus. Bull. Bot.*, 1: 293 (1988)—New Guinea (Papua New Guinea, W. Irian), New Britain, New Ireland.
vanimoensis *S.H. Sohmer in Bishop Mus. Bull. Bot.*, 1: 296 (1988)—New Guinea (Papua New Guinea).
vinkii *S.H. Sohmer in Bishop Mus. Bull. Bot.*, 1: 300 (1988)—New Guinea (W. Irian).
waiuensis *S.H. Sohmer in Bishop Mus. Bull. Bot.*, 1: 303 (1988)—New Guinea (Papua New Guinea).
womersleyi *S.H. Sohmer in Bishop Mus. Bull. Bot.*, 1: 307 (1988)—New Guinea (Papua New Guinea).
yapaensis *S.H. Sohmer in Bishop Mus. Bull. Bot.*, 1: 309 (1988)—New Guinea (Papua New Guinea, W. Irian).
yutajensis *J.A. Steyermark in Ann. Missouri Bot. Gard.*, 75(1): 348 (1988)—Venezuela.

PSYDRAX (Rubiac.).
nitidum *(Craib) K.M. Wong in Tree Fl. Malaya*, 4: 400 (1989): Canthium nitidum.

PTERNOPETALUM (Umbellif.).
tanakae
var. fulcrantum *Y.H. Zhang in Bull. Bot. Res. North-East. Forest. Inst.*, 9(3): 59 (1989)—China.
trifoliatum *R.H. Shan & F.T. Pu in Acta Phytotax. Sin.*, 27(1): 64 (1989)—China.

PTEROCARPUS (Legumin.).
marsupium
subsp. acuminatus *(Prain) K. Thothathri, Taxon. Revis. Tribe Dalbergieae in Indian Subcont.*: 227 (1987): P. marsupium var. acuminata.

PTEROCERAS (Orchidac.).
pricei *(Rolfe) L.V. Aver'yanov, Prelim. list Vietnam. Orchids*, 2: 120 (1988): Dendrocolla pricei.
pustulatum *K. Senghas in Orchidee*, 40(1): 5 (1989)—Philippine Is.

PTEROCYPSELA C. Shih in Acta Phytotax. Sin., 26(5): 385 (1988). *COMPOSITAE.*
elata *(Hemsl.) C. Shih in Acta Phytotax. Sin.*, 26(5): 385 (1988): Lactuca elata.

PTEROCYPSELA :–
formosana *(Maxim.) C. Shih in Acta Phytotax. Sin.*, 26(5): 389 (1988): Lactuca formosana.
indica *(L.) C. Shih in Acta Phytotax. Sin.*, 26(5): 387 (1988): Lactuca indica.
laciniata *(Houtt.) C. Shih in Acta Phytotax. Sin.*, 26(5): 388 (1988): Prenanthes laciniata.
raddeana *(Maxim.) C. Shih in Acta Phytotax. Sin.*, 26(5): 386 (1988): Lactuca raddeana.
sonchus *(Lévl. & Vant.) C. Shih in Acta Phytotax. Sin.*, 26(5): 388 (1988): Lactuca sonchus.
triangulata *(Maxim.) C. Shih in Acta Phytotax. Sin.*, 26(5): 386 (1988): Lactuca triangulata.

PTEROSTYLIS (Orchidac.).
nigricans *D.L. Jones & M.A. Clements in Austrobaileya*, 2(5): 550 (1988)—Australia (Queensland, New South Wales).

PTILOTUS (Amaranthac.).
barkeri *G. Benl in J. Adelaide Bot. Gard.*, 11(2): 195 (1989)—Australia (S. Australia).
crispus *G. Benl in Nuytsia*, 6(3): 319 (1988)—Australia (W. Australia).

PTYSSIGLOTTIS (Acanthac.).
isophylla *(C.B. Clarke) B. Hansen in Nordic J. Bot.*, 9(2): 214 (1989): Polytrema isophylla.
kunthiana *(Nees) B. Hansen in Nordic J. Bot.*, 9(2): 214 (1989): Asystasia kunthiana.

PUCCINELLIA (Gramin.).
neglecta *(Tzvel.) S.V. Bubnova in Spis. Rast. Gerb. Fl. SSSR*, 26(131–136): 40 (1988): P. borealis subsp. neglecta.

PUERARIA (Legumin.).
edulis
var. likiangensis *P.C. Li in Acta Bot. Yunnanica*, 11(3): 302 (1989)—China.
thunbergii
f. alba *M.H. Lee in Korean J. Pl. Taxon.*, 13(2): 109 (1983)—Korea.

PULSATILLA (Ranunculac.).
cernua
f. plumbea *J.X. Ji & Y.T. Zhao in Bull. Bot. Res. North-East. Forest. Inst.*, 9(4): 69 (1989)—China.
turczaninovii
f. albiflora *Y.Z. Zhao in Acta Sci. Nat. Univ. Intramongolicae*, 19(4): 657 (1988)—China.

PUYA (Bromeliac.).
antioquensis *L.B. Smith & R.W. Read in Rev. Acad. Colomb. Cienc. Exact. Fis. Nat.*, 65: 234 (1989)—Colombia.
horrida *L.B. Smith & R.W. Read in Rev. Acad. Colomb. Cienc. Exact. Fis. Nat.*, 65: 234 (1989)—Colombia.
solomonii *G.S. Varadarajan in J. Bromeliad Soc.*, 39(3): 121 (1989)—Bolivia.

PYCREUS (Cyperac.).
polystachyos
subsp. holosericeus *(Link) T. Koyama in Bishop Mus. Occas. Pap.*, 29: 125 (1989): Cyperus holosericeus.
subsp. riograndensis *D.A. Simpson in Kew Bull.*, 44(2): 281 (1989)—Brazil.

PYRENARIA (Theac.).
burmanica *T.K. Paul & M.P. Nayar in J. Jap. Bot.*, 64(1): 8 (1989)—Burma.

PYRGOPHYLLUM (Gagnep.) T.L. Wu & Z.Y. Chen in Acta Phytotax. Sin., 27(2): 126 (1989): Kaempferia subgen. Pyrgophyllum (Zingiberac.).
yunnanensis *(Gagnep.) T.L. Wu & Z.Y. Chen in Acta Phytotax. Sin.*, 27(2): 127 (1989): Kaempferia yunnanensis.

PYROLA (Ericac.).
asarifolia
subsp. incarnata *(DC.) E. Haber & H. Takahashi in Bot. Mag. Tokyo*, 101(1064): 492 (1988): P. rotundifolia var. incarnata.

PYRUS (Rosac.).
×amphigenea *Domin ex J. Dostálek in Folia Geobot. Phytotax.*, 24(1): 105 (1989)—Czechoslovakia. (P. communis × pyraster).
pyrifolia
var. talyschensis *V.N. Gladkova in Novosti Sist. Vÿssh. Rast.*, 25: 100 (1988)—U.S.S.R. (Caucasus).
ussuriensis
var. heterophylla *V.N. Gladkova in Novosti Sist. Vÿssh. Rast.*, 25: 102 (1988)—U.S.S.R. (Caucasus).

Q

QUADRANGULA M.G. Baumann-Bodenheim, Syst. Fl. Neu-Caledonien, 4: 102 (1988 publ. Jan. 1989), nom. nov., without replaced synonym ref.; et Syst. Fl. Neu-Caledonien, 5: 97 (May 1989), without replaced synonym ref.: Casuarina § Gymnostoma Poiss. (Casuarinac.).
chamaecyparis *(Poiss.) M.G. Baumann-Bodenheim, Syst. Fl. Neu-Caledonien*, 4: 102 (1988 publ. Jan. 1989), without basionym ref.; *et Syst. Fl. Neu-Caledonien*, 5: 97 (May 1989), without basionym page: Casuarina chamaecyparis.
deplancheana *(Miq.) M.G. Baumann-Bodenheim, Syst. Fl. Neu-Caledonien*, 4: 102 (1988 publ. Jan. 1989), without basionym ref.; *et Syst. Fl. Nau-Caledonien*, 5: 97 (May 1989), without basionym page: Casuarina deplancheana.

QUARARIBEA (Bombacac.).
aurantiocalyx *W.S. Alverson in Brittonia,* 41(1): 68 (1989)—Costa Rica, Panama.
costaricensis *W.S. Alverson in Brittonia,* 41(1): 70 (1989)—Costa Rica.
gomeziana *W.S. Alverson in Brittonia,* 41(1): 61 (1989)—Costa Rica, Panama.
pendula *W.S. Alverson in Brittonia,* 41(1): 64 (1989)—Costa Rica, Panama.
santaritensis *W.S. Alverson in Brittonia,* 41(1): 66 (1989)—Panama.

QUERCUS (Fagac.).
faginea
var. oscensis *P. Montserrat in Soc. Ech. Pl. Vasc. Eur. Bassin Médit. Bull.,* 22: 62 (1986–87 publ. 1988)—Spain.
frainetto *M. Tenore, Fl. Napol.* 1, Prodr. Supp. II: lxxii (?1813)—Italy.
gussonei *(Borzì) S. Brullo in Not. Fitosoc.,* 19(1): 205 (1984): Q. cerris var. gussonei.
semicarpifolia
var. miyabei *(Hayata) S.S. Ying, Coloured Illustr. Fl. Taiwan,* 3: 195 (1988): Q. spinosa var. miyabei.

R

RABDOSIA (Labiat.).
excisa
var. chiisanensis *(Nakai) G. Murata in Acta Phytotax. Geobot.,* 40(1–4): 100 (1989): Amethystanthus excisus var. chiisanensis.

RACEMOBAMBOS (Gramin.).
ciliata *(A. Camus) C.S. Chao & S.A. Renvoize in Kew Bull.,* 44(2): 365 (1989): Arundinaria ciliata.

RACOSPERMA (Legumin.).
juliferum
subsp. curvinervium *(Maiden) L. Pedley in Austrobaileya,* 2(5): 571 (1988): Acacia curvinervia.
latescens *(Benth.) L. Pedley in Austrobaileya,* 2(5): 571 (1988): Acacia latescens.
perryi *(Pedley) L. Pedley in Austrobaileya,* 2(5): 571 (1988): Acacia perryi.

RAFFLESIA (Rafflesiac.).
tengku-adlinii *K. Mat Salleh & A. Latiff in Blumea,* 34(1): 112 (1989)—Borneo (Sabah).

×**RAMBERLEA** J. Halda in Preslia, 61(2): 127 (1989). *GESNERIACEAE.* (Haberlea × Ramonda).
kistlerae *J. Halda in Preslia,* 61(2): 127 (1989)—Cultivation. (Haberlea rhodopensis × Ramonda myconii).

RAMOSMANIA (Rubiac.).
rodriguesii *D.D. Tirvengadum in Compte R. Séances Soc. Biogéogr.,* 65(1): 14 (1989)—Rodrigues Is.

RANUNCULUS (Ranunculac.).
sect. Parviflori *A.G. Elenevskiĭ & T.G. Derviz-Sokolova in Byull. Mosk. Obshch. Ispȳt. Prir., Biol.,* 94(1): 121 (1989).
alaschanicus *Y.Z. Zhao in Bull. Bot. Res. North-East. Forest. Inst.,* 9(1): 64 (1989)—China.
ashibetsuensis *G. Wiegleb in Acta Phytotax. Geobot.,* 39(4–6): 130 (1988)—Japan.
brutius
subsp. crimaeus *(Juz.) A.G. Elenevskiĭ in Byull. Mosk. Obshch. Ispȳt. Prir., Biol.,* 94(1): 114 (1989): R. crimaeus.
cuneilaminatus *W. Greuter & Burdet in Willdenowia,* 19(1): 45 (1989), nom. nov.: R. cuneatus *Boiss.*
intramongolicus *Y.Z. Zhao in Bull. Bot. Res. North-East. Forest. Inst.,* 9(1): 69 (1989)—China.
kurdicus *W. Greuter & Burdet in Willdenowia,* 19(1): 46 (1989), nom. nov.: R. diversifolius. *Boiss. & Kotschy.*
luminarius *Pignatti ex W. Greuter in Willdenowia,* 19(1): 46 (1989)—Italy.
lunaris *T. Brodtbeck in Bauhinia,* 9(1): 89 (1988)—Germany.
macrotis *T. Brodtbeck in Bauhinia,* 9(1): 85 (1988)—Switzerland, Germany.
marginatus
subsp. trachycarpus *(Fisch. & Mey.) A.G. Elenevskiĭ & T.G. Derviz-Sokolova in Byull. Mosk. Obshch. Ispȳt. Prir., Biol.,* 94(1): 117 (1989): R. trachycarpus.
nipponicus
var. ohtagawensis *G. Wiegleb in Acta Phytotax. Geobot.,* 39(4–6): 129 (1988)—Japan.
var. okayamensis *G. Wiegleb in Acta Phytotax. Geobot.,* 39(4–6): 129 (1988)—Japan.
pedatus
subsp. silvisteppaceus *(Dubovik) A.G. Elenevskiĭ & T.G. Derviz-Sokolova in Byull. Mosk. Obshch. Ispȳt. Prir., Biol.,* 94(1): 120 (1989): R. silvisteppaceus.
quinatus *T. Brodtbeck in Bauhinia,* 9(1): 83 (1988)—France, Switzerland.
sardous
subsp. philonotis *(Ehrh.) A.G. Elenevskiĭ & T.G. Derviz-Sokolova in Byull. Mosk. Obshch. Ispȳt. Prir., Biol.,* 94(1): 117 (1989): R. philonotis.
spicatus
subsp. aspromontanus *(Huter) W. Greuter & Burdet in Willdenowia,* 19(1): 46 (1989): R. aspromontanus.
subsp. maroccanus *(Cosson) W. Greuter & Burdet in Willdenowia,* 19(1): 46 (1989): R. cortusifolius var. maroccanus.
stellaris *T. Brodtbeck in Bauhinia,* 9(1): 92 (1988)—Germany, Switzerland.

RANUNCULUS :-
trichophyllus
var. kadzusensis *(Makino) G. Wiegleb in Acta Phytotax. Geobot.,* 39(4–6): 128 (1988): R. kadzusensis.
xinningensis *W.T. Wang in Bull. Bot. Res. North-East. Forest. Inst.,* 9(2): 10 (1989)—China.
yinshanensis *(Y.Z. Zhao) Y.Z. Zhao in Bull. Bot. Res. North-East. Forest. Inst.,* 9(1): 67 (1989): R. pulchellus var. yinshanensis.

RAOULIA (Compos.).
×petrimia *Kit Tan & R.J.D. McBeath in Quart. Bull. Alp. Gard. Soc.,* 56(4): 311 (1988)—New Zealand. (R. eximia × petriensis).

RAPATEA (Rapateac.).
chimantensis *J.A. Steyermark in Ann. Missouri Bot. Gard.,* 75(1): 313 (1988)—Venezuela.

RAPHIONACME (Asclepiadac.).
borenensis *H.J.T. Venter & M.G. Gilbert in Bot. J. Linn. Soc.,* 99(4): 401 (1989)—Ethiopia, Kenya, Somali Republic.
lobulata *H.J.T. Venter & R.L. Verhoeven in S. Afr. J. Bot.,* 54(6): 603 (1988)—South Africa (Cape Province).
lucens *H.J.T. Venter & R.L. Verhoeven in S. Afr. J. Bot.,* 54(6): 607 (1988)—South Africa (Natal), Mozambique.

RAUIA (Rutac.).
subtruncata *J.A. Steyermark in Ann. Missouri Bot. Gard.,* 75(1): 315 (1988)—Venezuela.

RAVENIOPSIS (Rutac.).
liesneri *J.A. Steyermark in Ann. Missouri Bot. Gard.,* 75(1): 317 (1988)—Venezuela.

REAUMURIA (Tamaricac.).
oxiana
var. persica *(Boiss.) M. Assadi in Fl. Iran (ed. M. Assadi et al.),* 1: 9 (1989): Eichwaldia persica.
sivasica *Kit Tan & B. Yildiz in Notes Roy. Bot. Gard. Edinburgh,* 45(3): 440 (1988 publ. 1989)—Turkey.

REBUTIA (Cactac.).
walteri *L. Diers in Kakt. And. Sukk.,* 40(8): 189 (1989)—Argentina.

RECORDOXYLON (Legumin.).
pulcherrimum *R. Barneby in Acta Bot. Venezuel.,* 14(3): 20 (1984)—Venezuela.

REICHARDIA (Compos.).
picroides
var. gracilis *(Schultz Bip.) O. de Bolòs & J. Vigo in Folia Bot. Misc.,* 6: 86 (1989): Picridium intermedium var. gracile.
var. robusta *(Willk.) O. de Bolòs & J. Vigo in Folia Bot. Misc.,* 6: 86 (1989): Picridium intermedium var. robustum.

RELDIA (Gesneriac.).
calcarata *L.P. Kvist & L.E. Skog in Nordic J. Bot.,* 8(6): 606 (1989)—Ecuador.
grandiflora *L.P. Kvist & L.E. Skog in Nordic J. Bot.,* 8(6): 607 (1989)—Colombia.
minutiflora *(L. Skog) L.P. Kvist & L.E. Skog in Nordic J. Bot.,* 8(6): 607 (1989): Nautilocalyx minutiflorus.
var. veraguensis *(Wiehl.) L.P. Kvist & L.E. Skog in Nordic J. Bot.,* 8(6): 609 (1989): R. veraguensis.
multiflora *L.P. Kvist & L.E. Skog in Nordic J. Bot.,* 8(6): 611 (1989)—Ecuador.

REMIJIA (Rubiac.).
sessilis *J.A. Steyermark in Ann. Missouri Bot. Gard.,* 75(3): 1084 (1988)—Venezuela.

RENANTHERELLA (Orchidac.).
histrionica
subsp. auyongii *(E.A. Christ.) K. Senghas in Schlechter Orchideen,* 1(21): 1327 (1988): Renanthera auyongii.

RENNELLIA (Rubiac.).
amoena *(Bremek.) J.T. Johansson in Blumea,* 34(1): 11 (1989): Didymoecium amoenum.
microcephala *(Ridl.) K.M. Wong in Tree Fl. Malaya,* 4: 405 (1989), with incorrect basionym: Myrioneuron microcephalum.

RESEDA (Resedac.).
alba
subsp. trigyna *(Batt.) W. Greuter & Burdet in Willdenowia,* 19(1): 47 (1989): R. gayana subsp. trigyna.

RETROPHYLLUM C.N. Page in Notes Roy. Bot. Gard. Edinburgh, 45(2): 379 (1988 publ. 1989). *PODOCARPACEAE.*
comptonii *(Buchh.) C.N. Page in Notes Roy. Bot. Gard. Edinburgh,* 45(2): 380 (1988 publ. 1989): Podocarpus comptonii.
minor *(Carr.) C.N. Page in Notes Roy. Bot. Gard. Edinburgh,* 45(2): 380 (1988 publ. 1989): Nageia minor.
piresii *(Silba) C.N. Page in Notes Roy. Bot. Gard. Edinburgh,* 45(2): 380 (1988 publ. 1989): Decussocarpus piresii.
rospigliosii *(Pilg.) C.N. Page in Notes Roy. Bot. Gard. Edinburgh,* 45(2): 380 (1988 publ. 1989): Podocarpus rospigliosii.
vitiense *(Seem.) C.N. Page in Notes Roy. Bot. Gard. Edinburgh,* 45(2): 380 (1988 publ. 1989): Podocarpus vitiensis.

RHAGADIOLUS (Compos.).
stellatus
subsp. edulis *(Gaertn.) O. de Bolòs & J. Vigo in Folia Bot. Misc.,* 6: 85 (1989): R. edulis.

RHAMNUS (Rhamnac.).
acuminata *B. Maguire & J.A. Steyermark in Mem. New York Bot. Gard.,* 51: 121 (1989)—Venezuela.

RHAMNUS :-
longipes *J.A. Steyermark in Ann. Missouri Bot. Gard.*, 75(3): 1066 (1988)—Venezuela.
marahuacensis *J. Steyermark & B. Maguire in Acta Bot. Venezuel.*, 14(3): 22 (1984)—Venezuela.
neblinensis *B. Maguire & J.A. Steyermark in Mem. New York Bot. Gard.*, 51: 119 (1989)—Venezuela.
psilocarpa *B. Maguire & J.A. Steyermark in Mem. New York Bot. Gard.*, 51: 119 (1989)—Venezuela.
serpyllacea *W. Greuter & Burdet in Willdenowia*, 19(1): 48 (1989), nom. nov.: R. serpyllifolia *Emberger & Maire*.
sipapoensis *J.A. Steyermark in Ann. Missouri Bot. Gard.*, 75(3): 1066 (1988)—Venezuela.

RHEUM (Polygonac.).
compactum
var. orientale *(Losinsk.) N.N. Tsvelev in Sosud. Rast. Sovet. Dal'nego Vostoka*, 4: 55 (1989): R. orientale.

RHODALSINE (Caryophyllac.).
geniculata
var. procumbens *(Vahl) A. Dubuis in Soc. Ech. Pl. Vasc. Eur. Bassin Médit. Bull.*, 22: 63 (1986–87 publ. 1988): Arenaria procumbens.

RHODAMNIA (Myrtac.).
pauciovulata *G.P. Guymer in Austrobaileya*, 2(5): 515 (1988)—Australia (Queensland).

RHODOCHITON (Scrophulariac.).
hintonii *(Elisens) D.A. Sutton, Revis. tribe Antirrhineae:* 508 (1988): Lophospermum hintonii.
nubicola *(Elisens) D.A. Sutton, Revis. tribe Antirrhineae:* 509 (1988): Lophospermum nubicola.

RHODODENDRON (Ericac.).
ser. Argyrophylla *H.H. Davidian, Rhododendron Species, 2 (Ser. Arboreum-Lacteum):* 309 (1989).
subser. Campanulata *H.H. Davidian, Rhododendron Species, 2 (Ser. Arboreum-Lacteum):* 309 (1989).
ser. Fulgensia *H.H. Davidian, Rhododendron Species, 2 (Ser. Arboreum-Lacteum):* 309 (1989).
ser. Griersoniana *H.H. Davidian, Rhododendron Species, 2 (Ser. Arboreum-Lacteum):* 309 (1989).
subser. Lanata *H.H. Davidian, Rhododendron Species, 2 (Ser. Arboreum-Lacteum):* 309 (1989).
ser. Parishia *H.H. Davidian, Rhododendron Species, 2 (Ser. Arboreum-Lacteum):* 310 (1989).
ser. Sherriffia *H.H. Davidian, Rhododendron Species, 2 (Ser. Arboreum-Lacteum):* 310 (1989).

RHODODENDRON :-
subser. Williamsiana *H.H. Davidian, Rhododendron Species, 2 (Ser. Arboreum-Lacteum):* 310 (1989).
anthosphaerum
var. eritimum *(Balf. f. & W.W. Sm.) H.H. Davidian, Rhododendron Species, 2 (Ser. Arboreum-Lacteum):* 261 (1989): R. eritimum.
bellum *H.P. Yang in Bull. Bot. Res. North-East. Forest. Inst.*, 9(1): 18 (1989)—China.
bhutanense *D.G. Long & Bowes Lyon in Notes Roy. Bot. Gard. Edinburgh*, 45(2): 327 (1988 publ. 1989)—Bhutan.
×charadzeae *A.P. Khokhryakov & Mazurenko in Byull. Glavn. Bot. Sada (Moscow)*, 152: 58 (1989)—U.S.S.R. (Caucasus). (R. smirnovii × ungernii).
delavayi
var. albotomentosum *H.H. Davidian, Rhododendron Species, 2 (Ser. Arboreum-Lacteum):* 308, 75 (1989)—Burma.
ebianense *M.Y. Fang in Rhodendron Notes Records*, 2: 23 (1988)—China.
fictolacteum
var. miniforme *H.H. Davidian, Rhododendron Species, 2 (Ser. Arboreum-Lacteum):* 308, 170 (1989)—China, Xizang.
glaucophyllum
subsp. tubiforme *(Cowan & Davidian) D.G. Long in Notes Roy. Bot. Gard. Edinburgh*, 45(2): 330 (1988 publ. 1989): R. glaucophyllum var. tubiforme.
heftii *H.H. Davidian, Rhododendron Species, 2 (Ser. Arboreum-Lacteum):* 307, 149 (1989)—Nepal.
heteroclitum *H.P. Yang in Bull. Bot. Res. North-East. Forest. Inst.*, 9(1): 17 (1989)—China.
kesangiae *D.G. Long & Rushforth in Notes Roy. Bot. Gard. Edinburgh*, 45(2): 331 (1988 publ. 1989)—Bhutan.
lagopus
var. niphophilum *(Yamazaki) T. Yamazaki in J. Jap. Bot.*, 63(12): 410 (1988): R. nudipes subsp. niphophyllum.
var. tokushimense *(Yamazaki) T. Yamazaki in J. Jap. Bot.*, 63(12): 410 (1988): R. nudipes var. tokushimense.
var. tsurugisanense *(Yamazaki) T. Yamazaki in J. Jap. Bot.*, 63(12): 410 (1988): R. nudipes var. tsurugisanense.
lanigerum
var. silvaticum *(Cowan) H.H. Davidian, Rhododendron Species, 2 (Ser. Arboreum-Lacteum):* 77 (1989): R. silvaticum.
macrosmithii *H.H. Davidian, Rhododendron Species, 2 (Ser. Arboreum-Lacteum):* 113 (1989), nom. nov.: R. smithii *Nutt. ex W.J. Hooker*.
mucronulatum
var. chejuense *H.H. Davidian, Rhododendron Species, 2 (Ser. Arboreum-Lacteum):* 309 (1989)—Korea.

RHODODENDRON :-
oblancifolium *M.Y. Fang in Rhodendron Notes Records,* 2: 25 (1988)—China.
periclymenoides
f. glandiferum *(Porter) L.J. Uttal in Sida,* 13(2): 169 (1988): Azalea nudiflora var. glandifera.
poluninii *H.H. Davidian, Rhododendron Species, 2 (Ser. Arboreum-Lacteum):* 307, 155 (1989)—Bhutan.
roxieanum
var. parvum *H.H. Davidian, Rhododendron Species, 2 (Ser. Arboreum-Lacteum):* 308 (1989)—China.
scyphocalyx
var. septendrionale *Tagg ex H.H. Davidian, Rhododendron Species, 2 (Ser. Arboreum-Lacteum):* 308 (1989)—Burma, China.
serpyllifolium
f. album *T. Yamazaki in J. Jap. Bot.,* 64(9): 280 (1989)—Japan.
tsariense
var. magnum *H.H. Davidian, Rhododendron Species, 2 (Ser. Arboreum-Lacteum):* 308, 157 (1989)—Bhutan.
var. trimoense *H.H. Davidian, Rhododendron Species, 2 (Ser. Arboreum-Lacteum):* 308, 157 (1989)—Xizang.
viscidum *C.Z. Guo & Z.H. Liu in Acta Bot. Yunnanica,* 10(4): 497 (1988)—China.

RHODOSTEMONODAPHNE (Laurac.).
steyermarkiana *(Allen) H. van der Werff in Ann. Missouri Bot. Gard.,* 76(2): 474 (1989): Ocotea steyermarkiana.

RHUS (Anacardiac.).
wilsonii
var. glabra *Y.T. Wu in Fl. Sichuanica,* 4: 135 (1988)—China.

RHYNCHOSIA (Legumin.).
megalocalyx *M. Thulin in Nordic J. Bot.,* 8(5): 480 (1989)—Somali Republic.

RHYNCHOSPORA (Cyperac.).
chinensis
subsp. spiciformis *(Hillebr.) T. Koyama in Bishop Mus. Occas. Pap.,* 29: 127 (1989): R. spicaeformis.
rosemariana *D.A. Simpson in Kew Bull.,* 44(2): 289 (1989)—Brazil.
rugosa
subsp. lavarum *(Gaud.) T. Koyama in Bishop Mus. Occas. Pap.,* 29: 127 (1989): R. lavarum.

RICHERIA (Euphorbiac.).
dressleri *G.L. Webster in Ann. Missouri Bot. Gard.,* 75(3): 1094 (1988)—Panama, Costa Rica.

ROBINIA (Legumin.).
halodendron *P.S. Pallas, Reise Russ. Reich.,* 2(2): 741 (1773), in Index Kewensis vol. 2, fasc. 4 as 'halodendrum'—Siberia.†

ROBIQUETIA (Orchidac.).
merrilliana *(Ames) E. Lückel, M. Wolff & J.J. Wood in Orchidee,* 40(3): 109 (1989): Sarcanthus merrillianus.

ROEGNERIA (Gramin.).
alashanica *Keng ex Y.L. Keng & S.L. Chen in J. Nanjing Univ. (Nat. Sci.),* 3: 73 (1963)—China.
aliena *Keng ex Y.L. Keng & S.L. Chen in J. Nanjing Univ. (Nat. Sci.),* 3: 31 (1963)—China.
altissima *Keng ex Y.L. Keng & S.L. Chen in J. Nanjing Univ. (Nat. Sci.),* 3: 53 (1963)—China.
anthosachnoides *Keng ex Y.L. Keng & S.L. Chen in J. Nanjing Univ. (Nat. Sci.),* 3: 65 (1963)—China.
aristiglumis *Y.L. Keng & S.L. Chen in J. Nanjing Univ. (Nat. Sci.),* 3: 55 (1963)—Xinjiang, China.
breviglumis *Keng ex Y.L. Keng & S.L. Chen in J. Nanjing Univ. (Nat. Sci.),* 3: 48 (1963)—China.
brevipes *Keng ex Y.L. Keng & S.L. Chen in J. Nanjing Univ. (Nat. Sci.),* 3: 49 (1963)—China.
calcicola *Keng ex Y.L. Keng & S.L. Chen in J. Nanjing Univ. (Nat. Sci.),* 3: 21 (1963)—China.
dolichathera *Keng ex Y.L. Keng & S.L. Chen in J. Nanjing Univ. (Nat. Sci.),* 3: 19 (1963)—China.
dura *(Keng) Keng ex Y.L. Keng & S.L. Chen in J. Nanjing Univ. (Nat. Sci.),* 3: 54 (1963): Brachypodium durum.
foliosa *Keng ex Y.L. Keng & S.L. Chen in J. Nanjing Univ. (Nat. Sci.),* 3: 32 (1963)—China.
geminata *Y.L. Keng & S.L. Chen in J. Nanjing Univ. (Nat. Sci.),* 3: 80 (1963)—China.
glaberrima *Y.L. Keng & S.L. Chen in J. Nanjing Univ. (Nat. Sci.),* 3: 72 (1963)—Xinjiang.
glaucifolia *Keng ex Y.L. Keng & S.L. Chen in J. Nanjing Univ. (Nat. Sci.),* 3: 57 (1963)—Xizang.
grandiglumis *Keng ex Y.L. Keng & S.L. Chen in J. Nanjing Univ. (Nat. Sci.),* 3: 82 (1963)—China.
grandis *Keng ex Y.L. Keng & S.L. Chen in J. Nanjing Univ. (Nat. Sci.),* 3: 45 (1963)—China.
hirsuta *Keng ex Y.L. Keng & S.L. Chen in J. Nanjing Univ. (Nat. Sci.),* 3: 84 (1963)—China.
humilis *Y.L. Keng & S.L. Chen in J. Nanjing Univ. (Nat. Sci.),* 3: 40 (1963)—China.
hybrida *Keng ex Y.L. Keng & S.L. Chen in J. Nanjing Univ. (Nat. Sci.),* 3: 18 (1963)—China.
japonensis *(Honda) Keng ex Y.L. Keng & S.L. Chen in J. Nanjing Univ. (Nat. Sci.),* 3: 63 (1963): Agropyron japonense.

ROEGNERIA :–
kokonorica *Keng ex Y.L. Keng & S.L. Chen in J. Nanjing Univ. (Nat. Sci.),* 3: 88 (1963)—China.
laxiflora *Keng ex Y.L. Keng & S.L. Chen in J. Nanjing Univ. (Nat. Sci.),* 3: 75 (1963)—China.
leiantha *Keng ex Y.L. Keng & S.L. Chen in J. Nanjing Univ. (Nat. Sci.),* 3: 42 (1963)—China.
leiotropis *Keng ex Y.L. Keng & S.L. Chen in J. Nanjing Univ. (Nat. Sci.),* 3: 58 (1963)—China.
longiglumis *Keng ex Y.L. Keng & S.L. Chen in J. Nanjing Univ. (Nat. Sci.),* 3: 83 (1963)—China.
melanthera *(Keng) Keng ex Y.L. Keng & S.L. Chen in J. Nanjing Univ. (Nat. Sci.),* 3: 78 (1963): Agropyron melantherum.
minor *Keng ex Y.L. Keng & S.L. Chen in J. Nanjing Univ. (Nat. Sci.),* 3: 71 (1963)—China.
mutica *Keng ex Y.L. Keng & S.L. Chen in J. Nanjing Univ. (Nat. Sci.),* 3: 87 (1963)—China.
nutans *(Keng) Keng ex Y.L. Keng & S.L. Chen in J. Nanjing Univ. (Nat. Sci.),* 3: 48 (1963): Agropyron nutans.
parvigluma *Keng ex Y.L. Keng & S.L. Chen in J. Nanjing Univ. (Nat. Sci.),* 3: 47 (1963)—China.
platyphylla *Keng ex Y.L. Keng & S.L. Chen in J. Nanjing Univ. (Nat. Sci.),* 3: 35 (1963)—Xinjiang.
puberula *Keng ex Y.L. Keng & S.L. Chen in J. Nanjing Univ. (Nat. Sci.),* 3: 20 (1963)—China.
pubicaulis *Keng ex Y.L. Keng & S.L. Chen in J. Nanjing Univ. (Nat. Sci.),* 3: 30 (1963)—China.
purpurascens *Keng ex Y.L. Keng & S.L. Chen in J. Nanjing Univ. (Nat. Sci.),* 3: 56 (1963)—China.
rigidula *Keng ex Y.L. Keng & S.L. Chen in J. Nanjing Univ. (Nat. Sci.),* 3: 77 (1963)—China.
serotina *Keng ex Y.L. Keng & S.L. Chen in J. Nanjing Univ. (Nat. Sci.),* 3: 50 (1963)—China.
sinica *Keng ex Y.L. Chen & S.L. Chen in J. Nanjing Univ. (Nat. Sci.),* 3: 33 (1963)—China.
stenachyra *Keng ex Y.L. Keng & S.L. Chen in J. Nanjing Univ. (Nat. Sci.),* 3: 79 (1963)—China.
stricta *Keng ex Y.L. Keng & S.L. Chen in J. Nanjing Univ. (Nat. Sci.),* 3: 68 (1963)—China.
sylvatica *Y.L. Keng & S.L. Chen in J. Nanjing Univ. (Nat. Sci.),* 3: 36 (1963)—Xinjiang.
thoroldiana *(Oliv.) Keng ex Y.L. Keng & S.L. Chen in J. Nanjing Univ. (Nat. Sci.),* 3: 79 (1963): Agropyron thoroldianum.
varia *Keng ex Y.L. Keng & S.L. Chen in J. Nanjing Univ. (Nat. Sci.),* 3: 70 (1963)—China.

ROEGNERIA :–
villosa
subsp. coerulea *B.A. Yurtsev in Bot. Zhurn.,* 74(1): 113 (1989)—U.S.S.R. (Soviet Far East).
viridula *Y.L. Keng & S.L. Chen in J. Nanjing Univ. (Nat. Sci.),* 3: 39 (1963)—Xinjiang.

ROGERSONANTHUS B. Maguire & B.M. Boom in Mem. New York Bot. Gard., 51: 3 (1989). *GENTIANACEAE.*
arboreus *(Britton) B. Maguire & B.M. Boom in Mem. New York Bot. Gard.,* 51: 8 (1989): Chelonanthus arboreus.
quelchii *(N.E. Brown) B. Maguire & B.M. Boom in Mem. New York Bot. Gard.,* 51: 4 (1989): Lisianthus quelchii.
tepuiensis *(Gleason) B. Maguire & B.M. Boom in Mem. New York Bot. Gard.,* 51: 4 (1989): Calolisianthus tepuiensis.

ROGIERA (Rubiac.).
aprica *(Lundell) A. Borhidi in Acta Bot. Hung.,* 33(3–4): 301 (1987): Rondeletia aprica.
backhousii *(Hook. f.) A. Borhidi in Acta Bot. Hung.,* 33(3–4): 301 (1987): Rondeletia backhousii.
brachystantha *(Standl. & Steyerm.) A. Borhidi in Acta Bot. Hung.,* 33(3–4): 301 (1987): Rondeletia brachystantha.
chiapasensis *(Standl.) A. Borhidi in Acta Bot. Hung.,* 33(3–4): 301 (1987): Rondeletia chiapasensis.

ROLLINIA (Annonac.).
andicola *P.J.M. Maas & L.Y.T. Westra in Proc. Kon. Nederl. Akad. Wetensch., C,* 92(3): 297 (1989)—Ecuador, Peru.
bahiensis *P.J.M. Maas & L.Y.T. Westra in Proc. Kon. Nederl. Akad. Wetensch., C,* 92(3): 299 (1989)—Brazil.
chrysocarpa *P.J.M. Maas & L.Y.T. Westra in Proc. Kon. Nederl. Akad. Wetensch., C,* 92(3): 300 (1989)—Peru.
ferruginea *(R.E. Fries) P.J.M. Maas & L.Y.T. Westra in Proc. Kon. Nederl. Akad. Wetensch., C,* 92(3): 301 (1989): Rolliniopsis ferruginea.
fosteri *P.J.M. Maas & L.Y.T. Westra in Proc. Kon. Nederl. Akad. Wetensch., C,* 92(3): 302 (1989)—Peru, Ecuador.
glomerulifera *P.J.M. Maas & L.Y.T. Westra in Proc. Kon. Nederl. Akad. Wetensch., C,* 92(3): 303 (1989)—Peru, Ecuador.
helosioides *P.J.M. Maas & L.Y.T. Westra in Proc. Kon. Nederl. Akad. Wetensch., C,* 92(3): 305 (1989)—Ecuador.
hispida *P.J.M. Maas & L.Y.T. Westra in Proc. Kon. Nederl. Akad. Wetensch., C,* 92(3): 307 (1989)—Peru, Ecuador.
mammifera *P.J.M. Maas & L.Y.T. Westra in Proc. Kon. Nederl. Akad. Wetensch., C,* 92(3): 308 (1989)—Peru, Brazil.
pachyantha *P.J.M. Maas & L.Y.T. Westra in Proc. Kon. Nederl. Akad. Wetensch., C,* 92(3): 310 (1989)—Colombia.

ROLLINIA :–
schunkei *P.J.M. Maas & L.Y.T. Westra in Proc. Kon. Nederl. Akad. Wetensch., C,* 92(3): 311 (1989)—Peru, Brazil.
ubatubensis *P.J.M. Maas & L.Y.T. Westra in Proc. Kon. Nederl. Akad. Wetensch., C,* 92(3): 313 (1989)—Brazil.
velutina *E.J. van Marle in Proc. Kon. Nederl. Akad. Wetensch., C,* 92(3): 314 (1989)—Venezuela, Colombia.

ROMULEA (Iridac.).
hartungii
subsp. coronensis *A. Abraham in Feddes Repert.,* 99(11–12): 505, 506 (1988), without latin descr.—Canary Is.
subsp. hierroensis *A. Abraham in Feddes Repert.,* 99(11–12): 505 (1988), without latin descr.—Canary Is.
subsp. montana *A. Abraham in Feddes Repert.,* 99(11–12): 505 (1988), without latin descr.—Canary Is.
subsp. pallida *A. Abraham in Feddes Repert.,* 99(11–12): 505 (1988), without latin descr.—Canary Is.
lactea *A. Abraham in Feddes Repert.,* 99(11–12): 506 (1988)—Canary Is.

ROSA (Rosac.).
altidaghestanica *Sh.A. Guseĭnov in Bot. Zhurn.,* 74(6): 898 (1989)—U.S.S.R. (Caucasus).
baitagensis *R.V. Kamelin & I.A. Gubanov in Byull. Mosk. Obshch. Ispȳt. Prir., Biol.,* 93(5): 110 (1988)—Mongolia.
×beanii *P.V. Heath in Taxon,* 38(1): 97 (1989), nom. nov.: R. anemoniflora *Fortune ex Lindley.*
cziragensis *Sh.A. Guseĭnov in Bot. Zhurn.,* 74(6): 893 (1989)—U.S.S.R. (Caucasus).
darginica *Sh.A. Guseĭnov in Bot. Zhurn.,* 74(6): 892 (1989)—U.S.S.R. (Caucasus).
subbuschiana *Sh.A. Guseĭnov in Bot. Zhurn.,* 74(6): 895 (1989)—U.S.S.R. (Caucasus).
tlaratensis *Sh.A. Guseĭnov in Bot. Zhurn.,* 74(6): 896 (1989)—U.S.S.R. (Caucasus).
usischensis *Sh.A. Guseĭnov in Bot. Zhurn.,* 74(6): 894 (1989)—U.S.S.R. (Caucasus).

ROSCOEA J. Smith, Exot. Bot., 2: 97, t. 108 (1806). *ZINGIBERACEAE.* †

ROSTRARIA (Gramin.).
smyrnacea *(Trinius) H. Scholz in Willdenowia,* 19(1): 109 (1989): Trisetum smyrnaceum.

ROSULARIA (Crassulac.).
muratdaghensis *Kit Tan in Notes Roy. Bot. Gard. Edinburgh,* 45(2): 367 (1988 publ. 1989)—Turkey.
tauricola *Kit Tan in Notes Roy. Bot. Gard. Edinburgh,* 45(2): 365 (1988 publ. 1989)—Turkey.

ROTALA (Lythrac.).
cookii *K.T. Joseph & V.V. Sivarajan in Pl. Syst. Evol.,* 159(1–2): 143 (1988)—India.

ROTALA :–
vasudevanii *K.T. Joseph & V.V. Sivarajan in Proc. Indian Acad. Sci., Pl. Sci.,* 99(3): 195 (1989)—India.

×ROTORARA hort. in Orchid Rev., 97(1147): centre page pull-out p. 9 (1989). *ORCHIDACEAE.* (Bollea × Cochleanthea × Kefersteinia).

ROTULA (Boraginac.).
pohlii *(Kuhlmann) E.F. Guimarães & L. Mautone in Rev. Brasil. Biol.,* 49(1): 65 (1989): Rhabdia pohlii.

ROUCHERIA (Linac.).
monsalveae *A.H. Gentry in Ann. Missouri Bot. Gard.,* 75(4): 1433 (1988 publ. 1989), as 'Rouchera'—Colombia.

RUBIA (Rubiac.).
alaica *M.G. Pakhomova in Opred. Rast. Sred. Azii,* 9: 368 (1987)—U.S.S.R. (Middle Asia).

RUBUS (Rosac.).
acutus *T. Braeucker, 292 Deutsche Rubus-Arten und Formen:* 36 (1882)—Germany.
aestivalis *E.H.L. Krause in Bot. Jahrb. Syst.,* 16, Beibl. 39: 1 (1893)—Europe.
albicans *F. Krašan in Verh. Zool.-Bot. Ges. Wien,* 15: 357 (1865)—Iran.
atlanticus *E.H.L. Krause in Bot. Jahrb. Syst.,* 16, Beibl. 39: 2 (1893), nom. illegit. superfl. incl. R. sprengelii Wiehe—Europe.
beckeri *T. Braeucker, 292 Deutsche Rubus-Arten und Formen:* 83 (1882)—Germany.
bohemicus *F. Krašan in Verh. Zool.-Bot. Ges. Wien,* 15: 359 (1865)—Locality not stated.
botryoides *T. Braeucker, 292 Deutsche Rubus-Arten und Formen:* 82 (1882)—Germany.
braunii *T. Braeucker, 292 Deutsche Rubus-Arten und Formen:* 72 (1882)—Germany.
compositus *T. Braeucker, 292 Deutsche Rubus-Arten und Formen:* 31 (1882)—Germany.
concolor *E.H.L. Krause in Bot. Jahrb. Syst.,* 16, Beibl. 39: 3 (1893), nom. illegit. superfl. incl. R. hirtus Waldst. & Kit.—Locality not stated.
conoides *T. Braeucker, 292 Deutsche Rubus-Arten und Formen:* 75 (1882)—Germany.
contractipes *H.E. Weber in Osnabrücker Naturwiss. Mitt.,* 14: 144 (1988)—Germany.
cordatifolius *(Rogers ex Riddelsd.) D.E. Allen in Watsonia,* 17(4): 437 (1989): R. dumnoniensis var. cordatifolius.
florulentus *T. Braeucker, 292 Deutsche Rubus-Arten und Formen:* 30 (1882)—Germany.
fockei *T. Braeucker, 292 Deutsche Rubus-Arten und Formen:* 39 (1882)—Germany.
galbanus *T. Braeucker, 292 Deutsche Rubus-Arten und Formen:* 30 (1882)—Germany.
glandulosocarpus *M.X. Nie in Bull. Bot. Res. North-East. Forest. Inst.,* 9(3): 43 (1989)—China.
hastiferus *H.E. Weber in Osnabrücker Naturwiss. Mitt.,* 14: 140 (1988)—Germany.

RUBUS :–
hirsutior *Fitschen ex H.E. Weber in Osnabrücker Naturwiss. Mitt.*, 14: 144 (1988)—Germany.
horstensis *Focke ex T. Braeucker, 292 Deutsche Rubus-Arten und Formen:* 32 (1882)—Germany.
macvaughianus *J. Rzedowski & G. Calderón de Rzedowski in Acta Bot. Mex.*, 5: 1 (1989)—Mexico.
×masakii *N. Naruhashi in J. Phytogeogr. Taxon.*, 37(1): 1 (1989)—Japan. (R. crataegifolius × hirsutus).
micranthus *O. Sendtner in Flora,* 39(13): 200 (1856)—Germany.
mucronatus *T. Braeucker, 292 Deutsche Rubus-Arten und Formen:* 90 (1882)—Germany.
palmatus
f. coronarius *H. Ohba in J. Jap. Bot.*, 64(5): 160 (1989)—Japan.
prahlii *Focke ex R. Hülsen in Verh. Bot. Ver. Prov. Brandenburg,* 40: xxxiii (1898 publ. 1899)—Germany.
pullus *T. Braeucker, 292 Deutsche Rubus-Arten und Formen:* 52 (1882)—Germany.
rectangulatus *T. Braeucker, 292 Deutsche Rubus-Arten und Formen:* 69 (1882)—Germany.
rhytidophyllus *H.E. Weber in Osnabrücker Naturwiss. Mitt.*, 14: 148 (1988)—Germany.
rotundifolius *F. Krašan in Verh. Zool.-Bot. Ges. Wien,* 15: 357 (1865)—Jugoslavia, France, Italy.
rubiginosus *T. Braeucker, 292 Deutsche Rubus-Arten und Formen:* 62 (1882)—Germany.
subobscurus *T. Braeucker, 292 Deutsche Rubus-Arten und Formen:* 53 (1882)—Germany.
tenellus *T. Braeucker, 292 Deutsche Rubus-Arten und Formen:* 30 (1882)—Germany.
thyrsigeriformis *(Sudre) D.E. Allen in Watsonia,* 17(4): 436 (1989): R. insericatus var. thyrsigeriformis.
umbrosus *T. Braeucker, 292 Deutsche Rubus-Arten und Formen:* 31 (1882)—Germany.
vastulus *T. Braeucker, 292 Deutsche Rubus-Arten und Formen:* 74 (1882)—Germany.
velaris *T. Braeucker, 292 Deutsche Rubus-Arten und Formen:* 82 (1882)—Germany.
visurgianus *H.E. Weber in Osnabrücker Naturwiss. Mitt.*, 14: 152 (1988)—Germany.

RUDGEA (Rubiac.).
corocoroensis *J.A. Steyermark in Ann. Missouri Bot. Gard.*, 75(1): 349 (1988)—Venezuela.

RUELLIA (Acanthac.).
calimensis *D.C. Wasshausen in Rev. Acad. Colomb. Cienc. Exact. Fis. Nat.*, 65: 249 (1989)—Colombia.

RUELLIA :–
cuatrecasasii *D.C. Wasshausen in Rev. Acad. Colomb. Cienc. Exact. Fis. Nat.*, 65: 251 (1989)—Colombia.
macrosolen *Lillo ex C. Ezcurra in Syst. Bot.*, 14(3): 297 (1989)—Argentina, Bolivia, Paraguay.
mirandana *T.P. Ramamoorthy & Y. Hornelas Uribe in Pl. Syst. Evol.*, 159(3–4): 161 (1988), nom. nov.: Blechum grandiflorum *Oersted.*
sarukhaniana *T.P. Ramamoorthy in Ann. Missouri Bot. Gard.*, 75(4): 1664 (1988 publ. 1989)—Mexico.
tuxtlensis *T.P. Ramamoorthy & Y. Hornelas Uribe in Pl. Syst. Evol.*, 159(3–4): 162 (1988)—Mexico.

RUMEX (Polygonac.).
acetosa
subsp. hibernicus *(Rech. f.) J.R. Akeroyd in Watsonia,* 17(4): 444 (1989): R. hibernicus.
crispus
subsp. littoreus *(Hardy) J.R. Akeroyd in Watsonia,* 17(4): 444 (1989): R. crispus var. littoreus.
subsp. uliginosus *(Le Gall) J.R. Akeroyd in Watsonia,* 17(4): 444 (1989): R. crispus var. uliginosus.
leptocaulis *J. Brandbyge & K.H. Rechinger in Nordic J. Bot.*, 9(2): 203 (1989)—Ecuador.

RUMFORDIA (Compos.).
exauriculata *B.L. Turner in Phytologia,* 65(6): 491 (1989)—Mexico.

RUNGIA (Acanthac.).
adnata *(Imlay) B. Hansen in Nordic J. Bot.*, 9(2): 211 (1989): Justicia adnata.
burmanica *(C.B. Clarke) B. Hansen in Nordic J. Bot.*, 9(2): 211 (1989): Justicia burmanica.
clauda *(R. Ben.) B. Hansen in Nordic J. Bot.*, 9(2): 211 (1989): Justicia clauda.
eberhardtii *(R. Ben.) B. Hansen in Nordic J. Bot.*, 9(2): 211 (1989): Justicia eberhardtii.
monetaria *(R. Ben.) B. Hansen in Nordic J. Bot.*, 9(2): 211 (1989): Justicia monetaria.
naoensis *B. Hansen in Nordic J. Bot.*, 9(2): 211 (1989)—Thailand.
oligoneura *(Imlay) B. Hansen in Nordic J. Bot.*, 9(2): 212 (1989): Justicia oligoneura.
purpurascens *(Ridl.) B. Hansen in Nordic J. Bot.*, 9(2): 212 (1989): Justicia purpurascens.
selangorensis *(C.B. Clarke) B. Hansen in Nordic J. Bot.*, 9(2): 213 (1989): Justicia selangorensis.
subtilifolia *(Imlay) B. Hansen in Nordic J. Bot.*, 9(2): 213 (1989): Justicia subtilifolia.
vegeta *(Ridl.) B. Hansen in Nordic J. Bot.*, 9(2): 213 (1989): Justicia vegeta.

S

SABAZIA (Compos.).
elata *(Canne) B.L. Turner in Phytologia,* 65(6): 495 (1989): Galinsoga elata.
glandulosa *(Canne) B.L. Turner in Phytologia,* 65(6): 495 (1989): Galinsoga glandulosa.

SABIA (Sabiac.).
schumanniana
subsp. longipes *(Rehd. & Wils.) C.Y. Chang in Fl. Sichuanica,* 4: 161 (1988): S. schumanniana var. longipes.

SABICEA (Rubiac.).
bariensis *J.A. Steyermark in Ann. Missouri Bot. Gard.,* 75(1): 350 (1988)—Venezuela.

SAIROCARPUS D.A. Sutton, Revis. tribe Antirrhineae: 461 (1988). *SCROPHULARIACEAE.*
breweri *(A. Gray) D.A. Sutton, Revis. tribe Antirrhineae:* 474 (1988): Antirrhinum breweri.
cornutus *(Bentham) D.A. Sutton, Revis. tribe Antirrhineae:* 468 (1988): Antirrhinum cornutum.
costatus *(Wiggins) D.A. Sutton, Revis. tribe Antirrhineae:* 472 (1988): Antirrhinum costatum.
coulterianus *(A.DC.) D.A. Sutton, Revis. tribe Antirrhineae:* 468 (1988): Antirrhinum coulterianum.
elmeri *(Rothm.) D.A. Sutton, Revis. tribe Antirrhineae:* 475 (1988): Antirrhinum elmeri.
kingii *(S. Watson) D.A. Sutton, Revis. tribe Antirrhineae:* 470 (1988): Antirrhinum kingii.
multiflorus *(Pennell) D.A. Sutton, Revis. tribe Antirrhineae:* 467 (1988): Antirrhinum multiflorum.
nuttallianus *(A.DC.) D.A. Sutton, Revis. tribe Antirrhineae:* 464 (1988): Antirrhinum nuttallianum.
pusillus *(T.S. Brandegee) D.A. Sutton, Revis. tribe Antirrhineae:* 466 (1988): Antirrhinum pusillum.
subcordatus *(A. Gray) D.A. Sutton, Revis. tribe Antirrhineae:* 477 (1988): Antirrhinum subcordatum.
vexillocalyculatus *(Kellogg) D.A. Sutton, Revis. tribe Antirrhineae:* 472 (1988): Antirrhinum vexillocalyculatum.
virga *(A. Gray) D.A. Sutton, Revis. tribe Antirrhineae:* 466 (1988): Antirrhinum virga.
watsonii *(Vasey & Rose) D.A. Sutton, Revis. tribe Antirrhineae:* 471 (1988): Antirrhinum watsonii.

SALIX (Salicac.).
myrtilloides
subsp. ustnerensis *N.M. Bol'shakov in Bot. Zhurn.,* 74(10): 1497 (1989)—U.S.S.R. (Siberia).
nigricans
subsp. alpicola *(Buser) E. Lautenschlager in Bauhinia,* 9(1): 10 (1988): S. nigricans var. alpicola.

SALOMONIA (Polygalac.).
ciliata
f. oblongifolia *(DC.) V.S. Raju in Indian J. Bot.,* 9(1): 75 (1986): S. oblongifolia.

SALPINCTIUM T.J. Edwards in S. Afr. J. Bot., 55(1): 7 (1989). *ACANTHACEAE.*
hirsutum *T.J. Edwards in S. Afr. J. Bot.,* 55(1): 9 (1989)—Swaziland.
natalense *(Clarke) T.J. Edwards in S. Afr. J. Bot.,* 55(1): 10 (1989): Asystasia natalensis.
stenosiphon *(Clarke) T.J. Edwards in S. Afr. J. Bot.,* 55(1): 10 (1989): Asystasia stenosiphon.

SALSOLA (Chenopodiac.).
baryosma
subsp. gaetula *(Maire) H. Freitag in Flora (Germany),* 183(1–2): 159 (1989), without basionym ref.: S. foetida var. gaetula.

SALVIA (Labiat.).
amethystina
subsp. ampelophylla *(Epling) J.R.I. Wood & R.M. Harley in Kew Bull.,* 44(2): 241 (1989): S. ampelophylla.
bogotensis
subsp. aratocensis *J.R.I. Wood & R.M. Harley in Kew Bull.,* 44(2): 260 (1989)—Colombia.
subsp. sochensis *J.R.I. Wood & R.M. Harley in Kew Bull.,* 44(2): 258 (1989)—Colombia.
subsp. suratensis *J.R.I. Wood & R.M. Harley in Kew Bull.,* 44(2): 260 (1989)—Colombia.
carnea
var. punicans *(Epling) J.R.I. Wood & R.M. Harley in Kew Bull.,* 44(2): 252 (1989): S. punicans.
var. sciaphila *J.R.I. Wood & R.M. Harley in Kew Bull.,* 44(2): 253 (1989)—Colombia.
charpini *G. Alziar in Biocosme Mésogéen,* 6(1–2): 79 (1989), nom. nov.: S. flocculosa *Epl. & Mathias.*
chicamochae *J.R.I. Wood & R.M. Harley in Kew Bull.,* 44(2): 253 (1989)—Colombia.
dabieshanensis *J.Q. He in Acta Bot. Yunnanica,* 11(4): 409 (1989)—China.
eplingiana *G. Alziar in Biocosme Mésogéen,* 5(3–4): 86 (1988), nom. nov.: S. tricuspis *Epling.*
falcata *J.R.I. Wood & R.M. Harley in Kew Bull.,* 44(2): 270 (1989)—Colombia.

SALVIA :–
herbanica *A. Santos & M. Fernández in Lazaroa,* 9: 52 (1986 publ. 1988)—Canary Is.
melaleuca
subsp. totensis *J.R.I. Wood & R.M. Harley in Kew Bull.,* 44(2): 234 (1989)—Colombia.
nubigena *J.R.I. Wood & R.M. Harley in Kew Bull.,* 44(2): 236 (1989)—Colombia.
pauciserrata
subsp. calocalicina *(Briq.) J.R.I. Wood & R.M. Harley in Kew Bull.,* 44(2): 245 (1989): S. calocalicina.
subsp. derasa *(Benth.) J.R.I. Wood & R.M. Harley in Kew Bull.,* 44(2): 244 (1989): S. derasa.
subsp. erythrocalicina *J.R.I. Wood & R.M. Harley in Kew Bull.,* 44(2): 245 (1989)—Colombia.
subsp. lasiocalicina *J.R.I. Wood & R.M. Harley in Kew Bull.,* 44(2): 245 (1989)—Colombia.
phlomoides
subsp. boissieri *(Noë) J.L. Rosúa & G. Blanca in Collect. Bot.,* 17(2): 230 (1988 publ. 1989): S. boissieri.
polystachya *A.J. Cavanilles, Icon.,* 1: 17 (1791), as 'polystachia'—Mexico.
rubescens
subsp. colombiana *(Epling) J.R.I. Wood & R.M. Harley in Kew Bull.,* 44(2): 232 (1989): S. colombiana.
subsp. dolichothrix *J.R.I. Wood & R.M. Harley in Kew Bull.,* 44(2): 229 (1989)—Colombia.
subsp. truxillensis *(Briq.) J.R.I. Wood & R.M. Harley in Kew Bull.,* 44(2): 231 (1989): S. truxillensis.
rufula
subsp. latens *(Benth.) J.R.I. Wood & R.M. Harley in Kew Bull.,* 44(2): 268 (1989): S. latens.
var. nutans *J.R.I. Wood & R.M. Harley in Kew Bull.,* 44(2): 268 (1989)—Colombia.
subsp. paezorum *J.R.I. Wood & R.M. Harley in Kew Bull.,* 44(2): 265 (1989)—Colombia.
sclareoides
var. baetica *(Boiss.) R. Figuerola in Collect. Bot.,* 17(2): 309 (1988 publ. 1989): S. baetica.
sphacelioides
subsp. trianae *J.R.I. Wood & R.M. Harley in Kew Bull.,* 44(2): 257 (1989)—Colombia.
stibali *G. Alziar in Biocosme Mésogéen,* 6(1–2): 79 (1989), nom. nov.: S. pauciflora *Stibal.*
uribei *J.R.I. Wood & R.M. Harley in Kew Bull.,* 44(2): 261 (1989)—Colombia.
whitehousei *G. Alziar in Biocosme Mésogéen,* 5(3–4): 86 (1988), nom. nov.: S. dolichantha *(Cory) Whitehouse.*

SANICULA (Umbellif.).
odorata *(Raf.) K.M. Pryer & L.R. Phillippe in Canad. J. Bot.,* 67(3): 703 (1989): Triclinium odoratum.

SANICULA :–
tienmuensis
var. pauciflora *R.H. Shan & F.T. Pu in Acta Phytotax. Sin.,* 27(1): 66 (1989)—China.

SANTALUM (Santalac.).
austrocaledonicum
var. minutum *N. Hallé in Fl. N. Caléd. & Dépend.,* 15: 110 (1988)—New Caledonia.
var. pilosulum *N. Hallé in Fl. N. Caléd. & Dépend.,* 15: 110 (1988)—New Caledonia.

SARCOCHILUS (Orchidac.).
sect. Calcares *K. Senghas in Schlechter Orchideen,* 1(22): 1390 (1989).
williamsianus *P.J. Kores in Allertonia,* 5(1): 187 (1989)—Fiji.

SARCODRIMYS (Baill.) M.G. Baumann-Bodenheim, Syst. Fl. Neu-Caledonien, 4: 102 (1988 publ. Jan. 1989), without basionym ref.; et Syst. Fl. Neu-Caledonien, 5: 97 (May 1989), without basionym ref.: Drimys § Sarcodrimys (Winterac.).
crassifolium *(Baill.) M.G. Baumann-Bodenheim, Syst. Fl. Neu-Caledonien,* 4: 102 (1988 publ. Jan. 1989), without basionym ref.; *et Syst. Fl. Neu-Caledonien,* 5: 97 (May 1989), with basionym ref., generic name invalid: Drimys crassifolium.

SASA (Gramin.).
subgen. Sasamorpha *(Nakai) C.H. Hu in Bamboo Res.,* 1985(2): 60 (1985): gen. Sasamorpha.
hubeiensis *(C.H. Hu) C.H. Hu in Bamboo Res.,* 1985(2): 60 (1985): Sasamorpha hubeiensis.
kurilensis
f. marmorea *(Koidz.) S. Suzuki in J. Jap. Bot.,* 64(9): 275 (1989): S. kurilensis var. marmorea.
masamuneana *(Makino) C.S. Chao & S.A. Renvoize in Kew Bull.,* 44(2): 368 (1989): Pleioblastus masamuneanus.
megalophylla
f. aureovariegata *S. Suzuki in J. Jap. Bot.,* 64(9): 275 (1989)—Japan.
qingyuanensis *(C.H. Hu) C.H. Hu in Bamboo Res.,* 1985(2): 62 (1985): Sasamorpha qingyuanensis.
rubrovaginata *C.H. Hu in Bamboo Res.,* 1985(2): 59 (1985)—China.

SATUREJA (Labiat.).
abyssinica
subsp. condensata *(Hedberg) S. Seybold in Stuttgarter Beitr. Naturk., A,* 421: 18 (1988): S. abyssinica var. condensata.
paradoxa *(Vatke) Engler ex S. Seybold in Stuttgarter Beitr. Naturk., A,* 421: 13 (1988): Calamintha paradoxa.
punctata
subsp. ovata *(Bentham) S. Seybold in Stuttgarter Beitr. Naturk., A,* 421: 31 (1988): Micromeria ovata.

SAURAUIA (Actinidiac.).
oroquensis *D.D. Soejarto in Brittonia,* 41(1): 28 (1989)—Colombia, Venezuela.

SAUSSUREA (Compos.).
bella *Y. Ling in Contrib. Inst. Bot. Nation. Acad. Peiping,* 6(2): 87 (1949)—China.†
gubanovii *R.V. Kamelin in Byull. Mosk. Obshch. Ispўt. Prir., Biol.,* 93(5): 113 (1988)—Mongolia.
handeliana *Y. Ling in Contrib. Inst. Bot. Nation. Acad. Peiping,* 6(2): 88 (1949), nom. nov.: S. modesta *Handel-Mazzetti.*
hwangshanensis *Y. Ling in Contrib. Inst. Bot. Nation. Acad. Peiping,* 6(2): 79 (1949)—China.
kokonorensis *Y. Ling in Contrib. Inst. Bot. Nation. Acad. Peiping,* 6(2): 90 (1949)—China.
nematolepis *Y. Ling in Contrib. Inst. Bot. Nation. Acad. Peiping,* 6(2): 67 (1949)—China.
nipponica
subsp. higomontana *(Honda) H.T. Im in J. Fac. Sci. Univ. Tokyo, Bot.,* 14(4): 261 (1989): S. higomontana.
subsp. yakusimensis *(Masamune) H.T. Im in J. Fac. Sci. Univ. Tokyo, Bot.,* 14(4): 263 (1989): S. yakusimensis.
oligocephala *(Ling) Y. Ling in Contrib. Inst. Bot. Nation. Acad. Peiping,* 6(2): 83 (1949): S. acrophila var. oligocephala.
oligolepis *Y. Ling in Contrib. Inst. Bot. Nation. Acad. Peiping,* 6(2): 81 (1949)—China.
paucijuga *Y. Ling in Contrib. Inst. Bot. Nation. Acad. Peiping,* 6(2): 88 (1949)—China.
reniformis *Y. Ling in Contrib. Inst. Bot. Nation. Acad. Peiping,* 6(2): 77 (1949)—China.
stenocephala *Y. Ling in Contrib. Inst. Bot. Nation. Acad. Peiping,* 6(2): 75 (1949)—China.
tenella *Y. Ling in Contrib. Inst. Bot. Nation. Acad. Peiping,* 6(2): 85 (1949)—China.
tenuicaulis *Y. Ling in Contrib. Inst. Bot. Nation. Acad. Peiping,* 6(2): 66 (1949)—China.
variiloba *Y. Ling in Contrib. Inst. Bot. Nation. Acad. Peiping,* 6(2): 71 (1949)—China.

SAXIFRAGA (Saxifragac.).
aristulata
var. earistulata *T.C. Ku in Bull. Bot. Res. North-East. Forest. Inst.,* 9(4): 14 (1989)—China.
aurantiaca
f. lanceolata *T.C. Ku in Bull. Bot. Res. North-East. Forest. Inst.,* 9(4): 18 (1989)—China.
balongshanensis *T.C. Ku in Bull. Bot. Res. North-East. Forest. Inst.,* 9(4): 7 (1989)—China.
biflora *T.C. Ku in Bull. Bot. Res. North-East. Forest. Inst.,* 9(4): 18 (1989)—China.

SAXIFRAGA :-
clavistamineoides *T.C. Ku in Bull. Bot. Res. North-East. Forest. Inst.,* 9(4): 19 (1989)—China.
cuneata
subsp. corbariensis *(Timb.-Lagr.) G. Mateo Sanz & M.B. Crespo Villalba in Fontqueria,* 24: 7 (1989): S. corbariensis.
cymiformis *T.C. Ku in Bull. Bot. Res. North-East. Forest. Inst.,* 9(4): 4 (1989)—China.
gongshanensis *T.C. Ku in Bull. Bot. Res. North-East. Forest. Inst.,* 9(4): 9 (1989)—China.
granulata
subsp. glaucescens *(Reuter) S. Rivas-Martínez, F. Fernández González & D. Sánchez-Mata in Opusc. Bot. Pharm. Complutensis,* 2: 116 (1986): S. glaucescens.
hypericoides
var. glabrescens *T.C. Ku in Bull. Bot. Res. North-East. Forest. Inst.,* 9(4): 3 (1989)—China.
f. longipetala *T.C. Ku in Bull. Bot. Res. North-East. Forest. Inst.,* 9(4): 2 (1989)—China.
kangdingensis *T.C. Ku in Bull. Bot. Res. North-East. Forest. Inst.,* 9(4): 9 (1989)—China.
latipetala *T.C. Ku in Bull. Bot. Res. North-East. Forest. Inst.,* 9(4): 6 (1989)—China.
lixianensis *T.C. Ku in Bull. Bot. Res. North-East. Forest. Inst.,* 9(4): 17 (1989)—China.
longipetala *T.C. Ku in Bull. Bot. Res. North-East. Forest. Inst.,* 9(4): 10 (1989)—China.
mandenovae *J. Soják in Čas. Nár. Muz. (Prague),* 154(2): 93 (1985 publ. 1988), nom. nov.: S. meyeri *Mandenova.*
microgyna
f. uniflora *T.C. Ku in Bull. Bot. Res. North-East. Forest. Inst.,* 9(4): 19 (1989)—China.
montana
f. densifolia *T.C. Ku in Bull. Bot. Res. North-East. Forest. Inst.,* 9(4): 13 (1989)—China.
f. oblongipetala *T.C. Ku in Bull. Bot. Res. North-East. Forest. Inst.,* 9(4): 13 (1989)—China.
obovatipetala *T.C. Ku in Bull. Bot. Res. North-East. Forest. Inst.,* 9(4): 16 (1989)—China.
oligophylla *T.C. Ku in Bull. Bot. Res. North-East. Forest. Inst.,* 9(4): 8 (1989)—China.
orogredensis *S. Rivas-Martínez, V. de la Fuente & D. Sánchez-Mata in Opusc. Bot. Pharm. Complutensis,* 4: 123 (1988)—Spain.
ovatipetala *T.C. Ku in Bull. Bot. Res. North-East. Forest. Inst.,* 9(4): 3 (1989)—China.
pauciflora *T.C. Ku in Bull. Bot. Res. North-East. Forest. Inst.,* 9(4): 14 (1989)—China.
peplidifolia
var. angustipetala *T.C. Ku in Bull. Bot. Res. North-East. Forest. Inst.,* 9(4): 2 (1989)—China.
platyloba *G. Mateo Sanz & M.B. Crespo Villalba in Fontqueria,* 24: 7 (1989)—Spain.
reflexa *T.C. Ku in Bull. Bot. Res. North-East. Forest. Inst.,* 9(4): 4 (1989)—China.

SAXIFRAGA :-
riagdomensis *T.C. Ku in Bull. Bot. Res. North-East. Forest. Inst.*, 9(4): 2 (1989)—Xizang.
rupestris *T.C. Ku in Bull. Bot. Res. North-East. Forest. Inst.*, 9(4): 7 (1989)—China.
spathulifolia *T.C. Ku in Bull. Bot. Res. North-East. Forest. Inst.*, 9(4): 11 (1989)—China.
wenchuanensis *T.C. Ku in Bull. Bot. Res. North-East. Forest. Inst.*, 9(4): 16 (1989)—China.
xiaojinensis *T.C. Ku in Bull. Bot. Res. North-East. Forest. Inst.*, 9(4): 5 (1989)—China.
zayuensis *T.C. Ku in Bull. Bot. Res. North-East. Forest. Inst.*, 9(4): 12 (1989)—Xizang.
f. angustipetala *T.C. Ku in Bull. Bot. Res. North-East. Forest. Inst.*, 9(4): 13 (1989)—China.
zhejiangensis *Z. Wei & Y.B. Chang in Bull. Bot. Res. North-East. Forest. Inst.*, 9(2): 33 (1989)—China.
zogangensis *T.C. Ku in Bull. Bot. Res. North-East. Forest. Inst.*, 9(4): 13 (1989)—Xizang.
var. pilosa *T.C. Ku in Bull. Bot. Res. North-East. Forest. Inst.*, 9(4): 14 (1989)—Xizang.

SCABIOSA (Dipsacac.).
sect. Olivierianae *K.H. Rechinger in Willdenowia*, 19(1): 139 (1989).
deserticola *K.H. Rechinger in Willdenowia*, 19(1): 142 (1989)—Iraq, Sinai Peninsula, Jordan, Syria, Saudi Arabia, Iran, U.S.S.R. (Middle Asia), Afghanistan, Pakistan.
transcaspica *K.H. Rechinger in Willdenowia*, 19(1): 148 (1989)—U.S.S.R. (Middle Asia, Caucasus), Iran.

SCABIOSIOPSIS K.H. Rechinger in Willdenowia, 19(1): 153 (1989). *DIPSACACEAE.*
enigmatica *K.H. Rechinger in Willdenowia*, 19(1): 153 (1989)—Iran.

SCAPHIUM (Sterculiac.).
velutinosum *A.J.G.H. Kostermans in Reinwardtia*, 10(5): 478 (1988)—Sumatra.

SCAPHOSEPALUM (Orchidac.).
sect. Distichium *C.A. Luer, Ic. Pleurothallid. V (Monogr. Syst. Bot. Missouri Bot. Gard. 26)*: 26 (1988).
sect. Leiocaulium *C.A. Luer, Ic. Pleurothallid. V (Monogr. Syst. Bot. Missouri Bot. Gard. 26)*: 26 (1988).
cimex *C.A. Luer & Hirtz in C.A. Luer, Ic. Pleurothallid. V (Monogr. Syst. Bot. Missouri Bot. Gard. 26)*: 46 (1988)—Ecuador.
fimbriatum *C.A. Luer & Hirtz in C.A. Luer, Ic. Pleurothallid. V (Monogr. Syst. Bot. Missouri Bot. Gard. 26)*: 57 (1988)—Ecuador.
swertiifolium
subsp. exiguum *C.A. Luer & Escobar in C.A. Luer, Ic. Pleurothallid. V (Monogr. Syst. Bot. Missouri Bot. Gard. 26)*: 91 (1988)—Colombia, Ecuador.

SCAPHOSEPALUM :-
triceratops *C.A. Luer & Andreetta in C.A. Luer, Ic. Pleurothallid. V (Monogr. Syst. Bot. Missouri Bot. Gard. 26)*: 96 (1988)—Ecuador.
xystra *C.A. Luer, Ic. Pleurothallid. V (Monogr. Syst. Bot. Missouri Bot. Gard. 26)*: 104 (1988)—Venezuela.

SCHEFFLERA (Araliac.).
albocapitata *M.J. & J.F.M. Cannon in Bull. Brit. Mus. (Nat. Hist.), Bot.*, 19: 54 (1989)—Panama.
aquaverensis *M.J. & J.F.M. Cannon in Bull. Brit. Mus. (Nat. Hist.), Bot.*, 19: 54 (1989)—Panama.
bifida *M.J. & J.F.M. Cannon in Bull. Brit. Mus. (Nat. Hist.), Bot.*, 19: 51 (1989)—Panama.
cabalionii *P.P. Lowry in Bull. Mus. Nation. Hist. Nat., B, Adansonia*, Ser. 4, 11(2): 125 (1989)—Vanuatu.
caduca *M.J. & J.F.M. Cannon in Bull. Brit. Mus. (Nat. Hist.), Bot.*, 19: 47 (1989)—Panama.
cartagoensis *M.J. & J.F.M. Cannon in Bull. Brit. Mus. (Nat. Hist.), Bot.*, 19: 54 (1989)—Costa Rica.
cicatricata *M.J. & J.F.M. Cannon in Bull. Brit. Mus. (Nat. Hist.), Bot.*, 19: 58 (1989)—Panama.
coclensis *M.J. & J.F.M. Cannon in Bull. Brit. Mus. (Nat. Hist.), Bot.*, 19: 32 (1989)—Panama.
elegantissima *(Veitch ex Masters) P.P. Lowry & D.G. Frodin in Baileya*, 23(1): 9 (1989): Aralia elegantissima.
instita *M.J. & J.F.M. Cannon in Bull. Brit. Mus. (Nat. Hist.), Bot.*, 19: 58 (1989)—Costa Rica.
jefensis *M.J. & J.F.M. Cannon in Bull. Brit. Mus. (Nat. Hist.), Bot.*, 19: 43 (1989)—Panama.
kerchoveiana *(Veitch ex W. Richards) D.G. Frodin & P.P. Lowry in Baileya*, 23(1): 9 (1989): Aralia kerchoveiana.
latiligulata *M.J. & J.F.M. Cannon in Bull. Brit. Mus. (Nat. Hist.), Bot.*, 19: 43 (1989)—Panama.
macphersonii *M.J. & J.F.M. Cannon in Bull. Brit. Mus. (Nat. Hist.), Bot.*, 19: 34 (1989)—Panama.
octostyla *M.J. & J.F.M. Cannon in Bull. Brit. Mus. (Nat. Hist.), Bot.*, 19: 38 (1989)—Panama.
panamensis *M.J. & J.F.M. Cannon in Bull. Brit. Mus. (Nat. Hist.), Bot.*, 19: 34 (1989)—Panama.
pubens *M.J. & J.F.M. Cannon in Bull. Brit. Mus. (Nat. Hist.), Bot.*, 19: 51 (1989)—Costa Rica.
reiniana *D.G. Frodin in Proc. Acad. Nat. Sci. Philadelphia*, 140(2): 337 (1988), nom. nov.: S. aromatica f. scandens *Hochreutiner*.

SCHEFFLERA :-
revolutissima *M.G. Baumann-Bodenheim, Syst. Fl. Neu-Caledonien,* 5: 77, 67 (1989), without latin descr. or type—New Caledonia.
rodriguesiana *Frodin ex M.J. & J.F.M. Cannon in Bull. Brit. Mus. (Nat. Hist.), Bot.,* 19: 61 (1989), nom. nov.: Didymopanax pittieri *Marchal.*
sapoensis *M.J. & J.F.M. Cannon in Bull. Brit. Mus. (Nat. Hist.), Bot.,* 19: 38 (1989)—Panama.
simplex *J.A. Steyermark & Holst in Ann. Missouri Bot. Gard.,* 75(3): 1082 (1988)—Venezuela.
vanuatua *P.P. Lowry in Bull. Mus. Nation. Hist. Nat., B, Adansonia,* Ser. 4, 11(2): 129 (1989)—Vanuatu.
veitchii *(Carrière) D.G. Frodin & P.P. Lowry in Baileya,* 23(1): 11 (1989): Aralia veitchii.
whitefoordiae *M.J. & J.F.M. Cannon in Bull. Brit. Mus. (Nat. Hist.), Bot.,* 19: 47 (1989)—Panama.
yutajensis *J.A. Steyermark & Holst in Ann. Missouri Bot. Gard.,* 75(3): 1082 (1988)—Venezuela.

SCHIEDEELLA (Orchidac.).
dodii *P. Burns-Balogh in Orchidee,* 40(5): 169 (1989)—Hispaniola (Dominican Republic).
fauci-sanguinea *(Dod) P. Burns-Balogh in Orchidee,* 40(5): 171 (1989), without basionym date: Spiranthes fauci-sanguinea.
nonantzin *(Glz. Tamayo ex McVaugh) González Tamayo ex M.A. Soto Arenas in Orquídea (México),* 11: 271 (1988): Spiranthes nonantzin.

SCHIZACHYRIUM (Gramin.).
mitchelliana *B.K. Simon in Austrobaileya,* 3(1): 90 (1989)—Australia (W. Australia).

SCHIZOGLOSSUM (Asclepiadac.).
amatolicum *O.M. Hilliard in Notes Roy. Bot. Gard. Edinburgh,* 45(2): 182 (1988 publ. 1989)—South Africa (Cape Province).
rubiginosum *O.M. Hilliard in Notes Roy. Bot. Gard. Edinburgh,* 45(2): 182 (1988 publ. 1989)—South Africa (Natal).

SCHIZOPETALON (Crucif.).
arcuatum *I.A. Al-Shehbaz in Harvard Pap. Bot.,* 1: 34 (1989)—Chile.
brachycarpum *I.A. Al-Shehbaz in Harvard Pap. Bot.,* 1: 38 (1989)—Chile.
corymbosum *I.A. Al-Shehbaz in Harvard Pap. Bot.,* 1: 41 (1989)—Chile.
maipoanum *P. Ravenna in Onira, Bot. Leafl.,* 1(4): 32 (1988), as 'Schyzopetalon'—Chile.

SCHOENOCAULON (Melanthiac.).
jaliscense
var. regulare *(Brinker) D. Frame in R. McVaugh, Fl. Novo-Galiciana,* 15: 266 (1989): S. regulare.

SCHOENOCAULON :-
lauricola *McVaugh ex D. Frame in R. McVaugh, Fl. Novo-Galiciana,* 15: 267 (1989)—Mexico.
pellucidum *D. Frame in R. McVaugh, Fl. Novo-Galiciana,* 15: 272 (1989)—Mexico.
tigrense *D. Frame in R. McVaugh, Fl. Novo-Galiciana,* 15: 274 (1989)—Mexico.

SCHOENOCEPHALIUM (Rapateac.).
sipapoanum *B. Maguire in Acta Bot. Venezuel.,* 14(3): 17 (1984)—Venezuela.

SCHOENOPLECTUS (Cyperac.).
lacustris
subsp. validus *(Vahl) T. Koyama in Bishop Mus. Occas. Pap.,* 29: 128 (1989): Scirpus validus.

SCHOEPFIA (Olacac.).
clarkii *J.A. Steyermark in Ann. Missouri Bot. Gard.,* 75(3): 1061 (1988)—Venezuela.

SCHWARTZIA (Marcgraviac.).
spiciflora *(A.L. Juss.) H.G. Bedell in R.A. Howard, Fl. Lesser Antilles (Dicot. 2),* 5: 309 (1989): Marcgravia spiciflora.

SCILLA (Hyacinthac.).
buekkensis *F. Speta in Naturk. Jahrb. Linz,* 22: 42 (1976)—Hungary, Czechoslovakia.
drunensis *(Speta) F. Speta in Naturk. Jahrb. Linz,* 22: 34 (1976): S. bifolia subsp. drunensis.
lus. albescens *Z. Kereszty in Bot. Közlem.,* 74–75(1–2): 69 (1987–88 publ. 1988)—Hungary.
lus. albida *Z. Kereszty in Bot. Közlem.,* 74–75(1–2): 69 (1987–88 publ. 1988)—Hungary.
subsp. buekkensis *(Speta) Z. Kereszty in Bot. Közlem.,* 74–75(1–2): 69 (1987–88 publ. 1988), without basionym ref.: S. buekkensis.
siberica *A.H. Haworth in H. Andrews, Botanist's Repository,* 6: t. 365 (1804)—Siberia.†
spetana *Z. Kereszty in Bot. Közlem.,* 74–75(1–2): 69 (1987–88 publ. 1988)—Hungary.
vindobonensis
lus. alba *(Beck) Z. Kereszty in Bot. Közlem.,* 74–75(1–2): 67 (1987–88 publ. 1988): S. bifolia β alba.
lus. albicans *Z. Kereszty in Bot. Közlem.,* 74–75(1–2): 68 (1987–88 publ. 1988)—Hungary.
subsp. borhidiana *Z. Kereszty in Bot. Közlem.,* 74–75(1–2): 68 (1987–88 publ. 1988)—Hungary.
lus. rosacea *Z. Kereszty in Bot. Közlem.,* 74–75(1–2): 67 (1987–88 publ. 1988), homonym see p. 68—Hungary.
lus. rosacea *Z. Kereszty in Bot. Közlem.,* 74–75(1–2): 68 (1987–88 publ. 1988), homonym see p. 67—Hungary.

SCILLA :–
lus. rubra *(Beck) Z. Kereszty in Bot. Közlem.*, 74–75(1–2): 67 (1987–88 publ. 1988): S. bifolia γ rubra.
var. transdanubialis *Z. Kereszty in Bot. Közlem.*, 74–75(1–2): 68 (1987–88 publ. 1988)—Hungary.

SCLEROCACTUS (Cactac.).
wetlandicus *F. Hochstätter in Succulenta,* 68(6): 123, 125 (1989)—U.S.A. (Utah).

SCORZONERA (Compos.).
subgen. Lasiospora *(Cass.) N.N. Tsvelev in Novosti Sist. Vÿssh. Rast.*, 25: 190 (1988): gen. Lasiospora.
laciniata
f. decumbens *(Guss.) O. de Bolòs & J. Vigo in Folia Bot. Misc.*, 6: 85 (1989): S. calcitrapifolia b. decumbens.
f. latifolia *(Gren. & Godr.) O. de Bolòs & J. Vigo in Folia Bot. Misc.*, 6: 85 (1989): Podospermum laciniatum var. latifolia.
f. subulata *(Lam.) O. de Bolòs & J. Vigo in Folia Bot. Misc.*, 6: 85 (1989): S. subulata.

SCROPHULARIA (Scrophulariac.).
bahorucana *T.A. Zanoni in Moscosoa*, 5: 108 (1989)—Hispaniola (Dominican Republic).

SEBASTIANIA (Euphorbiac.).
panamensis *G.L. Webster in Ann. Missouri Bot. Gard.*, 75(3): 1128 (1988)—Panama.

SECURIGERA (Legumin.).
carinata *P. Lassen in Willdenowia,* 19(1): 62 (1989)—Rhodes, Turkey.
cretica *(L.) P. Lassen in Willdenowia,* 19(1): 60 (1989): Coronilla cretica.
elegans *(Pančić) P. Lassen in Willdenowia,* 19(1): 60 (1989): Coronilla elegans.
globosa *(Lam.) P. Lassen in Willdenowia,* 19(1): 60 (1989): Coronilla globosa.
grandiflora *P. Lassen in Willdenowia,* 19(1): 60 (1989), nom. nov.: Coronilla grandiflora *Boiss.*
libanotica *(Boiss.) P. Lassen in Willdenowia,* 19(1): 60 (1989): Coronilla libanotica.
orientalis *(Mill.) P. Lassen in Willdenowia,* 19(1): 60 (1989): Coronilla orientalis.
parviflora *(Desv.) P. Lassen in Willdenowia,* 19(1): 60 (1989): Artrolobium parviflorum.
somalensis *(Thulin) P. Lassen in Willdenowia,* 19(1): 60 (1989): Coronilla somalensis.
varia *(L.) P. Lassen in Svensk Bot. Tidskr.*, 83(2): 86 (1989): Coronilla varia.

SEDUM (Crassulac.).
creticum
var. monocarpicum *H. 't Hart in Taxon,* 38(4): 650 (1989), nom. nov.: S. creticum *Boiss. & Heldr.*
himalense
subsp. taohoense *(Fu) Yu.P. Kozhevnikov in Bot. Zhurn.*, 74(4): 543 (1989): Rhodiola taohoensis.

SEDUM :–
jiulungshanense *Y.C. Ho in Bull. Bot. Res. North-East. Forest. Inst.*, 9(4): 31 (1989)—China.
ludlowii *(Ohba) Yu.P. Kozhevnikov in Bot. Zhurn.*, 74(4): 543 (1989): Rhodiola ludlowii.
nicaense
subsp. dianium *(O. Bolós) S. Rivas-Martínez in Opusc. Bot. Pharm. Complutensis,* 3: 88 (1986): S. sediforme subsp. dianium.
var. saguntinum *(O. Bolós) S. Rivas-Martínez in Opusc. Bot. Pharm. Complutensis,* 3: 88 (1986): S. sediforme var. saguntinum.
ohbae *Yu.P. Kozhevnikov in Bot. Zhurn.*, 74(4): 543 (1989), nom. nov.: Rhodiola angusta *Nakai.*
orbatum *R. Moran & J. Meyrán in Cact. Suc. Mex.*, 33(3): 55 (1988)—Cultivation.
pinnatifidum *(Boriss.) Yu.P. Kozhevnikov in Bot. Zhurn.*, 74(4): 543 (1989): Rhodiola pinnatifida.
roseum
subsp. arcticum *(Boriss.) Yu.P. Kozhevnikov in Bot. Zhurn.*, 74(4): 537 (1989): Rhodiola arctica.
serratum *(Ohba) Yu.P. Kozhevnikov in Bot. Zhurn.*, 74(4): 537 (1989): Rhodiola serrata.
sherriffii *(Ohba) Yu.P. Kozhevnikov in Bot. Zhurn.*, 74(4): 537 (1989): Rhodiola sherriffii.
tianmushanense *Y.C. Ho & F. Chai in Bull. Bot. Res. North-East. Forest. Inst.*, 9(4): 32 (1989)—China.

×SELENIPHYLCHIA F. Süpplie in Succulenta, 67(12): 258 (1988). *CACTACEAE.* (Epiphyllum × Nopalxochia × Selenicereus).

SEMIAQUILEGIA (Ranunculac.).
adoxoides
var. grandis *D.Q. Wang in Bull. Bot. Res. North-East. Forest. Inst.*, 9(4): 53 (1989)—China.
dauciformis *D.Q. Wang in Bull. Bot. Res. North-East. Forest. Inst.*, 9(4): 51 (1989)—China.

SEMIARUNDINARIA (Gramin.).
densiflora *(Rendle) T.H. Wen in J. Bamboo Res.*, 8(1): 24 (1989): Arundinaria densiflora.
f. villosa *(S.L. Chen & Yao) T.H. Wen in J. Bamboo Res.*, 8(1): 24 (1989): Brachystachyum densiflorum var. villosum.
sinica *T.H. Wen in J. Bamboo Res.*, 8(1): 13 (1989)—China.

SEMPERVIVUM (Crassulac.).
atropatanum *J.A.N. Parnell in Willdenowia,* 18(2): 419 (1989), as 'atropatanus'—Iran.

SENECIO (Compos.).
billieturneri *T.M. Barkley in Phytologia,* 67(3): 238 (1989)—Mexico.
boluangensis *H. Koyama in Acta Phytotax. Geobot.,* 39(4–6): 156 (1988)—Thailand.
erucifolius
subsp. grandidentatus *(Ledeb.) B. Nordenstam in Fl. Iran.,* 164: 76 (1989): S. grandidentatus.
flaccidus
var. durangensis *(Greenm.) T.M. Barkley in Phytologia,* 67(3): 237 (1989): S. durangensis.
glaucus
subsp. hyblaeus *S. Brullo in Boll. Accad. Gioenia Sci. Nat. Catania,* 18(325): 279 (1985)—Sicily.
hintonii *(H. Robins. & Brettell) J. Pruski & T.M. Barkley in Phytologia,* 67(3): 238 (1989): Roldana hintonii.
iranicus *B. Nordenstam in Fl. Iran.,* 164: 95 (1989)—Iran.
laetus
var. analogus *(DC.) R. Mathur in R.R. Rao et al., Fl. Ind. Enumerat. - Asterac.:* 69 (1988), as '(Hook. f.)' & without ref. to true basionym: Basionym not stated.
var. sisymbriiformis *(DC.) R. Mathur in R.R. Rao et al., Fl. Ind. Enumerat. - Asterac.:* 69 (1988), as 'sisymbriformis (Hook. f.)' & without ref. to true basionym: Basionym not stated.
var. spectabilis *(Hook. f.) R. Mathur in R.R. Rao et al., Fl. Ind. Enumerat. - Asterac.:* 69 (1988): S. chrysanthemoides var. spectabilis.
lasiocaulon *T.M. Barkley in Phytologia,* 67(3): 240 (1989)—Mexico.
marquezii *B.L. Turner in Brittonia,* 40(1): 82 (1988), as 'marquesii'; *see also Brittonia,* 41(1): 43 (1989)— Mexico.†
metepecus *B.L. Turner in Phytologia,* 66(6): 465 (1989)—Mexico.
multidentatus
var. huachucanus *(A. Gray) T.M. Barkley in Phytologia,* 67(3): 238 (1989): S. huachucanus.
namnaoensis *H. Koyama in Acta Phytotax. Geobot.,* 39(4–6): 156 (1988)—Thailand.
nesomiorum *B.L. Turner in Phytologia,* 66(6): 467 (1989)—Mexico.
pattersonii *B.L. Turner in Phytologia,* 65(5): 373 (1988)—Mexico.
paulsenii
subsp. khorasanicus *(Rech. f. & Aell.) B. Nordenstam in Fl. Iran.,* 164: 68 (1989): S. khorasanicus.
porphyresthes *T.M. Barkley in Phytologia,* 67(3): 243 (1989)—Mexico.
pseudopicridis *T.M. Barkley in Phytologia,* 67(3): 244 (1989)—Mexico.
scalaris
var. carmenensis *C.C. Freeman in Phytologia,* 67(3): 249 (1989)—Mexico.

SENECIO :-
var. parrasianus *(Greenm.) C.C. Freeman in Phytologia,* 67(3): 238 (1989): S. parrasianus.
spellenbergii *T.M. Barkley in Phytologia,* 67(3): 248 (1989)—U.S.A. (New Mexico).

SENNA (Legumin.).
appendiculata *(Vogel) J.H. Wiersema in Taxon,* 38(4): 652 (1989): Cassia appendiculata.
galegifolia *(L.) R. Barneby & A. Lourteig in Phytologia,* 65(5): 400 (1988): Cassia galegifolia.

SERAPIAS (Orchidac.).
nothosubsect. Asmed *H. Baumann & S. Künkele in Mitt. Arbeitskr. Heim. Orch. Baden-Württemberg,* 21(3): 743 (1989). (S. subsect. Mediopetalae × sect. Serapias).
nothosubsect. Aspla *H. Baumann & S. Künkele in Mitt. Arbeitskr. Heim. Orch. Baden-Württemberg,* 21(3): 742 (1989). (S. subsect. Platypetalae × sect. Serapias).
nothosubsect. Assteno *H. Baumann & S. Künkele in Mitt. Arbeitskr. Heim. Orch. Baden-Württemberg,* 21(3): 743 (1989). (S. sect. Serapias × subsect. Stenopetalae).
subsect. Mediopetalae *H. Baumann & S. Künkele in Mitt. Arbeitskr. Heim. Orch. Baden-Württemberg,* 21(3): 741 (1989).
nothosubsect. Menopetalae *H. Baumann & S. Künkele in Mitt. Arbeitskr. Heim. Orch. Baden-Württemberg,* 21(3): 742 (1989). (S. subsect. Mediopetalae × subsect. Stenopetalae).
nothosubsect. Pladiopetalae *H. Baumann & S. Künkele in Mitt. Arbeitskr. Heim. Orch. Baden-Württemberg,* 21(3): 742 (1989). (S. subsect. Mediopetalae × subsect. Platypetalae).
nothosubsect. Planopetalae *H. Baumann & S. Künkele in Mitt. Arbeitskr. Heim. Orch. Baden-Württemberg,* 21(3): 742 (1989). (S. subsect. Platypetalae × subsect. Stenopetalae).
subsect. Platypetalae *H. Baumann & S. Künkele in Mitt. Arbeitskr. Heim. Orch. Baden-Württemberg,* 21(3): 741 (1989).
nothosect. Serapiaria *H. Baumann & S. Künkele in Mitt. Arbeitskr. Heim. Orch. Baden-Württemberg,* 21(3): 741 (1989). (S. sect. Bilamellaria × sect. Serapias).
subsect. Stenopetalae *H. Baumann & S. Künkele in Mitt. Arbeitskr. Heim. Orch. Baden-Württemberg,* 21(3): 741 (1989).
cycladum *H. Baumann & S. Künkele in Mitt. Arbeitskr. Heim. Orch. Baden-Württemberg,* 21(3): 755 (1989)—Greece.
×cypria *H. Baumann & S. Künkele in Mitt. Arbeitskr. Heim. Orch. Baden-Württemberg,* 21(3): 812 (1989)—Cyprus. (S. bergonii × levantina).
×euxina *H. Baumann & S. Künkele in Mitt. Arbeitskr. Heim. Orch. Baden-Württemberg,* 21(3): 810 (1989)—Turkey. (S. bergonii × feldwegiana).

SERAPIAS :-
feldwegiana *H. Baumann & S. Künkele in Mitt. Arbeitskr. Heim. Orch. Baden-Württemberg,* 21(3): 763, 760 (1989)—Turkey.
×garganica *H. Baumann & S. Künkele in Mitt. Arbeitskr. Heim. Orch. Baden-Württemberg,* 21(3): 831 (1989)—Italy. (S. orientalis × vomeracea).
gracilis *C.A.J. Kreutz in Mitt. Arbeitskr. Heim. Orch. Baden-Württemberg,* 21(1): 116 (1989)—Spain, Portugal.
levantina *H. Baumann & S. Künkele in Mitt. Arbeitskr. Heim. Orch. Baden-Württemberg,* 21(3): 766 (1989)—Israel, Syria, Lebanon, Turkey.
lorenziana *H. Baumann & S. Künkele in Mitt. Arbeitskr. Heim. Orch. Baden-Württemberg,* 21(3): 775, 770 (1989)—Morocco.
orientalis
subsp. apulica *H. Baumann & S. Künkele in Mitt. Arbeitskr. Heim. Orch. Baden-Württemberg,* 21(3): 779 (1989)—Italy.
subsp. carica *H. Baumann & S. Künkele in Mitt. Arbeitskr. Heim. Orch. Baden-Württemberg,* 21(3): 782 (1989)—Turkey, Rhodes, East Aegean Is.
×oulmesiaca *H. Baumann & S. Künkele in Mitt. Arbeitskr. Heim. Orch. Baden-Württemberg,* 21(3): 821 (1989)—Morocco. (S. lingua × lorenziana).
×provincialis *H. Baumann & S. Künkele in Mitt. Arbeitskr. Heim. Orch. Baden-Württemberg,* 21(3): 819 (1989)—France. (S. cordigera × olbia).
vomeracea
subsp. longipetala *(Ten.) H. Baumann & S. Künkele in Mitt. Arbeitskr. Heim. Orch. Baden-Württemberg,* 21(3): 796 (1989): Helleborine longipetala.
×wettsteinii
nothosubsp. halicarnassia *H. Baumann & S. Künkele in Mitt. Arbeitskr. Heim. Orch. Baden-Württemberg,* 21(3): 815 (1989)—Turkey. (S. bergonii × orientalis subsp. carica).

SERJANIA (Sapindac.).
coradinii *M.S. Ferrucci & G. Vieira Somner in Bonplandia,* 6(3): 166 (1989)—Brazil.

SESAMOIDES (Resedac.).
clusii *(Sprengel) W. Greuter & Burdet in Willdenowia,* 19(1): 47 (1989): Reseda clusii.

SESAMUM (Pedaliac.).
calycinum
subsp. repens *(Engler & Gilg) U. Grabow-Seidensticker in Mitt. Inst. Allg. Bot. Hamburg,* 22: 232 (1988): S. repens.
parviflorum *U. Grabow-Seidensticker in Mitt. Inst. Allg. Bot. Hamburg,* 22: 232 (1988)—Zaïre.

SESBANIA (Legumin.).
leptocarpa
var. minimiflora *(Gillett) G.P. Lewis in Kirkia,* 13(1): 39 (1988 publ. 1989): S. mossambicensis subsp. minimiflora.
tetraptera
subsp. rogersii *(Phill. & Hutch.) G.P. Lewis in Kirkia,* 13(1): 50 (1988 publ. 1989): S. rogersii.

SICYOS (Cucurbitac.).
alba *(H. St. John) I.R.H. Telford in Phytologia,* 67(3): 210 (1989): Sarx alba.
anunu *(H. St. John) I.R.H. Telford in Phytologia,* 67(3): 211 (1989): S. hillebrandii var. anunu.
herbstii *(H. St. John) I.R.H. Telford in Phytologia,* 67(3): 210 (1989): Cladocarpa herbstii.

SIDA (Malvac.).
arenicola *S.T. Reynolds & A.E. Holland in Austrobaileya,* 2(5): 463 (1988)—Australia (N. Terr., W. Australia).
argillacea *A.E. Holland & S.T. Reynolds in Austrobaileya,* 2(5): 465 (1988)—Australia (Queensland, N. Terr.).
asterocalyx *S.T. Reynolds & A.E. Holland in Austrobaileya,* 2(5): 463 (1988)—Australia (Queensland, W. Australia).
brachypoda *A.E. Holland & S.T. Reynolds in Austrobaileya,* 2(5): 459 (1988)—Australia (Queensland, N. Terr., W. Australia).
canariensis *A.J. Cavanilles in Anales Ci. Nat.,* 3(7): 72 (1801)—Canary Is.
everistiana *S.T. Reynolds & A.E. Holland in Austrobaileya,* 2(5): 461 (1988)—Australia (Queensland, N. Terr.).
peruviana *Lamarck, Encycl.,* 1: 6 (1783)—Peru.
reflexa *Lamarck, Encycl.,* 1: 6 (1783)—Locality not stated.

SIDERITIS (Labiat.).
arborescens
var. antiatlantica *(Maire) A. Dobignard in Saussurea,* 19: 108 (1988 publ. 1989): S. ochroleuca var. antiatlantica.
funkiana
subsp. talaverana *O. Socorro, L. Cano & M.C. Espinar in Acta Bot. Malacitana,* 13: 168 (1988), nom. nov.: S. lagascana *Willk.*

SIGESBECKIA (Compos.).
pringlei *D.L. Schulz in Haussknechtia, Mitt. Thüring. Bot. Ges.,* 4: 28 (1988)—Mexico.
subsp. tuerckheimii *D.L. Schulz in Haussknechtia, Mitt. Thüring. Bot. Ges.,* 4: 29 (1988)—Hispaniola (Dominican Republic, Haiti).

SILENE (Caryophyllac.).
perlmanii *W.L. Wagner, D.R. Herbst & S.H. Sohmer in Bishop Mus. Occas. Pap.,* 29: 94 (1989)—Hawaiian Is.

SILPHIUM (Compos.).
wasiotensis *M.E. Medley in Sida,* 13(3): 285 (1989)—U.S.A. (Kentucky).

SIMIRA (Rubiac.).
ignicola *J.A. Steyermark in Ann. Missouri Bot. Gard.,* 75(3): 1086 (1988)—Venezuela.

SINARUNDINARIA (Gramin.).
sect. Chimonocalamus *(Hsueh & Yi) C.S. Chao & S.A. Renvoize in Kew Bull.,* 44(2): 353 (1989): gen. Chimonocalamus.
alpina *(K. Schum.) C.S. Chao & S.A. Renvoize in Kew Bull.,* 44(2): 361 (1989): Arundinaria alpina.
anceps *(Mitf.) C.S. Chao & S.A. Renvoize in Kew Bull.,* 44(2): 359 (1989): Arundinaria anceps.
debilis *(Thwaites) C.S. Chao & S.A. Renvoize in Kew Bull.,* 44(2): 355 (1989): Arundinaria debilis.
densifolia *(Munro) C.S. Chao & S.A. Renvoize in Kew Bull.,* 44(2): 354 (1989): Arundinaria densifolia.
elegans *(Kurz) C.S. Chao & S.A. Renvoize in Kew Bull.,* 44(2): 357 (1989): Arundinaria elegans.
falcata *(Nees) C.S. Chao & S.A. Renvoize in Kew Bull.,* 44(2): 357 (1989): Arundinaria falcata.
floribunda *(Thwaites) C.S. Chao & S.A. Renvoize in Kew Bull.,* 44(2): 356 (1989): Arundinaria floribunda.
gallatlyi *(Gamble) C.S. Chao & S.A. Renvoize in Kew Bull.,* 44(2): 354 (1989): Arundinaria gallatlyi.
griffithiana *(Munro) C.S. Chao & S.A. Renvoize in Kew Bull.,* 44(2): 353 (1989): Arundinaria griffithiana.
hirsuta *(Munro) C.S. Chao & S.A. Renvoize in Kew Bull.,* 44(2): 355 (1989): Arundinaria hirsuta.
hookeriana *(Munro) C.S. Chao & S.A. Renvoize in Kew Bull.,* 44(2): 358 (1989): Arundinaria hookeriana.
humbertii *(A. Camus) C.S. Chao & S.A. Renvoize in Kew Bull.,* 44(2): 362 (1989): Arundinaria humbertii.
intermedia *(Munro) C.S. Chao & S.A. Renvoize in Kew Bull.,* 44(2): 357 (1989): Arundinaria intermedia.
madagascariensis *(A. Camus) C.S. Chao & S.A. Renvoize in Kew Bull.,* 44(2): 362 (1989): Arundinaria madagascariensis.
maling *(Gamble) C.S. Chao & S.A. Renvoize in Kew Bull.,* 44(2): 356 (1989): Arundinaria maling.
microphylla *(Munro) C.S. Chao & S.A. Renvoize in Kew Bull.,* 44(2): 354 (1989): Arundinaria microphylla.
pantlingii *(Gamble) C.S. Chao & S.A. Renvoize in Kew Bull.,* 44(2): 359 (1989): Arundinaria pantlingii.

SINARUNDINARIA :-
polystachya *(Gamble) C.S. Chao & S.A. Renvoize in Kew Bull.,* 44(2): 359 (1989): Arundinaria polystachya.
rolloana *(Gamble) C.S. Chao & S.A. Renvoize in Kew Bull.,* 44(2): 355 (1989): Arundinaria rolloana.
schmidiana *(A. Camus) C.S. Chao & S.A. Renvoize in Kew Bull.,* 44(2): 360 (1989): Arundinaria schmidiana.
walkeriana *(Munro) C.S. Chao & S.A. Renvoize in Kew Bull.,* 44(2): 354 (1989): Arundinaria walkeriana.
wightiana *(Nees) C.S. Chao & S.A. Renvoize in Kew Bull.,* 44(2): 356 (1989): Arundinaria wightiana.

SINCLAIRIA (Compos.).
sect. Liabellum *(Rydb.) B.L. Turner in Phytologia,* 67(2): 201 (1989): gen. Liabellum.
subgen. Megaliabum *(Rydb.) B.L. Turner in Phytologia,* 67(2): 197 (1989): gen. Megaliabum.
angustissima *(A. Gray) B.L. Turner in Phytologia,* 67(2): 205 (1989): Liabum angustissimum.
cervina *(B.L. Robins.) B.L. Turner in Phytologia,* 67(2): 204 (1989): Liabum cervinum.
gentryi *(H. Robins.) B.L. Turner in Phytologia,* 67(2): 204 (1989): Liabellum gentryi.
glabra
var. hypoleuca *(Greenm.) B.L. Turner in Phytologia,* 67(2): 180 (1989): Liabum glabrum var. hypoleucum.
var. minor *B.L. Turner in Phytologia,* 67(2): 182 (1989)—Mexico.
hintoniorum *B.L. Turner in Phytologia,* 67(2): 201 (1989)—Mexico.
palmeri *(A. Gray) B.L. Turner in Phytologia,* 67(2): 203 (1989): Liabum palmeri.

SINOBAMBUSA (Gramin.).
sat *(Balansa) C.S. Chao & S.A. Renvoize in Kew Bull.,* 44(2): 366 (1989): Arundinaria sat.

SIPAPOANTHA B. Maguire & B.M. Boom in Mem. New York Bot. Gard., 51: 23 (1989). *GENTIANACEAE.*
ostrina *B. Maguire & B.M. Boom in Mem. New York Bot. Gard.,* 51: 26 (1989)—Venezuela.

SIPHONOGLOSSA (Acanthac.).
leptantha *(Nees) K.L. Immelman in Bothalia,* 19(2): 209 (1989): Gendarussa leptantha.
subsp. late-ovata *(C.B. Cl.) K.L. Immelman in Bothalia,* 19(2): 209 (1989): Justicia pulegioides subsp. late-ovata.
mexicana *R.A. Hilsenbeck in Madroño,* 36(3): 198 (1989)—Mexico.
nkandlaensis *K.L. Immelman in Bothalia,* 19(2): 209 (1989)—South Africa (Natal, Cape Province).

SIPHONOGLOSSA :–
ramosa
var. discolor *(S.F. Blake) R.A. Hilsenbeck in Madroño,* 36(3): 204 (1989): S. discolor.
var. hondurensis *(Standley & Steyerm.) R.A. Hilsenbeck in Madroño,* 36(3): 205 (1989): S. hondurensis.
sessilis
var. calcarea *(Leonard) R.A. Hilsenbeck in Madroño,* 36(3): 206 (1989): S. calcarea.

SISYRINCHIUM (Iridac.).
abietum *R. McVaugh, Fl. Novo-Galiciana,* 15: 314 (1989)—Mexico.
aquidaunicum *P. Ravenna in Onira, Bot. Leafl.,* 1(5): 44 (1988)—Brazil.
caeteanum *P. Ravenna in Onira, Bot. Leafl.,* 1(3): 31 (1988)—Brazil.
minense *P. Ravenna in Onira, Bot. Leafl.,* 1(2): 16 (1988)—Brazil.

SKIMMIA (Rutac.).
laureola
subsp. lancasteri *N.P. Taylor in Kew Mag.,* 6(4): 162 (1989), nom. nov., as 'subsp. nov.': S. melanocarpa *Rehder & Wilson.*
var. wattii *K. Narayanan Nair & M.P. Nayar in J. Bombay Nat. Hist. Soc.,* 85(3): 653 (1988 publ. 1989)—India.

SLOANEA (Elaeocarpac.).
bolivarensis *J.A. Steyermark in Ann. Missouri Bot. Gard.,* 75(4): 1572 (1988 publ. 1989)—Venezuela.
cataniapensis *J.A. Steyermark in Ann. Missouri Bot. Gard.,* 75(4): 1572 (1988 publ. 1989)—Venezuela.
davidsei *J.A. Steyermark in Ann. Missouri Bot. Gard.,* 75(4): 1575 (1988 publ. 1989)—Venezuela.
longiaristata *J.A. Steyermark in Ann. Missouri Bot. Gard.,* 75(4): 1577 (1988 publ. 1989)—Venezuela.
merevariensis *Pittier ex J.A. Steyermark in Ann. Missouri Bot. Gard.,* 75(4): 1579 (1988 publ. 1989)—Venezuela.
parvifructa *J.A. Steyermark in Ann. Missouri Bot. Gard.,* 75(4): 1579 (1988 publ. 1989)—Brazil.
sipapoana *J.A. Steyermark in Ann. Missouri Bot. Gard.,* 75(4): 1580 (1988 publ. 1989)—Venezuela.
steyermarkii
subsp. autanae *(Steyerm.) J.A. Steyermark in Ann. Missouri Bot. Gard.,* 75(4): 1583 (1988 publ. 1989): S. autanae.
subsp. jauaensis *(Steyerm.) J.A. Steyermark in Ann. Missouri Bot. Gard.,* 75(4): 1582 (1988 publ. 1989): S. jauaensis.
subpsilocarpa *J.A. Steyermark in Ann. Missouri Bot. Gard.,* 75(4): 1583 (1988 publ. 1989)—Venezuela.
wurdackii *J.A. Steyermark in Ann. Missouri Bot. Gard.,* 75(4): 1583 (1988 publ. 1989)—Venezuela.

SMELOWSKIA (Crucif.).
inopinata
subsp. pseudoalba *V.N. Voroshilov in Byull. Glavn. Bot. Sada (Moscow),* 149: 33 (1988)—U.S.S.R. (Soviet Far East).

SMYTHEA B.C. Seeman in Bonplandia, 10(5): 69, t. 9 (15 March 1862). *RHAMNACEAE.* †
pacifica *B.C. Seeman in Bonplandia,* 10(5): 69, t. 9 (15 March 1862)—Fiji.†

SOLANUM (Solanac.).
subsect. Estolonifera *J.G. Hawkes in Taxon,* 38(3): 490 (1989).
ser. Lignicaulia *J.G. Hawkes in Taxon,* 38(3): 490 (1989).
superser. Rotata *J.G. Hawkes in Taxon,* 38(3): 491 (1989).
ser. Simplicissima *C.M. Ochoa in Rev. Acad. Colomb. Cienc. Exact. Fis. Nat.,* 65: 321 (1989).
superser. Stellata *J.G. Hawkes in Taxon,* 38(3): 490 (1989).
abollatum *H. St. John, Native Hawaiian sp. Solanum (Hawaiian Pl. Stud. 82):* 5 (1988)—Hawaiian Is.
americanum
subsp. nodiflorum *(Jacq.) R.J.F. Henderson in Austrobaileya,* 2(5): 555 (1988): S. nodiflorum.
subsp. nutans *(R. Henderson) R.J.F. Henderson in Austrobaileya,* 2(5): 555 (1988): S. nodiflorum subsp. nutans.
angustior *H. St. John, Native Hawaiian sp. Solanum (Hawaiian Pl. Stud. 82):* 5 (1988)—Hawaiian Is.
bahianum *S. Knapp in Bull. Brit. Mus. (Nat. Hist.), Bot.,* 19: 109 (1989)—Brazil.
chillonanum *C.M. Ochoa in Phytologia,* 67(3): 235 (1989), nom. nov.: S. tenellum *Ochoa.*
concarense *A.T. Hunziker in Kurtziana,* 20: 190 (1989)—Argentina.
falconense *S. Knapp in Bull. Brit. Mus. (Nat. Hist.), Bot.,* 19: 103 (1989)—Venezuela.
forbesii *H. St. John, Native Hawaiian sp. Solanum (Hawaiian Pl. Stud. 82):* 6 (1988)—Hawaiian Is.
globosum *H. St. John, Native Hawaiian sp. Solanum (Hawaiian Pl. Stud. 82):* 7 (1988), nom. nov., as 'sp. nov.': S. incompletum var. glabratum *Hillebr.*
gonyrhachis *S. Knapp in Bull. Brit. Mus. (Nat. Hist.), Bot.,* 19: 105 (1989)—Bolivia.
honopuense *H. St. John, Native Hawaiian sp. Solanum (Hawaiian Pl. Stud. 82):* 9 (1988)—Hawaiian Is.
imbaburense *S. Knapp in Bull. Brit. Mus. (Nat. Hist.), Bot.,* 19: 94 (1989)—Ecuador.
lanaiense *H. St. John, Native Hawaiian sp. Solanum (Hawaiian Pl. Stud. 82):* 12 (1988)—Hawaiian Is.
lydgatei *H. St. John, Native Hawaiian sp. Solanum (Hawaiian Pl. Stud. 82):* 13 (1988)—Hawaiian Is.

SOLANUM :–
macdanielsii *H. St. John, Native Hawaiian sp. Solanum (Hawaiian Pl. Stud. 82):* 14 (1988)—Hawaiian Is.
nelsoni
var. vadosum *H. St. John, Native Hawaiian sp. Solanum (Hawaiian Pl. Stud. 82):* 18 (1988)—Hawaiian Is.
nesophilum *H. St. John, Native Hawaiian sp. Solanum (Hawaiian Pl. Stud. 82):* 18 (1988)—Hawaiian Is.
plowmanii *S. Knapp in Bull. Brit. Mus. (Nat. Hist.), Bot.,* 19: 104 (1989)—Peru.
popolo *H. St. John, Native Hawaiian sp. Solanum (Hawaiian Pl. Stud. 82):* 19 (1988)—Hawaiian Is.
pubinervosum *H. St. John, Native Hawaiian sp. Solanum (Hawaiian Pl. Stud. 82):* 21 (1988)—Hawaiian Is.
restingae *S. Knapp in Bull. Brit. Mus. (Nat. Hist.), Bot.,* 19: 109 (1989)—Brazil.
ruizii *S. Knapp in Bull. Brit. Mus. (Nat. Hist.), Bot.,* 19: 91 (1989), nom. nov.: S. lanceolatum *Ruiz López & Pavón.*
santosii *S. Knapp in Bull. Brit. Mus. (Nat. Hist.), Bot.,* 19: 106 (1989)—Brazil.
semilobatum *H. St. John, Native Hawaiian sp. Solanum (Hawaiian Pl. Stud. 82):* 23 (1988), nom. nov., as 'sp. nov.': S. incompletum var. mauiense *Hillebr.*
simplicissimum *C.M. Ochoa in Rev. Acad. Colomb. Cienc. Exact. Fis. Nat.,* 65: 321 (1989)—Peru.
spirale
var. tetrasepalum *H. Chu in Acta Sci. Nat. Univ. Sunyatseni,* 28(2): 67 (1989)—China.
toliaraea *W.G. D'Arcy & A. Rakotozafy in Ann. Missouri Bot. Gard.,* 76(1): 351 (1989)—Madagascar.
tonsum *H. St. John, Native Hawaiian sp. Solanum (Hawaiian Pl. Stud. 82):* 24 (1988)—Hawaiian Is.

SOLDANELLA (Primulac.).
pusilla
subsp. alpicola *(F.K. Meyer) J. Chrtek in Cas. Nár. Muz. (Prague),* 154(3–4): 170 (1985 publ. 1989): S. alpicola.
subsp. pirinica *(F.K. Meyer) J. Chrtek in Cas. Nár. Muz. (Prague),* 154(3–4): 170 (1985 publ. 1989): S. pirinica.

SOLIDAGO (Compos.).
ericamerioides *G.L. Nesom in Phytologia,* 67(2): 143 (1989)—Mexico.
gypsophila *G.L. Nesom in Phytologia,* 67(2): 142 (1989)—Mexico.
hintoniorum *G.L. Nesom in Phytologia,* 67(2): 145 (1989)—Mexico.
macvaughii *G.L. Nesom in Phytologia,* 67(4): 301 (1989)—Mexico.

SOLIDAGO :–
virgaurea
subsp. fallit-tirones *(Font Quer) S. Rivas-Martínez, F. Fernández González & D. Sánchez-Mata in Opusc. Bot. Pharm. Complutensis,* 2: 118 (1986): S. virgaurea var. fallit-tirones.
wrightii
var. orientalis *G.L. Nesom in Phytologia,* 67(2): 147 (1989)—Mexico.

SONCHUS (Compos.).
tenerrimus
subvar. dianae *(Lacaita ex Willk.) O. de Bolòs & J. Vigo in Folia Bot. Misc.,* 6: 86 (1989), without exact basionym date: S. dianae.
subvar. pectinatus *(DC.) O. de Bolòs & J. Vigo in Folia Bot. Misc.,* 6: 86 (1989): S. pectinatus.

SONDOTTIA P.S. Short in Muelleria, 7(1): 113 (1989). *COMPOSITAE.*
connata *(W.V. Fitzg.) P.S. Short in Muelleria,* 7(1): 114 (1989): Angianthus connatus.
glabrata *P.S. Short in Muelleria,* 7(1): 115 (1989)—Australia (W. Australia).

SONERILA (Melastomatac.).
kanniyakumariana *R. Gopalan & A.N. Henry in J. Bombay Nat. Hist. Soc.,* 86(1): 82 (1989)—India.

SORBARIA (Rosac.).
tomentosa
var. angustifolia *(Wenzig) K. Rahn in Nordic J. Bot.,* 8(6): 561 (1989): Spiraea sorbifolia var. angustifolia.

SORBUS (Rosac.).
croceocarpa *P.D. Sell in Watsonia,* 17(4): 392 (1989)—Great Britain.
lanpingensis *L.T. Lu in Bull. Bot. Res. North-East. Forest. Inst.,* 9(2): 51 (1989)—China.
reducta
var. pubescens *L.T. Lu in Bull. Bot. Res. North-East. Forest. Inst.,* 9(2): 52 (1989)—China.

SPATHOGLOTTIS (Orchidac.).
smithii *P.J. Kores in Allertonia,* 5(1): 134 (1989)—Fiji.

SPERANSKIA (Euphorbiac.).
yunnanensis *S.M. Hwang in Bull. Bot. Res. North-East. Forest. Inst.,* 9(4): 38 (1989)—China.

SPERMACOCE (Rubiac.).
bahamensis *(Britton) R.A. Howard in Phytologia,* 65(4): 287 (1988): Borreria bahamensis.
berteroana *R.A. Howard in Phytologia,* 65(4): 287 (1988), nom. nov.: Borreria scandens *DC.*
bisepala *B. Verdcourt in Fl. Zambes.,* 5(1): 172 (1989)—Zimbabwe.

SPERMACOCE :-
brittonii *(Standley) R.A. Howard in Phytologia,* 65(4): 288 (1988): Borreria brittonii.
capillaris *(Correll) R.A. Howard in Phytologia,* 65(4): 288 (1988): Borreria capillaris.
dussii *(Standley) R.A. Howard in Phytologia,* 65(4): 288 (1988): Borreria dussii.
felis-insulae *(Correll) R.A. Howard in Phytologia,* 65(4): 288 (1988): Borreria felis-insulae.
inaguensis *(Britton) R.A. Howard in Phytologia,* 65(4): 288 (1988): Borreria inaguensis.
princeae
var. mwinilungae *B. Verdcourt in Fl. Zambes.,* 5(1): 180 (1989)—Zambia.
samfya *B. Verdcourt in Fl. Zambes.,* 5(1): 186 (1989)—Zambia.
savannarum *(Britton) R.A. Howard in Phytologia,* 65(4): 288 (1988): Borreria savannarum.
schlechteri *K. Schum. ex B. Verdcourt in Fl. Zambes.,* 5(1): 180 (1989)—Mozambique.
spermacocina *(K. Schum.) D.M. Bridson & C. Puff in Kew Bull.,* 44(1): 138 (1989): Paragophyton spermacocinum.
thymifolia *(Griseb.) R.A. Howard in Phytologia,* 65(4): 288 (1988): Borreria thymifolia.

SPHAEROBAMBOS S. Dransfield in Kew Bull., 44(3): 428 (1989). *GRAMINEAE.*
hirsuta *S. Dransfield in Kew Bull.,* 44(3): 428 (1989)—Borneo (Sabah).
philippinensis *(Gamble) S. Dransfield in Kew Bull.,* 44(3): 432 (1989): Guadua philippinensis.
subtilis *S. Dransfield in Kew Bull.,* 44(3): 432 (1989)—Celebes.

SPHEDAMNOCARPUS (Malpighiac.).
pruriens
subsp. galphimiifolius *(Juss.) P.D. de Villiers & D.J. Botha in Fl. S. Afr.,* 18(3): 67 (1986): Acridocarpus galphimiifolius.

SPIRADICLIS (Rubiac.).
arunachalensis *D.B. Deb & R.C. Rout in Candollea,* 44(1): 225 (1989)—India, Assam, China.
seshagirii *D.B. Deb & R.C. Rout in Candollea,* 44(1): 227 (1989)—India.

SPIRAEA (Rosac.).
daochengensis *L.T. Lu in Bull. Bot. Res. North-East. Forest. Inst.,* 9(2): 49 (1989), as 'daochegensis'—China.

SPIRANTHES (Orchidac.).
infernalis *C.J. Sheviak in Rhodora,* 91(867): 226 (1989)—U.S.A. (Nevada).

SPONDIANTHUS (Euphorbiac.).
preussii
subsp. glaber *(Engl.) J. Léonard & J. Nkounkou in Bull. Jard. Bot. Nation. Belg.,* 59(1–2): 143 (1989): Spondianthus glaber.

SPOROBOLUS (Gramin.).
exilis *(Trin.) B. Balansa in L. Morot, Journal de Botanique,* 4: 164 (1890): Vilfa exilis.
temomairemensis *E.J. Judziewicz & P.M. Peterson in Syst. Bot.,* 14(4): 525 (1989)—Brazil, Surinam.

STACHYOPSIS (Labiat.).
canescens *(Rgl.) T.A. Adylov & M. Tulyaganova in Opred. Rast. Sred. Azii,* 9: 114 (1987): Phlomis oblongata var. canescens.

STACHYS (Labiat.).
baytopiorum *Kit Tan & B. Yildiz in Notes Roy. Bot. Gard. Edinburgh,* 45(3): 450 (1988 publ. 1989)—Turkey.
herrerana *J. Rzedowski & G. Calderón de Rzedowski in Acta Bot. Mex.,* 3: 1 (1988)—Mexico.
horvaticii *K. Micevski in Acta Bot. Croat.,* 28: 452 (1969)—Jugoslavia.
langmaniae *J. Rzedowski & G. Calderón de Rzedowski in Acta Bot. Mex.,* 3: 2 (1988)—Mexico.
sivasica *Kit Tan & B. Yildiz in Notes Roy. Bot. Gard. Edinburgh,* 45(3): 449 (1988 publ. 1989)—Turkey.
vulnerabilis *J. Rzedowski & G. Calderón de Rzedowski in Acta Bot. Mex.,* 3: 3 (1988)—Mexico.
willemsei *Kit Tan & Hedge in Notes Roy. Bot. Gard. Edinburgh,* 45(3): 451 (1988 publ. 1989)—Turkey.

STANHOPEA (Orchidac.).
avicula *R.L. Dressler in Amer. Orchid Soc. Bull.,* 58(9): 885 (1989)—Panama.
×herrenhusana *R. Jenny in Orchidee,* 40(1): 32 (1989)—Colombia. (S. reichenbachiana × tricornis).

STAUNTONIA (Lardizabalac.).
brunoniana *(Decne) W.B. Hemsley in Hook. Ic. Pl.:* t. 2843 (1907): Parvatia brunoniana.

STEGOLEPIS (Rapateac.).
minor *J.A. Steyermark in Ann. Missouri Bot. Gard.,* 75(1): 315 (1988)—Venezuela.

STELIS (Orchidac.).
magnicava *D.D. Dod in Moscosoa,* 5: 233 (1989)—Hispaniola (Haiti).

STELLARIA (Caryophyllac.).
parviumbellata *Y.Z. Zhao in Acta Sci. Nat. Univ. Intramongolicae,* 20(2): 226 (1989), as 'parvi-umbellata'—China.
viridifolia *(Khokhr.) A.P. Khokhryakov, Analiz Fl. Kolÿmskogo Nagor'ya:* 10 (1989): S. fischerana subsp. viridifolia.

STELLILABIUM (Orchidac.).
boylei *J.T. Atwood in Ic. Pl. Trop.,* 14 (Orch. Costa Rica, pt. 1): t. 1392 (1989)—Costa Rica.

STELLILABIUM :-
bullpenense *J.T. Atwood in Ic. Pl. Trop.*, 14 (Orch. Costa Rica, pt. 1): t. 1393 (1989)—Costa Rica.
campbelliorum *J.T. Atwood in Ic. Pl. Trop.*, 14 (Orch. Costa Rica, pt. 1): t. 1394 (1989), as 'campbellorum'—Costa Rica.
monteverdense *J.T. Atwood in Ic. Pl. Trop.*, 14 (Orch. Costa Rica, pt. 1): t. 1395 (1989)—Costa Rica.

STENOGYNE (Labiat.).
campanulata *S.G. Weller & A. Sakai in Bishop Mus. Occas. Pap.*, 29: 101 (1989)—Hawaiian Is.

STENOMESSON (Amaryllidac.).
sect. Adnata *P. Ravenna in Onira, Bot. Leafl.*, 1(2): 20 (1988).
cuzcoense *(Varg.) P. Ravenna in Onira, Bot. Leafl.*, 1(2): 20 (1988): Urceolina cuzcoensis.
gasteroides *P. Ravenna in Onira, Bot. Leafl.*, 1(2): 18 (1988)—Peru.
parvulum *P. Ravenna in Onira, Bot. Leafl.*, 1(2): 17 (1988)—Peru.
sunchubambae *P. Ravenna in Onira, Bot. Leafl.*, 1(2): 17 (1988)—Peru.
weberbaueri *(Varg.) P. Ravenna in Onira, Bot. Leafl.*, 1(2): 20 (1988): Urceolina weberbaueri.

STENOTUS (Compos.).
pulvinatus *(R. Moran) G.L. Nesom in Phytologia*, 67(2): 113 (1989): Haplopappus pulvinatus.

STEPHANIA (Menispermac.).
dentifolia *H.S. Lo & M. Yang in Guihaia*, 8(4): 313 (1988)—China.
lincangensis *H.S. Lo & M. Yang in Guihaia*, 8(4): 311 (1988)—China.
macrantha *H.S. Lo & M. Yang in Guihaia*, 8(4): 309 (1988)—China.
officinarum *H.S. Lo & M. Yang in Guihaia*, 8(4): 310 (1988)—China.
yunnanensis
var. trichocalyx *H.S. Lo & M. Yang in Guihaia*, 8(4): 313 (1988)—China.

STERCULIA (Sterculiac.).
acerifolia *A. Cunningham ex G. Don, Gen. Syst.*, 1: 517 (1831)V)—Australia. Y1

STERNBERGIA (Amaryllidac.).
vernalis *(Miller) R. Gorer & J.H. Harvey in Plantsman*, 10(4): 204 (1989): Amaryllis vernalis.

STEYERBROMELIA L.B. Smith in Acta Bot. Venezuel., 14(3): 8 (1984), without type. *BROMELIACEAE.*
deflexa *L.B. Smith & Robinson in Acta Bot. Venezuel.*, 14(3): 12 (1984), generic name not validly publ.—Venezuela.

STEYERBROMELIA :-
discolor *L.B. Smith & Robinson in Acta Bot. Venezuel.*, 14(3): 8 (1984), generic name not validly publ.—Venezuela.

STIGMAPHYLLON (Malpighiac.).
bahiense *C. Anderson in Syst. Bot.*, 14(4): 509 (1989)—Brazil.
blanchetii *C. Anderson in Syst. Bot.*, 14(4): 511 (1989)—Brazil.
cavernulosum *C. Anderson in Syst. Bot.*, 14(4): 513 (1989)—Brazil.

STILLINGIA (Euphorbiac.).
parvifolia *I.M. Sánchez Vega, A. Sagástegui Alva & M.J. Huft in Ann. Missouri Bot. Gard.*, 75(4): 1666 (1988 publ. 1989)—Peru.

STIMPSONIA (Primulac.).
chamaedryoides
f. rubriflora *J.Z. Shao in Bull. Bot. Res. North-East. Forest. Inst.*, 9(3): 73 (1989)—China.

STIPA (Gramin.).
ayacuchensis *O. Tovar in Opusc. Bot. Pharm. Complutensis*, 4: 85 (1988)—Peru.
dasyvaginata
subsp. longiglumis *H. Scholz in Willdenowia*, 19(1): 131 (1989)—Algeria.
gabesensis *B. Moraldo, M. Raffaelli & C. Ricceri in Candollea*, 44(1): 78 (1989)—Tunisia.
iberica
subsp. austroiberica *H. Scholz in Willdenowia*, 19(1): 129 (1989), as 'austro-iberica'—Spain.
isoldeae *H. Scholz in Willdenowia*, 19(1): 127 (1989)—Greece.
kelibiae *B. Moraldo, M. Raffaelli & C. Ricceri in Candollea*, 44(1): 80 (1989)—Tunisia.
keniensis *(Pilger) H. Freitag in Davis & Hedge Festschrift (ed. Kit Tan)*: 122 (1989): Oryzopsis keniensis.
subsp. somaliensis *H. Freitag in Davis & Hedge Festschrift (ed. Kit Tan)*: 124 (1989)—Somali Republic.
kralifii *B. Moraldo, M. Raffaelli & C. Ricceri in Candollea*, 44(1): 80 (1989)—Tunisia.
mandavillei *H. Freitag in Davis & Hedge Festschrift (ed. Kit Tan)*: 118 (1989)—Oman.
manrakica *Yu.A. Kotukhov in Bot. Zhurn.*, 74(3): 414 (1989)—U.S.S.R. (Middle Asia).
mayeri *J.O. Martinovský in Acta Bot. Croat.*, 30: 145 (1971)—Jugoslavia.
soukupii *O. Tovar in Opusc. Bot. Pharm. Complutensis*, 4: 90 (1988)—Peru.

STIXIS (Capparac.).
manipurensis *D.B. Deb & R.C. Rout in J. Bombay Nat. Hist. Soc.*, 86(1): 86 (1989)—India.

STOIBRAX (Umbellif.).
didymum *(Sonder) B.L. Burtt in Davis & Hedge Festschrift (ed. Kit Tan):* 145 (1989): Ptychotis didyma.
hanotei *(Br.-Bl. & Maire) B.L. Burtt in Davis & Hedge Festschrift (ed. Kit Tan):* 146 (1989): Tragiopsis hanotei.
involucratum *(Maire) B.L. Burtt in Davis & Hedge Festschrift (ed. Kit Tan):* 146 (1989): Trachyspermum involucratum.
pomelianum *(Maire) B.L. Burtt in Davis & Hedge Festschrift (ed. Kit Tan):* 146 (1989): Trachyspermum pomelianum.

STOMATOCHAETA (Compos.).
acuminata *J.F. Pruski in Brittonia,* 41(1): 35 (1989)—Venezuela.

STOMATOSTEMMA (Asclepiadac.).
pendulina *H.J.T. Venter & D.V. Field in Bot. J. Linn. Soc.,* 99(4): 398 (1989)—Mozambique.

STRAVADIUM (Lecythidac.).
neo-caledonicum *(Vieill.) M.G. Baumann-Bodenheim, Syst. Fl. Neu-Caledonien,* 5: 76 (1989): Barringtonia neo-caledonica.

STREBLUS (Morac.).
sect. Ampalis *(Bojer) C.C. Berg in Proc. Kon. Nederl. Akad. Wetensch., C,* 91(4): 358 (1988): gen. Ampalis.
dimepate *(Bureau) C.C. Berg in Proc. Kon. Nederl. Akad. Wetensch., C,* 91(4): 358 (1988): Pachytrophe dimepate.
usambarensis *(Engler) C.C. Berg in Proc. Kon. Nederl. Akad. Wetensch., C,* 91(4): 357 (1988): Sloetiopsis usambarensis.

STREPTANTHUS (Crucif.).
brachiatus
subsp. hoffmanii *R.W. Dolan & L.F. LaPré in Madroño,* 36(1): 36 (1989)—U.S.A. (California).
morrisonii
subsp. kruckebergii *R.W. Dolan & L.F. LaPré in Madroño,* 36(1): 39 (1989)—U.S.A. (California).

STREPTOCALYX (Bromeliac.).
colombianus
var. laxus *E. Gross in Abh. Akad. Wiss. Lit. Mainz, Math.-Naturwiss. Klasse, Trop. Subtrop. Pflanzenwelt,* 65: 72 (1988)—Colombia.

STREPTOCARPUS (Gesneriac.).
bindseili *E. Fischer in Feddes Repert.,* 99(9–10): 387 (1988)—Rwanda.

STREPTOCHAETA (Gramin.).
spicata
subsp. ecuatoriana *E.J. Judziewicz & T.R. Soderstrom in Smithsonian Contrib. Bot.,* 68: 39 (1989)—Ecuador.

STRYCHNOS (Loganiac.).
nicaraguensis *M.J. Huft in Ann. Missouri Bot. Gard.,* 75(1): 383 (1988)—Nicaragua.

STRYPHNODENDRON (Legumin.).
porcatum *D.A. Neill & E.M. de Lamare Occhioni in Ann. Missouri Bot. Gard.,* 76(1): 357 (1989)—Ecuador.

STYLOGYNE (Myrsinac.).
bracteolata *(Lundell) J.J. Pipoly in Ernstia,* 53: 2 (1989): Cybianthus bracteolatus.
lasseri *(Lundell) J.J. Pipoly in Ernstia,* 53: 4 (1989): Ardisia lasseri.
membranacea *J.J. Pipoly in Ernstia,* 53: 6 (1989)—Venezuela.

STYLOSANTHES (Legumin.).
guianensis
var. pauciflora *M. Brandão, N.M.S. Costa & R. Schultze-Kraft in Prog. Resum. Soc. Bot. Brasil, XXXVI Congr. Nac. Bot., Curitiba:* 323 (1985), without latin descr. or type—Brazil, Colombia.

STYRAX (Styracac.).
bashanensis *S.Z. Qu & K.Y. Wang in Bull. Bot. Res. North-East. Forest. Inst.,* 9(1): 27 (1989)—China.

SULCOREBUTIA (Cactac.).
augustinii *G. Hentzschel in Succulenta,* 68(7–8): 147 (1989)—Bolivia.
mariana *H. Swoboda in Succulenta,* 68(1): 4 (1989)—Bolivia.

SUNDACARPUS (Buchh. & Gray) C.N. Page in Notes Roy. Bot. Gard. Edinburgh, 45(2): 378 (1988 publ. 1989): Podocarpus sect. Sundacarpus (Podocarpac.).
amara *(Blume) C.N. Page in Notes Roy. Bot. Gard. Edinburgh,* 45(2): 378 (1988 publ. 1989): Podocarpus amara.

SVENKOELTZIA P. Burns-Balogh in Orchidee, 40(1): 12 (1989). *ORCHIDACEAE.*
congestiflora *(L.O. Williams) P. Burns-Balogh in Orchidee,* 40(1): 12 (1989): Spiranthes congestiflora.

SWERTIA (Gentianac.).
brownii *J. Shah in Pakistan J. Forest.,* 38(2): 82 (1988), nom. nov.: S. calycina *Brown.*
laticalyx *J. Shah in Pakistan J. Forest.,* 38(2): 80 (1988), as 'latycalyx'—China.
nepalensis *J. Shah in Pakistan J. Forest.,* 38(2): 79 (1988)—Nepal.
pauciflora *(Dalz.) J. Shah in Pakistan J. Forest.,* 38(2): 81 (1988): Ophelia pauciflora.
scottii *J. Shah in Pakistan J. Forest.,* 38(2): 80 (1988)—Ethiopia.
taylorii *J. Shah in Pakistan J. Forest.,* 38(2): 79 (1988)—Xizang.

SWERTIA :-
teres *(G. Don) J. Shah in Pakistan J. Forest.*, 38(2): 81 (1988): Ophelia teres.
umbellata *(Hook.) J. Shah in Pakistan J. Forest.*, 38(2): 82 (1988): Ophelia umbellata.
woodii *J. Shah in Pakistan J. Forest.*, 38(2): 80 (1988)—Yemen, Saudi Arabia.

SYCOPSIS (Hamamelidac.).
sinensis
subsp. formosana *(Kanehira) S.S. Ying, Coloured Illustr. Fl. Taiwan,* 3: 101 (1988): Distylium formosanum.

SYMBOLANTHUS (Gentianac.).
camanensis *B. Maguire & B.M. Boom in Mem. New York Bot. Gard.*, 51: 17 (1989)—Venezuela.
elisabethae
subsp. occidentalis *B. Maguire & B.M. Boom in Mem. New York Bot. Gard.*, 51: 17 (1989)—Venezuela.
huachamacariensis *J.A. Steyermark in Ann. Missouri Bot. Gard.*, 75(3): 1083 (1988)—Venezuela.
yaviensis *J.A. Steyermark in Ann. Missouri Bot. Gard.*, 75(3): 1084 (1988)—Venezuela.
yutajensis *B. Maguire & B.M. Boom in Mem. New York Bot. Gard.*, 51: 17 (1989)—Venezuela.

SYMPLOCOS (Symplocac.).
incahuasensis *A. Sagástegui-Alva & M.O. Dillon in Brittonia,* 41(1): 32 (1989)—Peru.
panamensis *G. McPherson in Ann. Missouri Bot. Gard.*, 75(1): 375 (1988)—Panama.
pulchra
subsp. coriacea *R. Gopalan & A.N. Henry in J. Econ. Taxon. Bot.*, 12(2): 481 (1988 publ. 1989)—India.
simaoensis *Y.Y. Qian in Bull. Bot. Res. North-East. Forest. Inst.*, 9(1): 41 (1989)—China.

SYNOTIS (Compos.).
borii *(Raizada) R. Mathur in R.R. Rao et al., Fl. Ind. Enumerat. - Asterac.:* 74 (1988): Senecio borii.
cappa
var. parishii *(Hook. f.) H. Koyama in Acta Phytotax. Geobot.*, 39(4–6): 161 (1988): Senecio densiflorus var. parishii.
jowaiensis *(Balakr.) R. Mathur in R.R. Rao et al., Fl. Ind. Enumerat. - Asterac.:* 74 (1988): Senecio jowaiensis.
phupeakensis *H. Koyama in Acta Phytotax. Geobot.*, 39(4–6): 162 (1988)—Thailand.

SYRINGA (Oleac.).
chuanxiensis *S.Z. Qu & X.L. Chen in Bull. Bot. Res. North-East. Forest. Inst.*, 9(3): 39 (1989)—China.
microphylla
var. flavoanthera *X.L. Chen in Bull. Bot. Res. North-East. Forest. Inst.*, 9(3): 41 (1989)—China.

SYRINGA :-
var. giraldiana *(Schneid.) S.Z. Qu & X.L. Chen in Bull. Bot. Res. North-East. Forest. Inst.*, 9(3): 41 (1989): S. giraldiana.
protolaciniata *P.S. Green & M.C. Chang in Kew Mag.*, 6(3): 121 (1989)—China.

SYSTENOTHECA J.L. Reveal & C.B. Hardham in Phytologia, 66(2): 85 (1989). *POLYGONACEAE.*
vortriedei *(Brandegee) J.L. Reveal & C.B. Hardham in Phytologia,* 66(2): 85 (1989): Chorizanthe vortriedei.

T

TABEBUIA (Bignoniac.).
bullata *A.H. Gentry in Moscosoa,* 5: 134 (1989)—Hispaniola (Dominican Republic).
ovatifolia *I. de Vattimo in Rodriguésia,* 38–40(64–66): 1 (1986–1988 publ. 1989)—Brazil.
vinosa *A.H. Gentry in Moscosoa,* 5: 136 (1989)—Hispaniola (Dominican Republic).
zanonii *A.H. Gentry in Moscosoa,* 5: 138 (1989)—Hispaniola (Dominican Republic).

TACCARUM (Arac.).
cardenasianum *J. Bogner in Willdenowia,* 19(1): 195 (1989)—Bolivia.

TACINGA (Cactac.).
braunii *E. Esteves Pereira in Kakt. And. Sukk.*, 40(6): 135 (1989)—Brazil.

TACOANTHUS (Acanthac.).
pearcei *H.E. Baillon in Bull. Soc. Linn. Paris,* 2: 832 (1890)—Bolivia.†

TAENIOPHYLLUM (Orchidac.).
confusum *P.J. Kores & L. Jonsson in Allertonia,* 5(1): 208 (1989)—Fiji.
smithii *P.J. Kores & L. Jonsson in Allertonia,* 5(1): 211 (1989)—Fiji.

TALISIA (Sapindac.).
amaruayana *J.A. Steyermark in Ann. Missouri Bot. Gard.*, 75(3): 1070 (1988)—Venezuela.
caudata *J.A. Steyermark in Ann. Missouri Bot. Gard.*, 75(3): 1072 (1988)—Venezuela.
glandulifera *J.A. Steyermark in Ann. Missouri Bot. Gard.*, 75(3): 1072 (1988)—French Guiana, Venezuela.
heterodoxa *J.A. Steyermark in Ann. Missouri Bot. Gard.*, 75(3): 1073 (1988)—Venezuela.
pentantha *J.A. Steyermark in Ann. Missouri Bot. Gard.*, 75(3): 1075 (1988)—Venezuela.
sancarlosiana *J.A. Steyermark in Ann. Missouri Bot. Gard.*, 75(3): 1075 (1988)—Venezuela.

TAMARIX (Tamaricac.).
gansuensis *H.Z. Zhang ex P.Y. Zhang & M.T. Liu in Acta Bot. Bor.-Occid. Sin.,* 8(4): 259 (1988)—China.
jintaensis *P.Y. Zhang & M.T. Liu in Acta Bot. Bor.-Occid. Sin.,* 8(4): 260 (1988), as 'jintaenaia' but see pp. 259 & 261—China.
sachensis *P.Y. Zhang & M.T. Liu in Acta Bot. Bor.-Occid. Sin.,* 8(4): 262 (1988)—Xinjiang.
tarimensis *P.Y. Zhang & M.T. Liu in Acta Bot. Bor.-Occid. Sin.,* 8(4): 263 (1988)—Xinjiang.

TAMONEA (Verbenac.).
boxiana *(Moldenke) R.A. Howard in Phytologia,* 65(4): 288 (1988): Ghinea boxiana.

TANACETUM (Compos.).
corymbosum
var. tenuifolium *(Willd.) D. Podlech in Mitt. Bot. Staatssamml. München,* 27: 70 (1988): Pyrethrum tenuifolium.

TANAKEA (Saxifragac.).
omeiensis
var. nanchuanensis *W.T. Wang in Bull. Bot. Res. North-East. Forest. Inst.,* 9(1): 8 (1989)—China.

TAPEINOSTEMON (Gentianac.).
breweri *J. Steyermark & B. Maguire in Acta Bot. Venezuel.,* 14(3): 24 (1984)—Venezuela.

TAPLINIA N.S. Lander in Nuytsia, 7(1): 37 (1989). *COMPOSITAE.*
saxatilis *N.S. Lander in Nuytsia,* 7(1): 38 (1989)—Australia (W. Australia).

TAPURA (Dichapetalac.).
magnifolia *G.T. Prance in Brittonia,* 40(4): 445 (1988)—Ecuador.

TARAXACUM (Compos.).
amansii *J. Hofstra in Bull. Soc. Bot. France, Lett. Bot.,* 135(4–5): 377 (1988 publ. 1989)—France.
aquitanum *J. Hofstra in Bull. Soc. Bot. France, Lett. Bot.,* 135(4–5): 377 (1988 publ. 1989)—France.
declivicola *J. Kirschner, C.E. Sonck & J. Štěpánek in Ann. Bot. Fenn.,* 26(3): 345 (1989)—Greece.
extimum *J. Kirschner, C.E. Sonck & J. Štěpánek in Ann. Bot. Fenn.,* 26(3): 343 (1989)—Greece.
multisinuatum *J. Kirschner, C.E. Sonck & J. Štěpánek in Ann. Bot. Fenn.,* 26(3): 345 (1989)—Greece.
noterophilum *J. Kirschner, C.E. Sonck & J. Štěpánek in Ann. Bot. Fenn.,* 26(3): 345 (1989)—Greece.
panhellenicum *C.E. Sonck in Ann. Bot. Fenn.,* 26(1): 51 (1989)—Greece.

TARAXACUM :-
praegracilens *C.E. Sonck in Ann. Bot. Fenn.,* 26(1): 52 (1989), nom. nov.: T. pseudogracilens *Sonck.*

TARENNA (Rubiac.).
mussaendoides *(Craib) K.M. Wong in Tree Fl. Malaya,* 4: 411 (1989): Randia mussaendoides.

TASMANNIA (Winterac.).
glaucifolia *J.B. Williams in Austral. J. Bot.,* 36(4): 396 (1988)—Australia (New South Wales).

TASSADIA (Asclepiadac.).
ivoniae *G. Morillo in Ernstia,* 51: 13 (1989), as 'ivonae'—Venezuela.
kamaensis *(Morillo) G. Morillo in Ernstia,* 51: 13 (1989): Cynanchum kamaense.
marahuacensis *G. Morillo in Ernstia,* 51: 13 (1989)—Venezuela.
medinae *(Morillo) G. Morillo in Ernstia,* 51: 14 (1989): Cynanchum medinae.

TAVERNIERA (Legumin.).
echinata *V. Mozaffarian in Iranian J. Bot.,* 4(1): 64 (1988), as 'Tavernieria'—Iran.

TEPHROSERIS (Compos.).
integrifolia
subsp. vindelicorum *B. Krach in Mitt. Bot. Staatssamml. München,* 27: 76 (1988)—Germany.

TEPHROSIA (Legumin.).
karkarensis *M. Thulin in Nordic J. Bot.,* 8(5): 463 (1989)—Somali Republic.
purpurea
var. yunnanensis *Z. Wei in Acta Phytotax. Sin.,* 27(1): 76 (1989)—China.
scopulata *M. Thulin in Nordic J. Bot.,* 8(5): 464 (1989)—Somali Republic.

TERNIOPSIS H.C. Chao in Contrib. Inst. Bot. Nation. Acad. Peiping, 6(1): 2 (1948 publ. 1949). *PODOSTEMACEAE.*
sessilis *H.C. Chao in Contrib. Inst. Bot. Nation. Acad. Peiping,* 6(1): 4 (1948 publ. 1949)—China.

TERNSTROEMIA (Theac.).
aracae *B.M. Boom in Brittonia,* 41(2): 136 (1989)—Brazil.
campinicola *B.M. Boom in Brittonia,* 41(2): 138 (1989)—Venezuela, Colombia, Brazil.
prancei *B.M. Boom in Brittonia,* 41(2): 140 (1989)—Brazil.

TESSMANNIANTHUS (Melastomatac.).
carinatus *F. Almeda in Ann. Missouri Bot. Gard.,* 76(1): 2 (1989)—Panama.
gordonii *F. Almeda in Ann. Missouri Bot. Gard.,* 76(1): 4 (1989)—Panama.
quadridomius *J.J. Wurdack in Rev. Acad. Colomb. Cienc. Exact. Fis. Nat.,* 65: 247 (1989)—Colombia.

TETRACHYRON (Compos.).
websteri *(Wussow & Urbatsch) B.L. Turner in Phytologia,* 65(3): 241 (1988): T. orizabaensis var. websteri.

TETRAGASTRIS (Burserac.).
occhionii *(Rizzini) D.C. Daly in Brittonia,* 41(1): 23 (1989): Protium occhionii.

TETRAPOLLINIA B. Maguire & B.M. Boom in Mem. New York Bot. Gard., 51: 31 (1989). *GENTIANACEAE.*
caerulescens *(Aublet) B. Maguire & B.M. Boom in Mem. New York Bot. Gard.,* 51: 31 (1989): Lisianthus caerulescens.

TETRARRHENA (Gramin.).
turfosa *N.G. Walsh in Muelleria,* 7(1): 95 (1989)—Australia (Victoria, New South Wales).

TETRASTIGMA (Vitac.).
crenatum *B.R. Jackes in Austrobaileya,* 3(1): 150 (1989)—Australia (Queensland).
nilagiricum *(Miq.) B.V. Shetty in Kew Bull.,* 44(3): 477 (1989): Cissus nilagirica.
petraeum *B.R. Jackes in Austrobaileya,* 3(1): 153 (1989)—Australia (Queensland).
thorsborneorum *B.R. Jackes in Austrobaileya,* 3(1): 155 (1989)—Australia (Queensland).

TETRORCHIDIUM (Euphorbiac.).
costaricense *M.J. Huft in Ann. Missouri Bot. Gard.,* 75(3): 1112 (1988)—Costa Rica, Panama.
microphyllum *M.J. Huft in Ann. Missouri Bot. Gard.,* 75(3): 1110 (1988)—Panama.

TEUCRIUM (Labiat.).
subsect. Botrys *A. Kästner in Biocosme Mésogéen,* 6(1–2): 76 (1989).
subsect. Canadensia *A. Kästner in Biocosme Mésogéen,* 6(1–2): 75 (1989).
subsect. Chamaedrys *A. Kästner in Biocosme Mésogéen,* 6(1–2): 71 (1989).
subsect. Cretica *A. Kästner in Biocosme Mésogéen,* 6(1–2): 71 (1989).
subsect. Fruticantia *A. Kästner in Biocosme Mésogéen,* 6(1–2): 70 (1989).
subsect. Isotriodon *A. Kästner in Biocosme Mésogéen,* 6(1–2): 73 (1989).
subsect. Marum *A. Kästner in Biocosme Mésogéen,* 6(1–2): 73 (1989).
subsect. Polium *(Miller) A. Kästner in Biocosme Mésogéen,* 6(1–2): 72 (1989): gen. Polium.
subsect. Pycnobotrys *(Benth.) A. Kästner in Biocosme Mésogéen,* 6(1–2): 74 (1989): T. sect. Pycnobotrys.
subsect. Scordium *A. Kästner in Biocosme Mésogéen,* 6(1–2): 75 (1989).
subsect. Scorodonia *A. Kästner in Biocosme Mésogéen,* 6(1–2): 74 (1989).
subsect. Spinularia *(Boiss.) A. Kästner in Biocosme Mésogéen,* 6(1–2): 75 (1989): T. sect. Spinularia.

TEUCRIUM :-
subsect. Stachyobotrys *(Benth.) A. Kästner in Biocosme Mésogéen,* 6(1–2): 75 (1989): T. sect. Stachyobotrys.
×estevei *F. Alcaraz, P. Sánchez-Gómez & J.S. Carrión in Lazaroa,* 9: 26 (1986 publ. 1988)—Spain. (T. capitatum subsp. gracillimum × libanitis).
glomeratum *A.J. Cavanilles in Anales Ci. Nat.,* 3(7): 62 (1801)—Canary Is.
homotrichum *(Font Quer) S. Rivas-Martínez in Opusc. Bot. Pharm. Complutensis,* 3: 88 (1986): T. carthaginense var. homotrichum.
luteum
subsp. aureiforme *(Rouy) S. Rivas-Martínez in Opusc. Bot. Pharm. Complutensis,* 3: 88 (1986): T. polium var. aureiforme.
subsp. lusitanicum *(Schreber) S. Rivas-Martínez in Opusc. Bot. Pharm. Complutensis,* 3: 88 (1986): T. polium subsp. lusitanicum.
taylorii
subsp. muticum *Yu.L. Menitskiĭ in Fl. Armenii,* 8: 18 (1987)—U.S.S.R. (Caucasus).

THALICTRUM (Ranunculac.).
simplex
subsp. rhodopaeum *(Rech. f.). Panov ex S. Kožuharov & A. Petrova in Ann. Bot. Fenn.,* 25(4): 390 (1988): T. rhodopaeum.

THAUMASTOCHLOA (Gramin.).
heteromorpha *B.K. Simon in Austrobaileya,* 3(1): 90 (1989)—Australia (Queensland).

THELYMITRA (Orchidac.).
fragrans *D.L. Jones & M.A. Clements in Austrobaileya,* 2(5): 550 (1988)—Australia (Queensland, New South Wales).

THEMEDA (Gramin.).
sect. Primothemedae *S.L. Chen & T.D. Zhuang in Bull. Bot. Res. North-East. Forest. Inst.,* 9(2): 55 (1989).
chinensis *(A. Camus) S.L. Chen & T.D. Zhuang in Bull. Bot. Res. North-East. Forest. Inst.,* 9(2): 59 (1989): T. ciliata subsp. chinensis.
trichiata *S.L. Chen & T.D. Zhuang in Bull. Bot. Res. North-East. Forest. Inst.,* 9(2): 58 (1989)—China.
unica *S.L. Chen & T.D. Zhuang in Bull. Bot. Res. North-East. Forest. Inst.,* 9(2): 56 (1989)—China.
yuanmounensis *S.L. Chen & T.D. Zhuang in Bull. Bot. Res. North-East. Forest. Inst.,* 9(2): 60 (1989)—China.
yunnanensis *S.L. Chen & T.D. Zhuang in Bull. Bot. Res. North-East. Forest. Inst.,* 9(2): 58 (1989)—China.

THIBAUDIA (Ericac.).
falconensis *J.L. Luteyn & M.L. Lebron-Luteyn in Acta Bot. Venezuel.,* 14(1): 167 (1983)—Venezuela.

THINOPYRUM (Gramin.).
sect. Lophopyrum *(Löve) D.R. Dewey in Gene Manipulat. in Pl. Improv. (ed. J.P. Gustafson):* 274 (1984), without basionym ref.: Lophopyrum sect. Lophopyrum.
sect. Trichophorae *(Nevski) D.R. Dewey in Gene Manipulat. in Pl. Improv. (ed. J.P. Gustafson):* 275 (1984): Agropyron ser. Trichophorae.
curvifolium *(Lange) D.R. Dewey in Gene Manipulat. in Pl. Improv. (ed. J.P. Gustafson):* 274 (1984): Agropyrum curvifolium.
elongatum *(Host) D.R. Dewey in Gene Manipulat. in Pl. Improv. (ed. J.P. Gustafson):* 274 (1984): Triticum elongatum.
flaccidifolium *(Boiss. & Heldr.) M. Moustakas in Pl. Syst. Evol.,* 161(3–4): 150 (1988 publ. 1989): Agropyron scirpeum var. flaccidifolium.
gentryi *(Melderis) D.R. Dewey in Gene Manipulat. in Pl. Improv. (ed. J.P. Gustafson):* 275 (1984): Agropyron gentryi.
podperae *(Nabelek) D.R. Dewey in Gene Manipulat. in Pl. Improv. (ed. J.P. Gustafson):* 275 (1984): Agropyron podperae.
scirpeum *(K. Presl) D.R. Dewey in Gene Manipulat. in Pl. Improv. (ed. J.P. Gustafson):* 275 (1984): Agropyron scirpeum.

THIOLLIEREA (Rubiac.).
campanulata *(Brongn.) M.G. Baumann-Bodenheim, Syst. Fl. Neu-Caledonien,* 5: 77 (1989): Grisia campanulata.
fritillarioides *(Schltr.) M.G. Baumann-Bodenheim, Syst. Fl. Neu-Caledonien,* 4: 103 (1988 publ. Jan. 1989), without basionym ref.; *et Syst. Fl. Neu-Caledonien,* 5: 98 (May 1989), with basionym ref.: Bikkia fritillarioides.
macrophylla *(Brongn.) M.G. Baumann-Bodenheim, Syst. Fl. Neu-Caledonien,* 4: 102 (1988 publ. Jan. 1989), without basionym ref.; *et Syst. Fl. Neu-Caledonien,* 5: 98 (May 1989), with basionym ref.: Grisia macrophylla.

THLADIANTHA (Cucurbitac.).
nudiflora
var. macrocarpa *Z. Zhang in Bull. Bot. Res. North-East. Forest. Inst.,* 9(4): 45 (1989)—China.
var. membranacea *Z. Zhang in Bull. Bot. Res. North-East. Forest. Inst.,* 9(4): 45 (1989)—China.

THRASYA (Gramin.).
stricta *A.G. Burman in Acta Bot. Venezuel.,* 14(4): 75 (1987)—Venezuela.

THUNBERGIA (Acanthac.).
bicolor *(Wight) K.S. Manilal & C.R. Suresh in J. Econ. Taxon. Bot.,* 5(2): 422 (1984): Schmidia bicolor.

THYMUS (Labiat.).
borgiae *S. Rivas-Martínez, A. Molina & G. Navarro in Opusc. Bot. Pharm. Complutensis,* 4: 118 (1988)—Spain.
godayanus *S. Rivas-Martínez, A. Molina & G. Navarro in Opusc. Bot. Pharm. Complutensis,* 4: 116 (1988)—Spain.
izcoi *S. Rivas-Martínez, A. Molina & G. Navarro in Opusc. Bot. Pharm. Complutensis,* 4: 114 (1988)—Spain.
kotschyanus
subsp. migricus *(Klok. & Shost.) Yu.L. Menitskiĭ in Fl. Armenii,* 8: 152 (1987): T. migricus.
subsp. pseudocollinus *Yu.L. Menitskiĭ in Fl. Armenii,* 8: 154 (1987)—U.S.S.R. (Caucasus).
neurophyllus *(Rech. f.) K.H. Rechinger in Ann. Bot. Fenn.,* 26(2): 169 (1989): Thymbra neurophylla.
subnervosus *A.I. Vvedenskiĭ, M.M. Nabiev & M. Tulyaganova in Opred. Rast. Sred. Azii,* 9: 361 (1987)—U.S.S.R. (Middle Asia).

TIBETIA (Legumin.).
forrestii *(Ali) P.C. Li in Acta Bot. Yunnanica,* 11(3): 287 (1989): Gueldenstaedtia forrestii.
liangshanensis *P.C. Li in Acta Bot. Yunnanica,* 11(3): 287 (1989)—China.
tongolensis
f. coelestis *(Diels) P.C. Li in Acta Bot. Yunnanica,* 11(3): 286 (1989): Astragalus coelestis.

TIGRIDIA (Iridac.).
chrysantha *Cruden & S.J. Walker ex R. McVaugh, Fl. Novo-Galiciana,* 15: 337 (1989)—Mexico.

TILLANDSIA (Bromeliac.).
confinis
var. minor *J. Rutschmann & W. Rauh in Abh. Akad. Wiss. Lit. Mainz, Math.-Naturwiss. Klasse, Trop. Subtrop. Pflanzenwelt,* 65: 47 (1988)—Ecuador.
dexteri *H.E. Luther in Selbyana,* 11: 54 (1989)—Costa Rica.
elvirae-grossiae *W. Rauh in Abh. Akad. Wiss. Lit. Mainz, Math.-Naturwiss. Klasse, Trop. Subtrop. Pflanzenwelt,* 65: 49 (1988)—Peru.
funckiana
var. recurvifolia *A. Blass ex W. Rauh in J. Bromeliad Soc.,* 39(3): 116 (1989)—Venezuela.
genseri *W. Rauh in Abh. Akad. Wiss. Lit. Mainz, Math.-Naturwiss. Klasse, Trop. Subtrop. Pflanzenwelt,* 65: 54 (1988)—Argentina.
guasamayensis *C.A. Palací & A.J. Gilmartin in Syst. Bot.,* 14(2): 152 (1989)—Argentina.

TILLANDSIA :-
guelzii *W. Rauh in Abh. Akad. Wiss. Lit. Mainz, Math.-Naturwiss. Klasse, Trop. Subtrop. Pflanzenwelt*, 65: 55 (1988)—Argentina.
harrisii *R. Ehlers in Bromelie*, 3: 34 (1987)—Guatemala.
huarazensis *Ehlers & Till in Bromelie*, 1988(2): 21 (1988)—Peru.
ilseana *W. Till, H. Halbritter & E. Zecher in J. Bromeliad Soc.*, 39(4): 153 (1989)—Mexico.
oropezana *L. Hromadnik in Bromelie*, 1989(1): 14 (1989)—Bolivia.
pentasticha *W. Rauh & R. Wülfinghoff in Abh. Akad. Wiss. Lit. Mainz, Math.-Naturwiss. Klasse, Trop. Subtrop. Pflanzenwelt*, 65: 59 (1988)—Mexico.
pfeufferi *W. Rauh in Abh. Akad. Wiss. Lit. Mainz, Math.-Naturwiss.. Klasse, Trop. Subtrop. Pflanzenwelt*, 65: 61 (1988)—Argentina.
porongoensis *L. Hromadnik & P. Schneider in Haussknechtia, Mitt. Thüring. Bot. Ges.*, 4: 39 (1988)—Argentina.
praschekii *R. Ehlers & K. Willinger in Bromelie*, 1989(2): 36 (1989)—Cuba.
pucaraensis *R. Ehlers in Bromelie*, 1989(1): 9 (1989)—Peru.
turquinensis *K. Willinger & Michálek in J. Bromeliad Soc.*, 39(1): 8 (1989)—Cuba.

TOCOCA (Melastomatac.).
croatii *F. Almeda in Proc. Calif. Acad. Sci.*, Ser. 4, 46(5): 148 (1989)—Panama.

TODDALIA (Rutac.).
asiatica
var. parva *Z.M. Tan in Bull. Bot. Res. North-East. Forest. Inst.*, 9(2): 47 (1989), without type—China.

TOFIELDIA (Melanthiac.).
sphaerocephala *(Khokhr.) A.P. Khokhryakov, Analiz Fl. Kolymskogo Nagor'ya:* 18 (1989): T. coccinea subsp. sphaerocephala.

TONGOLOA (Umbellif.).
rubronervis *S.L. Liu in Acta Phytotax. Sin.*, 27(1): 69 (1989)—China.
zhongdianensis *S.L. Liu in Acta Phytotax. Sin.*, 27(1): 68 (1989)—China.

TOONA (Meliac.).
ciliata
var. australis *(F. Muell.) K.N. Bahadur, Monogr. Genus Toona (Meliac.):* 78 (1988): Cedrela australis.
var. candollei *K.N. Bahadur, Monogr. Genus Toona (Meliac.):* 99 (1988)—India, China.
var. grandifoliola *(C. DC.) K.N. Bahadur, Monogr. Genus Toona (Meliac.):* 91 (1988): Cedrela microcarpa var. grandifoliola.
var. haslettii *(Haines) K.N. Bahadur, Monogr. Genus Toona (Meliac.):* 92 (1988): Cedrela toona var. haslettii.

TOONA :-
var. hexandra *(Wall.) K.N. Bahadur, Monogr. Genus Toona (Meliac.):* 93 (1988): Cedrela hexandra.
var. kingii *(C.DC.) K.N. Bahadur, Monogr. Genus Toona (Meliac.):* 100 (1988): Cedrela kingii.
var. listeri *(C.DC.) K.N. Bahadur, Monogr. Genus Toona (Meliac.):* 94 (1988): Cedrela toona var. listeri.
var. mollis *(Hand.-Mazz.) K.N. Bahadur, Monogr. Genus Toona (Meliac.):* 101 (1988): Cedrela mollis.
var. parviflora *(Benth.) K.N. Bahadur, Monogr. Genus Toona (Meliac.):* 94 (1988): Cedrela toona var. parviflora.
var. pilistila *(C.DC.) K.N. Bahadur, Monogr. Genus Toona (Meliac.):* 95 (1988): Cedrela toona var. pilistila.
var. pubinervis *(C.DC.) K.N. Bahadur, Monogr. Genus Toona (Meliac.):* 96 (1988): Cedrela toona var. pubinervis.
subsp. velutina *(DC.) K.N. Bahadur, Monogr. Genus Toona (Meliac.):* 97 (1988): Cedrela velutina.
var. velutina *(DC.) K.N. Bahadur, Monogr. Genus Toona (Meliac.):* 104 (1988): Cedrela velutina.
var. vestita *(C.T. White) K.N. Bahadur, Monogr. Genus Toona (Meliac.):* 106 (1988): Cedrela toona var. vestita.
microcarpa
var. pilipetala *(C.DC.) K.N. Bahadur, Monogr. Genus Toona (Meliac.):* 113 (1988): Cedrela toona var. pilipetala.
var. sahnii *K.N. Bahadur, Monogr. Genus Toona (Meliac.):* 114 (1988)—India, Bhutan.
sinensis
var. denticulata *(Chev.) K.N. Bahadur, Monogr. Genus Toona (Meliac.):* 118 (1988): T. microcarpa var. denticulata.
var. glabra *(C.DC.) K.N. Bahadur, Monogr. Genus Toona (Meliac.):* 120 (1988): Cedrela glabra.
var. grandifolia *(Chev.) K.N. Bahadur, Monogr. Genus Toona (Meliac.):* 121 (1988): T. microcarpa var. grandifolia.
var. puberula *(C.DC.) K.N. Bahadur, Monogr. Genus Toona (Meliac.):* 122 (1988): Cedrela serrata var. puberula.
sureni
var. celebica *(Koorders) K.N. Bahadur, Monogr. Genus Toona (Meliac.):* 133 (1988): Cedrela celebica.
var. cochinchinensis *(Pierre) K.N. Bahadur, Monogr. Genus Toona (Meliac.):* 135 (1988): T. febrifuga var. cochinchinensis.
var. inodora *(Hassk.) K.N. Bahadur, Monogr. Genus Toona (Meliac.):* 136 (1988): Cedrela inodora.
var. philippinensis *(Elm.) K.N. Bahadur, Monogr. Genus Toona (Meliac.):* 138 (1988): T. philippinensis.

TOONA :-
var. teysmannii *(Hassk.) K.N. Bahadur, Monogr. Genus Toona (Meliac.):* 143 (1988): Cedrela teysmannii.
ternatensis *(Miq.) K.N. Bahadur, Monogr. Genus Toona (Meliac.):* 144 (1988): Cedrela febrifuga f. ternatensis.

TORILIS (Umbellif.).
nodosa
subsp. nemoralis *S. Brullo in Colloqu. Phytosociolog.,* 12: 76 (1983); *see Pubbl. Ist. Bot. Univ. Catania* (1985)— Sicily.

TORULINIUM (Cyperac.).
odoratum
subsp. auriculatum *(Nees & Meyen ex Kunth) T. Koyama in Bishop Mus. Occas. Pap.,* 29: 125 (1989): Cyperus auriculatus.

TOULICIA (Sapindac.).
anomala *J.A. Steyermark in Ann. Missouri Bot. Gard.,* 75(3): 1076 (1988)—Venezuela.

TOURNEFORTIA (Boraginac.).
brantii *J.S. Miller in Ann. Missouri Bot. Gard.,* 76(2): 619 (1989)—Colombia.
spicata *J.S. Miller in Ann. Missouri Bot. Gard.,* 76(2): 619 (1989)—Colombia.

TRACHYMENE (Umbellif.).
clivicola *Boyland & A.E. Holland in Austrobaileya,* 3(1): 135 (1989)—Australia (Queensland).
hookeri *(Domin) A.E. Holland in Austrobaileya,* 3(1): 135 (1989): Didiscus procumbens var. hookeri.

TRAGIA (Euphorbiac.).
correae *M.J. Huft in Ann. Missouri Bot. Gard.,* 75(3): 1107 (1988)—Panama.

TRAGOPOGON (Compos.).
dubius
var. decipiens *(Chaub. ex Noulet) O. de Bolòs & J. Vigo in Folia Bot. Misc.,* 6: 85 (1989): T. majus var. decipiens.
var. major *(Jacq.) O. de Bolòs & J. Vigo in Folia Bot. Misc.,* 6: 85 (1989): T. major.

TRAPA (Trapac.).
bicornis
var. coreanus *Y.H. Chung & H.K. Choi in Korean J. Pl. Taxon.,* 17(1): 50 (1987), without latin descr. or type—Korea.

TREMA (Ulmac.).
tomentosa
var. viridis *(Planchon) H.J. Hewson in Fl. Australia,* 3: 190, 9 (1989): Sponia viridis.

TREMACRON (Gesneriac.).
aurantiacum *K.Y. Pan in Acta Phytotax. Sin.,* 26(6): 431 (1988)—China.
obliquifolium *K.Y. Pan in Acta Phytotax. Sin.,* 26(6): 429 (1988)—China.

TREMACRON :-
urceolatum *K.Y. Pan in Acta Phytotax. Sin.,* 26(6): 429 (1988)—China.

TRIADENUM (Guttif.).
tubulosum
var. walteri *(Gmel.) T.S. Cooperrider in Castanea,* 54(1): 10 (1989): Hypericon walteri.
virginicum
var. fraseri *(Spach) T.S. Cooperrider in Castanea,* 54(1): 9 (1989): Elodea fraseri.

TRIAINOLEPIS (Rubiac.).
sancta *B. Verdcourt in Fl. Zambes.,* 5(1): 50 (1989)—Mozambique.

TRIAS (Orchidac.).
bonaccordensis *C. Sathish Kumar in Blumea,* 34(1): 105 (1989)—India.
crassifolia *(Thw. ex Trimen) C. Sathish Kumar in Blumea,* 34(1): 108 (1989): Bulbophyllum crassifolium.

TRICHILIA (Meliac.).
lovettii *M. Cheek in Kew Bull.,* 44(3): 457 (1989)—Tanzania.

TRICHODIUM (Gramin.).
capillaris *(L.) A.W. Roth, Novae Plantarum Species:* 41 (1821): Agrostis capillaris.

TRICHOGLOTTIS (Orchidac.).
gibbosicalcar *(Seid.) K. Senghas in Schlechter Orchideen,* 1(21): 1315 (1988): Staurochilus gibbosicalcar.
ramosa *(Lindl.) K. Senghas in Schlechter Orchideen,* 1(21): 1315 (1988): Saccolabium ramosum.

TRICHOSANTHES (Cucurbitac.).
dafangensis *N.G. Ye & S.J. Li in Acta Phytotax. Sin.,* 26(6): 465 (1988), without type; et *in Acta Phytotax. Sin.,* 27(2): 153 (1989), with type—China.

TRICYRTIS (Convallariac.).
formosana
var. grandiflora *S.S. Ying, Coloured Illustr. Fl. Taiwan,* 3: 619 (1988)—Taiwan.

TRIDACTYLE (Orchidac.).
bicaudata
subsp. rupestris *H.P. Linder in Kew Bull.,* 44(2): 318 (1989)—South Africa (Cape Province, Natal).

TRIEENEA O.M. Hilliard in Notes Roy. Bot. Gard. Edinburgh, 45(3): 489 (1988 publ. 1989). *SCROPHULARIACEAE.*
elsiae *O.M. Hilliard in Notes Roy. Bot. Gard. Edinburgh,* 45(3): 489 (1988 publ. 1989)—South Africa (Cape Province).
frigida *O.M. Hilliard in Notes Roy. Bot. Gard. Edinburgh,* 45(3): 489 (1988 publ. 1989)—South Africa (Cape Province).

TRIEENEA :–
glutinosa *(Schltr.) O.M. Hilliard in Notes Roy. Bot. Gard. Edinburgh,* 45(3): 489 (1988 publ. 1989): Phyllopodium glutinosum.
lanciloba *O.M. Hilliard in Notes Roy. Bot. Gard. Edinburgh,* 45(3): 489 (1988 publ. 1989)—South Africa (Cape Province).
lasiocephala *O.M. Hilliard in Notes Roy. Bot. Gard. Edinburgh,* 45(3): 490 (1988 publ. 1989)—South Africa (Cape Province).
laxiflora *O.M. Hilliard in Notes Roy. Bot. Gard. Edinburgh,* 45(3): 490 (1988 publ. 1989)—South Africa (Cape Province).
longipedicellata *O.M. Hilliard in Notes Roy. Bot. Gard. Edinburgh,* 45(3): 490 (1988 publ. 1989)—South Africa (Cape Province).
schlechteri *(Hiern) O.M. Hilliard in Notes Roy. Bot. Gard. Edinburgh,* 45(3): 490 (1988 publ. 1989): Phyllopodium schlechteri.
taylorii *O.M. Hilliard in Notes Roy. Bot. Gard. Edinburgh,* 45(3): 491 (1988 publ. 1989)—South Africa (Cape Province).

TRIFOLIUM (Legumin.).
banaticum *(Heuffel) J. Májovský in Acta Fac. Rerum Nat. Univ. Comen., Bot.,* 35: 6 (1988): T. medium var. banaticum.
gordejevii *(Kom.) N.S. Pavlova in Sosud. Rast. Sovet. Dal'nego Vostoka,* 4: 330 (1989): Medicago gordejevii.

TRIGONOSTEMON (Euphorbiac.).
chinensis
f. fungii *(Merr.) Y.T. Chang in Acta Phytotax. Sin.,* 27(2): 149 (1989): T. fungii.
filipes *Y.T. Chang & X.L. Mo in Acta Phytotax. Sin.,* 27(2): 149 (1989)—China.

TRIKERAIA (Gramin.).
pappiformis *(Keng) P.C. Kuo & S.L. Lu ex J.L. Yang in Fl. Sichuanica,* 5(2): 197 (1988): Stipa pappiformis.

TRILLIUM (Trilliac.).
taiwanense *S.S. Ying in J. Jap. Bot.,* 64(5): 154 (1989)—Taiwan.

TRIMEZIA (Iridac.).
sect. Calolirion *P. Ravenna in Onira, Bot. Leafl.,* 1(1): 15 (1988).
sect. Juncella *P. Ravenna in Onira, Bot. Leafl.,* 1(1): 15 (1988).
sect. Neomarica *(Sprag.) P. Ravenna in Onira, Bot. Leafl.,* 1(1): 14 (1988), without basionym page: gen. Neomarica.
sect. Platyella *P. Ravenna in Onira, Bot. Leafl.,* 1(1): 15 (1988).
bauensis *P. Ravenna in Onira, Bot. Leafl.,* 1(1): 9 (1988)—Brazil.
bicolor *P. Ravenna in Onira, Bot. Leafl.,* 1(1): 5 (1988)—Brazil.
caerulea
subsp. eximia *P. Ravenna in Onira, Bot. Leafl.,* 1(1): 13 (1988)—Brazil.
subsp. rigida *P. Ravenna in Onira, Bot. Leafl.,* 1(1): 13 (1988)—Brazil.

TRIMEZIA :–
caeteana *P. Ravenna in Onira, Bot. Leafl.,* 1(1): 7 (1988)—Brazil.
capitellata *P. Ravenna in Onira, Bot. Leafl.,* 1(1): 2 (1988)—Brazil.
concinna *P. Ravenna in Onira, Bot. Leafl.,* 1(1): 6 (1988)—Brazil.
cristaliensis *P. Ravenna in Onira, Bot. Leafl.,* 1(1): 1 (1988)—Brazil.
fibrosa *P. Ravenna in Onira, Bot. Leafl.,* 1(1): 5 (1988)—Brazil.
fluminensis *P. Ravenna in Onira, Bot. Leafl.,* 1(1): 10 (1988)—Brazil.
itamarajuensis *P. Ravenna in Onira, Bot. Leafl.,* 1(1): 3 (1988)—Brazil.
juncifolia
subsp. aurantiaca *P. Ravenna in Onira, Bot. Leafl.,* 1(1): 12 (1988)—Brazil.
subsp. cristallicola *P. Ravenna in Onira, Bot. Leafl.,* 1(1): 11 (1988)—Brazil.
subsp. diamantinensis *P. Ravenna in Onira, Bot. Leafl.,* 1(1): 11 (1988)—Brazil.
subsp. parviflora *P. Ravenna in Onira, Bot. Leafl.,* 1(1): 12 (1988)—Brazil.
subsp. speciosa *P. Ravenna in Onira, Bot. Leafl.,* 1(1): 10 (1988)—Brazil.
nitida *P. Ravenna in Onira, Bot. Leafl.,* 1(1): 8 (1988)—Brazil.
paradoxa *P. Ravenna in Onira, Bot. Leafl.,* 1(1): 4 (1988)—Brazil.
portosecurensis *P. Ravenna in Onira, Bot. Leafl.,* 1(1): 7 (1988)—Brazil.
rotundata *P. Ravenna in Onira, Bot. Leafl.,* 1(1): 2 (1988)—Brazil.

TRIMORPHA (Compos.).
acris
var. asteroides *(Andrz. ex Besser) G.L. Nesom in Phytologia,* 67(1): 64 (1989): Erigeron asteroides.
var. debilis *(A. Gray) G.L. Nesom in Phytologia,* 67(1): 64 (1989): Erigeron acris var. debilis.
var. kamtschatica *(DC.) G.L. Nesom in Phytologia,* 67(1): 64 (1989): Erigeron kamtschaticus.
elata *(Hook.) G.L. Nesom in Phytologia,* 67(1): 64 (1989): Erigeron alpinus γ elatus.
lonchophylla *(Hook.) G.L. Nesom in Phytologia,* 67(1): 64 (1989): Erigeron lonchophyllus.
scotteri *(B. Boivin) G.L. Nesom in Phytologia,* 67(1): 64 (1989): Erigeron scotteri.

TRIPHORA (Orchidac.).
miserrima *(Cogn.) J. Acuña Galé, Cat. Descr. Orquideas Cubanas (Bol. Estac. Exper. Agron. Santiago de la Vegas, 60):* 18 (1938 publ. 1939), as 'Thiphora miserrima': Pogonia miserrima.

TRIPTEROSPERMUM (Gentianac.).
sect. Capsulifera *J. Murata in J. Fac. Sci. Univ. Tokyo, Bot.,* 14(4): 286 (1989).

TRIPTEROSPERMUM :–
alutaceofolium *(T.S. Liu & C.C. Kuo) J. Murata in J. Fac. Sci. Univ. Tokyo, Bot.*, 14(4): 301 (1989): T. taiwanense var. alutaceofolium.
australe *J. Murata in J. Fac. Sci. Univ. Tokyo, Bot.*, 14(4): 300 (1989)—China, Vietnam.
cordifolioides *J. Murata in J. Fac. Sci. Univ. Tokyo, Bot.*, 14(4): 299 (1989)—China.
involubile *N. Yonezawa in J. Phytogeogr. Taxon.*, 36(1): 9 (1988)—Japan.
luteoviride *(C.B. Clarke) J. Murata in J. Fac. Sci. Univ. Tokyo, Bot.*, 14(4): 303 (1989): Crawfurdia luteoviridis.
luzonense *(Vidal) J. Murata in J. Fac. Sci. Univ. Tokyo, Bot.*, 14(4): 305 (1989): Crawfurdia luzonensis.
sumatranum *J. Murata in J. Fac. Sci. Univ. Tokyo, Bot.*, 14(4): 298 (1989)—Sumatra.

TRITICUM (Gramin.).
aestivum
var. uralicum *L.V. Semenova in Sborn. Nauch. Tr. Prikl. Bot. Genet. Selek.*, 117: 127 (1988)—Cultivation.
var. vavilovianum *R.A. Udachin & I.Sh. Shakhmedov in Sborn. Nauch. Tr. Prikl. Bot. Genet. Selek.*, 117: 129 (1988)—Cultivation.
var. vigorovii *L.V. Semenova in Sborn. Nauch. Tr. Prikl. Bot. Genet. Selek.*, 117: 127 (1988)—Cultivation.
turanicum
var. quasinotabile *R.A. Udachin & S.A. Potokina in Sborn. Nauch. Tr. Prikl. Bot. Genet. Selek.*, 117: 128 (1988)—Cultivation.

TROCHODENDRON (Trochodendrac.)
aralioides *P.F. Siebold & J.G. Zuccarini, Fl. Japonica*, 1: 84, t. 39 & 40 (1839)—Japan.†

TROGLOPHYTON (Compos.).
acocksianum *O.M. Hilliard in Fl. S. Afr.*, 33(7:2): 42 (1983)—South Africa (Cape Province).
capillaceum
subsp. diffusum *(DC.) O.M. Hilliard in Fl. S. Afr.*, 33(7:2): 40 (1983): Helichrysum capillaceum var. diffusum.
elsiae *O.M. Hilliard in Fl. S. Afr.*, 33(7:2): 40 (1983)—South Africa (Cape Province).
leptomerum *O.M. Hilliard in Fl. S. Afr.*, 33(7:2): 45 (1983)—South Africa (Cape Province).
tenellum *O.M. Hilliard in Fl. S. Afr.*, 33(7:2): 44 (1983)—South Africa (Cape Province).

TROPHIS (Morac.).
sect. Echinocarpa *C.C. Berg in Proc. Kon. Nederl. Akad. Wetensch., C*, 91(4): 353 (1988).
sect. Malaisia *(Blanco) C.C. Berg in Proc. Kon. Nederl. Akad. Wetensch., C*, 91(4): 354 (1988): gen. Malaisia.

TROPHIS :–
sect. Olmedia *(Ruiz & Pavón) C.C. Berg in Proc. Kon. Nederl. Akad. Wetensch., C*, 91(4): 354 (1988): gen. Olmedia.
borbonica *(Duchartre) C.C. Berg in Proc. Kon. Nederl. Akad. Wetensch., C*, 91(4): 355 (1988): Maillardia borbonica.
caucana *(Pittier) C.C. Berg in Proc. Kon. Nederl. Akad. Wetensch., C*, 91(4): 354 (1988): Olmedia caucana.
montana *(Lénadri) C.C. Berg in Proc. Kon. Nederl. Akad. Wetensch., C*, 91(4): 355 (1988): Maillardia montana.

TUBEROLABIUM (Orchidac.).
papuanum *(Schltr.) J.J. Wood ex B. Lewis & P. Cribb, Orchids of Vanuatu:* 150 (1989): Saccolabium papuanum.

TULIPA (Liliac.).
hungarica
subsp. undulatifolia *(Roman) N. Roman & A. Beldie in Fl. Republ. Social. Român.*, 13: 52 (1976): T. hungarica var. undulatifolia.
subpraestans *A.I. Vvedenskii in Opred. Rast. Sred. Azii*, 2: 318 (1971), as 'subpreastans' in I.K. Supp. 16 in error—U.S.S.R. (Middle Asia).†

TURBINICARPUS (Cactac.).
krainzianus
f. minima *G. Frank in Succulenta*, 68(2): 40 (1989), without type—Mexico.

TURRAEA (Meliac.).
barbata *B.T. Styles & F. White in Bull. Jard. Bot. Nation. Belg.*, 59(1–2): 257 (1989)—Kenya.
cornucopia *B.T. Styles & F. White in Bull. Jard. Bot. Nation. Belg.*, 59(1–2): 257 (1989)—Kenya.
kimbozensis *M. Cheek in Kew Bull.*, 44(3): 465 (1989)—Tanzania.
kokwaroana *B.T. Styles & F. White in Bull. Jard. Bot. Nation. Belg.*, 59(1–2): 258 (1989)—Kenya.
mombassana
subsp. cuneata *(Gürke) B.T. Styles & F. White in Bull. Jard. Bot. Nation. Belg.*, 59(1–2): 258 (1989): T. cuneata.

TURUKHANIA (Legumin.).
ruthenica *(L.) N.S. Pavlova in Sosud. Rast. Sovet. Dal'nego Vostoka*, 4: 321 (1989): Trigonella ruthenica.
schischkinii *(Vass.) N.S. Pavlova in Sosud. Rast. Sovet. Dal'nego Vostoka*, 4: 319 (1989): Trigonella schischkinii.

TUXTLA J.L. Villaseñor & J.L. Strother in Syst. Bot., 14(4): 537 (1989). *COMPOSITAE.*
pittieri *(Greenman) J.L. Villaseñor & J.L. Strother in Syst. Bot.*, 14(4): 537 (1989): Zexmenia pittieri.

TWEEDIA (Asclepiadac.).
andina *(Phil.) G.H. Rua in Parodiana,* 5(2): 406 (1989): Oxypetalum andinum.

TYLECODON (Crassulac.).
albiflorus *P. Bruyns in S. Afr. J. Bot.,* 55(3): 333 (1989)—South Africa (Cape Province).
atropurpureus *P. Bruyns in S. Afr. J. Bot.,* 55(3): 332 (1989)—South Africa (Cape Province).
bayeri *E. van Jaarsveld in Aloe,* 25(2): 31 (1988)—South Africa (Cape Province).
ellaphieae *E.J. van Jaarsveld in Flow. Pl. Afr.,* 50(2): t. 1983 (1989)—South Africa (Cape Province).
pusillus *P. Bruyns in S. Afr. J. Bot.,* 55(3): 335 (1989)—South Africa (Cape Province).
sulphureus
var. armianus *E.J. van Jaarsveld in Flow. Pl. Afr.,* 50(2): t. 1984 (1989)—South Africa (Cape Province).

U

ULEX (Legumin.).
cantabricus *M.J. Álvarez Martínez, M.A. Fernández Casado, J.A. Fernández Prieto, H.S. Nava Fernández & M.L. Vera de la Puente in Candollea,* 43(2): 493 (1988)—Spain.

ULMUS (Ulmac.).
japonica
var. suberosa *(Turcz.) S.D. Zhao in Fl. Liaoningica, chief ed. S.X. Li,* 1: 268 (1988): U. pumila var. suberosa.
×mesocarpa *M. Kim & S. Lee in Korean J. Pl. Taxon.,* 19(1): 44 (1989), contrary to Art. H.5.2 ICBN (1988)—Korea. (U. davidiana var. japonica × macrocarpa).
minor
subsp. angustifolia *(Weston) C.A. Stace in Watsonia,* 17(4): 442 (1989): U. campestris var. angustifolia.
subsp. sarniensis *(Loudon) C.A. Stace in Watsonia,* 17(4): 442 (1989): U. campestris var. sarniensis.

URBANODENDRON (Laurac.).
bahiense *(Meissn.) J.G. Rohwer in Bot. Jahrb.,* 110(2): 166 (1988): Mespilodaphne bahiensis.
macrophyllum *J.G. Rohwer in Bot. Jahrb.,* 110(2): 166 (1988)—Brazil.

URGINEA (Hyacinthac.).
nana *S.O. Oyewole in Ann. Missouri Bot. Gard.,* 76(2): 623 (1989)—Nigeria.

UROSPATHELLA G.S. Bunting in Phytologia, 65(5): 391 (1988). *ARACEAE.*
wurdackii *(Bunting) G.S. Bunting in Phytologia,* 65(5): 391 (1988): Cyrtosperma wurdackii.

URTICA (Urticac.).
portosanctana *J.R. Press in Bocagiana,* 123: 4 (1988), nom. nov.: U. subincisa var. floribunda *Wedd.*
sondenii *(Simm.) Avror. ex D.V. Gel'tman in Novosti Sist. Vÿssh. Rast.,* 25: 76 (1988): U. dioica var. sondenii.

UTRICULARIA (Lentibulariac.).
sect. Australes *P. Taylor, Genus Utricularia (Kew Bull. Addit. Ser. 14):* 179 (1989).
sect. Choristothecae *P. Taylor, Genus Utricularia (Kew Bull. Addit. Ser. 14):* 508 (1989).
sect. Mirabiles *P. Taylor, Genus Utricularia (Kew Bull. Addit. Ser. 14):* 503 (1989).
sect. Steyermarkia *P. Taylor, Genus Utricularia (Kew Bull. Addit. Ser. 14):* 517 (1989).
letestui *P. Taylor, Genus Utricularia (Kew Bull. Addit. Ser. 14):* 401 (1989)—Central African Republic.
malabarica *M.K. Janarthanam & A.N. Henry in J. Bombay Nat. Hist. Soc.,* 86(1): 84 (1989)—India.

V

VACCINIUM (Ericac.).
bracteatum
var. rubellum *P.S. Hsu, J.X. Qiu, S.F. Huang & Y. Zhang in Acta Bot. Yunnanica,* 11(3): 319 (1989)—China.

VALERIANA (Valerianac.).
sect. Amplophus *(Raf.) B. Eriksen in Nordic J. Bot.,* 9(2): 186 (1989): gen. Amplophus.
sect. Astrephia *(Dufr.) B. Eriksen in Nordic J. Bot.,* 9(2): 185 (1989): gen. Astrephia.
sect. Belonanthus *(Graebn.) B. Eriksen in Nordic J. Bot.,* 9(2): 186 (1989): gen. Belonanthus.
sect. Stangea *(Graebn.) B. Eriksen in Nordic J. Bot.,* 9(2): 186 (1989): gen. Stangea.
alypifolia
subsp. argenteomarginata *B. Eriksen in Fl. Ecuador,* 34(186): 23 (1989)—Ecuador.
bryophila *F.R. Barrie in Sida,* 13(3): 266 (1989)—Mexico.
cernua *B. Eriksen in Fl. Ecuador,* 34(186): 20 (1989)—Ecuador.

VALERIANA :-
dageletiana
f. integra *(Nakai ex Maekawa) S.H. Park in J. Korean Pl. Taxon.*, 3(1–2): 26 (1971): V. dageletiana var. integra.
var. verticillata *S.H. Park in J. Korean Pl. Taxon.*, 3(1–2): 26 (1971)—Korea.
decussata
subsp. polemonioides *(H.B.K.) B. Eriksen in Fl. Ecuador*, 34(186): 52 (1989): V. polemonioides.
gallinae *F.R. Barrie in Sida*, 13(3): 269 (1989)—Mexico.
henrici *(Graebn.) B. Eriksen in Fl. Ecuador*, 34(186): 17 (1989): Stangea henrici.
protenta *B. Eriksen in Fl. Ecuador*, 34(186): 24 (1989)—Ecuador, Peru.
punctata
subsp. trichophylla *B. Eriksen in Fl. Ecuador*, 34(186): 23 (1989)—Ecuador.
secunda *B. Eriksen in Fl. Ecuador*, 34(186): 32 (1989)—Ecuador.

VANTANEA (Humiriac.).
depleta *G. McPherson in Ann. Missouri Bot. Gard.*, 75(3): 1148 (1988)—Panama.

VELLEREOPHYTON (Compos.).
felinum *O.M. Hilliard in Fl. S. Afr.*, 33(7:2): 35 (1983)—South Africa (Cape Province).
gracillimum *O.M. Hilliard in Fl. S. Afr.*, 33(7:2): 33 (1983)—South Africa (Cape Province).
niveum *O.M. Hilliard in Fl. S. Afr.*, 33(7:2): 34 (1983)—South Africa (Cape Province).
pulvinatum *O.M. Hilliard in Fl. S. Afr.*, 33(7:2): 35 (1983)—South Africa (Cape Province).

VERATRUM (Melanthiac.).
lobelianum
subsp. misae *(Širjaev) M. Šourková in Novit. Bot.*, 1988(4): 65 (1988): V. album var. misae.

VERBASCUM (Scrophulariac.).
×nydeggeri *A. Huber-Morath in Bauhinia*, 9(1): 40 (1988)—Turkey. (V. biledschikianum × lasianthum).
scoparium *V. Mozaffarian in Iranian J. Bot.*, 4(1): 66 (1988)—Iran.
xanthophoeniceum
var. albiflorum *D. Voliotis in Bot. Chronika*, 8(1–2): 36 (1988)—Greece.

VERBENA (Verbenac.).
arequipense *S.M. Botta in Darwiniana*, 28(1–4): 237 (1987 publ. 1988)—Peru.

VERBESINA (Compos.).
howardiana *J. Olsen in Phytologia*, 67(1): 107 (1989)—Hispaniola (Dominican Republic).
jacksonii *B.L. Turner in Phytologia*, 65(6): 493 (1989)—Mexico.
sinaloensis *B.L. Turner in Phytologia*, 66(6): 498 (1989)—Mexico.

VERNONIA (Compos.).
imeriensis *V.M. Badillo in Ernstia*, 53: 12 (1989)—Venezuela.
obtusa
var. bartlettii *(S.B. Jones) B.L. Turner in Phytologia*, 66(5): 462 (1989): V. duncanii subsp. bartlettii.
var. parkeri *(S.B. Jones) B.L. Turner in Phytologia*, 66(5): 462 (1989): V. obtusa subsp. parkeri.
proctorii *L.E. Urbatsch in Syst. Bot.*, 14(4): 589 (1989)—Puerto Rico.
teyucuarensis *A.L. Cabrera in Darwiniana*, 28(1–4): 187 (1987 publ. 1988)—Argentina.
triflosculosa
var. palmeri *(Rose) B.L. Turner in Phytologia*, 66(5): 464 (1989): V. palmeri.

VERONICA (Scrophulariac.).
×cantiana *D. McClintock in Plantsman*, 10(4): 214 (1989)—Cultivation. (V. dabneyi × officinalis).
cetikiana *A. Öztürk in Davis & Hedge Festschrift (ed. Kit Tan)*: 60 (1989)—Turkey.
gentianoides
var. alpina *A. Öztürk & M.A. Fischer in Davis & Hedge Festschrift (ed. Kit Tan)*: 58 (1989)—Turkey.
subsp. glacialis *(Nábělek) A. Öztürk & M.A. Fischer in Davis & Hedge Festschrift (ed. Kit Tan)*: 58 (1989): V. gentianoides var. glacialis.
kopgecidiensis *A. Öztürk & M.A. Fischer in Davis & Hedge Festschrift (ed. Kit Tan)*: 54 (1989)—Turkey.

VERONICASTRUM (Scrophulariac.).
borissovae *(Czer.) J. Soják in Sborn. Nár. Muz. Praze, B*, 39(1): 59 (1983): Leptandra borissovae.
tubiflorum *(Fisch. & Mey.) J. Soják in Sborn. Nár. Muz. Praze, B*, 39(1): 59 (1983): Veronica tubiflora.

VETIVERIA (Gramin.).
rigida *B.K. Simon in Austrobaileya*, 3(1): 95 (1989)—Australia (Queensland).

VEXILLABIUM (Orchidac.).
yakusimense
var. nakaianum *(F. Maekawa) T.B. Lee in J. Korean Pl. Taxon.*, 6(1–2): 18 (1975): V. nakaianum.

VICIA (Legumin.).
abbreviata
subsp. balansae *(Boiss.) A.K. Stankevich in Sborn. Nauch. Tr. Prikl. Bot. Genet. Selek.*, 117: 89 (1988): V. balansae.
cusnae *B. Foggi & C. Ricceri in Webbia*, 43(1): 25 (1989)—Italy.
eristalioides *N. Maxted in Notes Roy. Bot. Gard. Edinburgh*, 45(3): 453 (1988 publ. 1989)—Turkey.

VICIA :–
onobrychioides
subsp. alborosea *A. Dobignard in Saussurea,* 19: 100 (1988 publ. 1989)—Morocco.
sativa
subsp. consobrina *(Pomel) W. Greuter & Burdet in Willdenowia,* 19(1): 34 (1989): V. consobrina.
woroschilovii *N.S. Pavlova in Sosud. Rast. Sovet. Dal'nego Vostoka,* 4: 303 (1989), nom. nov.: V. pallida *Turcz.*

VIGNA (Legumin.).
subgen. Catiang *(DC.) V. Jaaska & V. Jaaska in Pl. Syst. Evol.,* 159(3–4): 157 (1988): Dolichos sect. Catiang.
benuensis *R.S. Pasquet & R. Maréchal in Canad. J. Bot.,* 67(4): 949 (1989)—Cameroun.
nyangensis *R. Mithen in Kew Bull.,* 44(1): 175 (1989)—Zimbabwe, Angola.

VIGUIERA (Compos.).
eriophora
var. poblano *(Panero & Schilling) B.L. Turner in Phytologia,* 66(5): 460 (1989): V. eriophora subsp. poblano.
excelsa
var. pachycephala *(DC.) B.L. Turner in Phytologia,* 66(5): 460 (1989): Tithonia pachycephala.
gentryi *B.L. Turner in Phytologia,* 66(5): 458 (1989)—Mexico.
iltisii *B.L. Turner in Phytologia,* 66(6): 502 (1989)—Mexico.
multiflora
var. macrocephala *(Heiser) B.L. Turner in Phytologia,* 66(5): 460 (1989): Heliomeris multiflora var. macrocephala.
nesomii *B.L. Turner in Phytologia,* 67(4): 339 (1989)—Mexico.
pinnatiloba
var. megaphylla *Butterwick ex B.L. Turner in Phytologia,* 66(5): 461 (1989)—Mexico.
torresii *B.L. Turner in Phytologia,* 66(5): 456 (1989)—Mexico.

VINCETOXICUM (Asclepiadac.).
balfourianum *(Schltr.) C.Y. Wu & D.Z. Li in Acta Bot. Yunnanica,* 11(1): 46 (1989): Cynanchum forrestii var. balfourianum.
biondioides *(W.T. Wang) C.Y. Wu & D.Z. Li in Acta Bot. Yunnanica,* 11(1): 48 (1989): Cynanchum biondioides.
chekiangense *(M. Cheng) C.Y. Wu & D.Z. Li in Acta Bot. Yunnanica,* 11(1): 48 (1989): Cynanchum chekiangense.
forrestii *(Schltr.) C.Y. Wu & D.Z. Li in Acta Bot. Yunnanica,* 11(1): 46 (1989): Cynanchum forrestii.
var. stenolobum *(Tsiang & Zhang) C.Y. Wu & D.Z. Li in Acta Bot. Yunnanica,* 11(1): 47 (1989): Cynanchum forrestii var. stenolobum.

VINCETOXICUM :–
glaucescens *(Decne.) C.Y. Wu & D.Z. Li in Acta Bot. Yunnanica,* 11(1): 48 (1989): Pentasachme glaucescens.
gracilipes *(Tsiang & Zhang) C.Y. Wu & D.Z. Li in Acta Bot. Yunnanica,* 11(1): 48 (1989): Cynanchum gracilipes.
hancockianum *(Maxim.) C.Y. Wu & D.Z. Li in Acta Bot. Yunnanica,* 11(1): 47 (1989): V. mongolicum β.? hancockianum.
hydrophilum *(Tsiang & Zhang) C.Y. Wu & D.Z. Li in Acta Bot. Yunnanica,* 11(1): 47 (1989): Cynanchum hydrophilum.
limprichtii *(Schltr.) C.Y. Wu & D.Z. Li in Acta Bot. Yunnanica,* 11(1): 47 (1989): Cynanchum limprichtii.
muliense *(Tsiang) C.Y. Wu & D.Z. Li in Acta Bot. Yunnanica,* 11(1): 46 (1989): Cynanchum muliense.
paniculatum *(Bunge) C.Y. Wu & D.Z. Li in Acta Bot. Yunnanica,* 11(1): 48 (1989): Asclepias paniculata.
riparium *(Tsiang & Zhang) C.Y. Wu & D.Z. Li in Acta Bot. Yunnanica,* 11(1): 47 (1989): Cynanchum riparium.
spirale *(Forssk.) D.Z. Li in Acta Bot. Yunnanica,* 11(1): 48 (1989): Asclepias spiralis.
stauntonii *(Decne.) C.Y. Wu & D.Z. Li in Acta Bot. Yunnanica,* 11(1): 47 (1989): Pentasacme stauntonii.
steppicola *(Hand.-Mazz.) C.Y. Wu & D.Z. Li in Acta Bot. Yunnanica,* 11(1): 47 (1989): Cynanchum steppicola.
taihangense *(Tsiang & Zhang) C.Y. Wu & D.Z. Li in Acta Bot. Yunnanica,* 11(1): 48 (1989): Cynanchum taihangense.
verticillatum
var. arenicola *(Tsiang & Zhang) C.Y. Wu & D.Z. Li in Acta Bot. Yunnanica,* 11(1): 47 (1989): Cynanchum verticillatum var. arenicola.

VIOLA (Violac.).
brevistipulata
f. albescens *E. Hama, Violets of Japan:* 181 (1987)—Japan.
chamissoniana
subsp. robusta *(Hillebr.) W.L. Wagner, D.R. Herbst & S.H. Sohmer in Bishop Mus. Occas. Pap.,* 29: 117 (1989): V. robusta.
subsp. tracheliifolia *(Ging.) W.L. Wagner, D.R. Herbst & S.H. Sohmer in Bishop Mus. Occas. Pap.,* 29: 117 (1989): V. tracheliifolia.
confusa
f. leucantha *E. Hama, Violets of Japan:* 181 (1987)—Japan.
faurieana
f. albifaurieana *E. Hama, Violets of Japan:* 181 (1987)—Japan.
hancockii
var. fangshanensis *J.W. Wang in Acta Phytotax. Sin.,* 27(3): 202 (1989)—China.

VIOLA :–
mandshurica
f. albiikedaeana *E. Hama, Violets of Japan:* 181 (1987)—Japan.
sacchalinensis
f. albialpina *E. Hama, Violets of Japan:* 181 (1987)—Japan.
f. chionantha *E. Hama, Violets of Japan:* 181 (1987)—Japan.
saxatilis
subsp. curtisii *(E. Foster) J. Kirschner & V. Skalický in Preslia,* 61(4): 317 (1989): V. curtisii.
subsp. polychroma *(A. Kerner) J. Kirschner & V. Skalický in Preslia,* 61(4): 317 (1989): V. polychroma.
sieboldii
f. albiflora *E. Hama, Violets of Japan:* 181 (1987)—Japan.
f. leucantha *E. Hama, Violets of Japan:* 181 (1987)—Japan.
violacea
f. leucantha *E. Hama, Violets of Japan:* 181 (1987)—Japan.
yezoensis
var. hopeiensis *(J.W. Wang & T.G. Ma) J.W. Wang & J. Yang in Acta Phytotax. Sin.,* 27(3): 202 (1989): V. hopeiensis.

VITEX (Verbenac.).
rotundifolia
f. albiflora *S.S. Ying, Colour. Illustr. Pl. Taiwan, Pt. 3; Colour. Illustr. Ligneous Pl. Taiwan,* 1: 504 (1987)—Taiwan.
taihangensis *L.B. Guo & S.Q. Zhou in Bull. Bot. Res. North-East. Forest. Inst.,* 9(4): 61 (1989)—China.

VITIS (Vitac.).
jinzhainensis *X.S. Shen in Acta Phytotax. Sin.,* 27(4): 304 (1989)—China.
shimenensis *W.T. Wang in Bull. Bot. Res. North-East. Forest. Inst.,* 9(1): 77 (1989)—China.
yuenlingensis *W.T. Wang in Bull. Bot. Res. North-East. Forest. Inst.,* 9(1): 75 (1989)—China.

VOANIOALA J. Dransfield in Kew Bull., 44(2): 192 (1989). *PALMAE.*
gerardii *J. Dransfield in Kew Bull.,* 44(2): 195 (1989)—Madagascar.

VOTOMITA (Melastomatac.).
roraimensis *T. Morley in Ann. Missouri Bot. Gard.,* 76(2): 439 (1989)—Brazil.

VRIESEA (Bromeliac.).
belloi *E.M.C. Leme in Bradea,* 5(16): 172 (1989)—Brazil.
debilis *E.M.C. Leme in Bradea,* 5(16): 173 (1989)—Brazil.
declinata *E.M.C. Leme in Bradea,* 5(16): 174 (1989)—Brazil.
odorata *E.M.C. Leme in Bradea,* 5(16): 175 (1989)—Brazil.

VRIESEA :–
rastrensis *E.M.C. Leme in Bradea,* 5(16): 176 (1989)—Brazil.

W

WAHLENBERGIA (Campanulac.).
caryophylloides *P.J. Smith in Nuytsia,* 7(1): 63 (1989)—Australia (Queensland, N. Terr., W. Australia).

WATSONIA (Iridac.).
subsect. Angustae *P. Goldblatt, Genus Watsonia (Ann. Kirstenbosch Bot. Gard., 19):* 109 (1989).
subsect. Brevibractea *P. Goldblatt, Genus Watsonia (Ann. Kirstenbosch Bot. Gard., 19):* 87 (1989).
sect. Gladioloides *P. Goldblatt, Genus Watsonia (Ann. Kirstenbosch Bot. Gard., 19):* 52 (1989).
subsect. Grandibractea *P. Goldblatt, Genus Watsonia (Ann. Kirstenbosch Bot. Gard., 19):* 99 (1989).
amabilis *P. Goldblatt, Genus Watsonia (Ann. Kirstenbosch Bot. Gard., 19):* 41 (1989)—South Africa (Cape Province).
amatolae *P. Goldblatt, Genus Watsonia (Ann. Kirstenbosch Bot. Gard., 19):* 58 (1989)—South Africa (Cape Province).
bella *N.E. Brown ex P. Goldblatt, Genus Watsonia (Ann. Kirstenbosch Bot. Gard., 19):* 71 (1989)—Swaziland, South Africa (Transvaal).
borbonica
subsp. ardernei *(Sander) P. Goldblatt, Genus Watsonia (Ann. Kirstenbosch Bot. Gard., 19):* 35 (1989): W. ardernei.
canaliculata *P. Goldblatt, Genus Watsonia (Ann. Kirstenbosch Bot. Gard., 19):* 83 (1989)—South Africa (Natal).
confusa *P. Goldblatt, Genus Watsonia (Ann. Kirstenbosch Bot. Gard., 19):* 79 (1989), without latin descr.—South Africa (Natal, Cape Province).
elsiae *P. Goldblatt, Genus Watsonia (Ann. Kirstenbosch Bot. Gard., 19):* 97 (1989)—South Africa (Cape Province).
inclinata *P. Goldblatt, Genus Watsonia (Ann. Kirstenbosch Bot. Gard., 19):* 67 (1989)—South Africa (Natal, Cape Province).
minima *P. Goldblatt, Genus Watsonia (Ann. Kirstenbosch Bot. Gard., 19):* 95 (1989)—South Africa (Cape Province).
mtamvunae *P. Goldblatt, Genus Watsonia (Ann. Kirstenbosch Bot. Gard., 19):* 74 (1989)—South Africa (Natal, Cape Province).

WATSONIA :–
paucifolia *P. Goldblatt, Genus Watsonia (Ann. Kirstenbosch Bot. Gard., 19):* 56 (1989)—South Africa (Cape Province).
pondoensis *P. Goldblatt, Genus Watsonia (Ann. Kirstenbosch Bot. Gard., 19):* 69 (1989)—South Africa (Natal, Cape Province).
pulchra *N.E. Brown ex P. Goldblatt, Genus Watsonia (Ann. Kirstenbosch Bot. Gard., 19):* 81 (1989)—Swaziland, South Africa (Natal, Transvaal).
rourkei *P. Goldblatt, Genus Watsonia (Ann. Kirstenbosch Bot. Gard., 19):* 132 (1989)—South Africa (Cape Province).

WEDELIA (Compos.).
aggregata *(Greenm.) B.L. Turner in Phytologia,* 65(5): 348 (1988): Aspilia (?) aggregata.
chihuahuana *B.L. Turner in Phytologia,* 65(5): 348 (1988)—Mexico.
cronquistii *B.L. Turner in Phytologia,* 65(5): 351 (1988)—Mexico.
gentryi *B.L. Turner in Phytologia,* 66(6): 500 (1989)—Mexico.
gonzaleziarum *B.L. Turner in Phytologia,* 65(5): 349 (1988), as 'gonzalezorum'—Mexico.
greenmanii *B.L. Turner in Phytologia,* 65(5): 351 (1988), nom. nov., with incorrect replaced synonym ref.: Aspilia stenophylla *Greenman.*
hintoniorum *B.L. Turner in Phytologia,* 65(5): 352 (1988)—Mexico.
keilii *B.L. Turner in Phytologia,* 65(5): 353 (1988)—Mexico.
purpurea *(Greenm.) B.L. Turner in Phytologia,* 65(5): 354 (1988): Aspilia purpurea.
scabra *(Cav.) B.L. Turner in Phytologia,* 65(5): 354 (1988): Buphthalmum scabrum.
tehuantepecana *B.L. Turner in Phytologia,* 65(5): 355 (1988)—Mexico.

WEINMANNIA (Cunoniac.).
neblinensis *B. Maguire & J.A. Steyermark in Mem. New York Bot. Gard.,* 51: 117 (1989)—Venezuela.

WHITFORDIODENDRON (Legumin.).
filipes
var. tomentosum *Z. Wei in Acta Phytotax. Sin.,* 27(1): 75 (1989)—China.

WIGANDIA (Hydrophyllac.).
urens
f. africana *(Brand) B. Verdcourt in Fl. Trop. E. Afr., Hydrophyllac.:* 1 (1989): W. kunthii f. africana.

WIKSTROEMIA (Thymelaeac.).
oahuensis
var. palustris *(Hochr.) B. Peterson in Bishop Mus. Occas. Pap.,* 29: 116 (1989): W. palustris.

WILDEMANIODOXA (Sapotac.).
laurentii
var. lundensis *(Cavaco) L. Liben in Bull. Jard. Bot. Nation. Belg.,* 58(3–4): 556 (1988): Chrysophyllum laurentii var. lundense.

WITHERINGIA (Solanac.).
filipes *Alain H. Liogier in Phytologia,* 65(6): 421 (1989)—Hispaniola (Dominican Republic).

WITTROCKIA (Bromeliac.).
paradoxa *E.M.C. Leme in Bradea,* 5(16): 170 (1989)—Brazil.
spiralipetala *E.M.C. Leme in Bradea,* 5(16): 171 (1989)—Brazil.

WURDACKANTHUS (Gentianac.).
frigidus *(Swartz) B. Maguire & B.M. Boom in Mem. New York Bot. Gard.,* 51: 9 (1989): Lisianthus frigidus.

X

XANTHOCEPHALUM (Compos.).
eradiatum *(M. Lane) G.L. Nesom in Phytologia,* 66(6): 485 (1989): X. gymnospermoides var. eradiatum.

XANTHOSIA (Umbellif.).
bungei *G.J. Keighery in Nordic J. Bot.,* 8(5): 445 (1989)—Australia (W. Australia).

XYRIS (Xyridac.).
diaphanobracteata *R. Kral & M. das Graças L. Wanderley in Ann. Missouri Bot. Gard.,* 75(1): 352 (1988)—Brazil.
dissitifolia *R. Kral & M. das Graças L. Wanderley in Ann. Missouri Bot. Gard.,* 75(1): 353 (1988)—Brazil.
itambensis *R. Kral & M. das Graças L. Wanderley in Ann. Missouri Bot. Gard.,* 75(1): 355 (1988)—Brazil.
jataiana *R. Kral & M. das Graças L. Wanderley in Ann. Missouri Bot. Gard.,* 75(1): 357 (1988)—Brazil.
kundelungensis *B. Brylska in Bull. Jard. Bot. Nation. Belg.,* 58(3–4): 549 (1988)—Zaïre.
lutescens *R. Kral & M. das Graças L. Wanderley in Ann. Missouri Bot. Gard.,* 75(1): 359 (1988)—Brazil.
obcordata *R. Kral & M. das Graças L. Wanderley in Ann. Missouri Bot. Gard.,* 75(1): 361 (1988)—Brazil.
pranceana *R. Kral & M. das Graças L. Wanderley in Ann. Missouri Bot. Gard.,* 75(1): 364 (1988)—Brazil.
sceptrifera *R. Kral & M. das Graças L. Wanderley in Ann. Missouri Bot. Gard.,* 75(1): 366 (1988)—Brazil.

XYRIS :–
seubertii
var. espinhacae *R. Kral & M. das Graças L. Wanderley in Ann. Missouri Bot. Gard.*, 75(1): 368 (1988)—Brazil.
villosicarinata *R. Kral & M. das Graças L. Wanderley in Ann. Missouri Bot. Gard.*, 75(1): 370 (1988)—Brazil.

Y

YEATESIA (Acanthac.).
mabryi *R.A. Hilsenbeck in Syst. Bot.*, 14(3): 436 (1989)—Mexico.
platystegia *(Torrey) R.A. Hilsenbeck in Syst. Bot.*, 14(3): 435 (1989): Tetramerium platystegium.

YUSHANIA (Gramin.).
canoviridis *G.H. Ye & Z.P. Wang in Acta Phytotax. Sin.*, 27(3): 228 (1989)—China.
glauca *T.P. Yi & T.L. Long in J. Bamboo Res.*, 8(2): 33 (1989)—China.
menghaiensis *T.P. Yi in Acta Bot. Yunnanica*, 10(4): 441 (1988)—China.

Z

ZAMIA (Zamiac.).
cremnophila *A.P. Vovides, B. Schutzman & B. Dehgan in Bot. Gaz.*, 149(3): 351 (1988)—Mexico.
soconuscensis *B. Schutzman, A.P. Vovides & B. Dehgan in Bot. Gaz.*, 149(3): 347 (1988)—Mexico.
standleyi *B. Schutzman in Syst. Bot.*, 14(2): 214 (1989)—Honduras.

ZANDERA D.L. Schulz in Haussknechtia, Mitt. Thüring. Bot. Ges., 4: 32 (1988). *COMPOSITAE.*
andersoniae *(Turner) D.L. Schulz in Haussknechtia, Mitt. Thüring. Bot. Ges.*, 4: 33 (1988): Sigesbeckia andersoniae.
blakei *(McVaugh & Lask.) D.L. Schulz in Haussknechtia, Mitt. Thüring. Bot. Ges.*, 4: 33 (1988): Trigonospermum blakei.
hartmanii *(Turner) D.L. Schulz in Haussknechtia, Mitt. Thüring. Bot. Ges.*, 4: 34 (1988): Sigesbeckia hartmanii.

ZANTEDESCHIA (Arac.).
odorata *P.L. Perry in S. Afr. J. Bot.*, 55(4): 447 (1989)—South Africa (Cape Province).

ZANTHOXYLUM (Rutac.).
acanthopodium
var. oligotrichum *Z.M. Tan in Bull. Bot. Res. North-East. Forest. Inst.*, 9(2): 43 (1989)—China.
dissitum
var. spinulosum *Z.M. Tan in Bull. Bot. Res. North-East. Forest. Inst.*, 9(2): 44 (1989)—China.

ZAPOTECA (Legumin.).
subgen. Aculeata *H.M. Hernández in Ann. Missouri Bot. Gard.*, 76(3): 855 (1989).
subgen. Amazonica *H.M. Hernández in Ann. Missouri Bot. Gard.*, 76(3): 856 (1989).
subgen. Nervosa *H.M. Hernández in Ann. Missouri Bot. Gard.*, 76(3): 808 (1989).
aculeata *(Spruce ex Benth.) H.M. Hernández in Ann. Missouri Bot. Gard.*, 76(3): 855 (1989): Calliandra aculeata.
alinae *H.M. Hernández in Ann. Missouri Bot. Gard.*, 76(3): 813 (1989)—Mexico.
amazonica *(Benth.) H.M. Hernández in Ann. Missouri Bot. Gard.*, 76(3): 858 (1989): Calliandra amazonica.
andina *H.M. Hernández in Ann. Missouri Bot. Gard.*, 76(3): 836 (1989)—Ecuador, Peru, Bolivia.
caracasana
subsp. weberbaueri *(Harms) H.M. Hernández in Ann. Missouri Bot. Gard.*, 76(3): 831 (1989): Calliandra weberbaueri.
costaricensis *(Britton & Rose) H.M. Hernández in Ann. Missouri Bot. Gard.*, 76(3): 836 (1989): Anneslia costaricensis.
filipes *(Benth.) H.M. Hernández in Ann. Missouri Bot. Gard.*, 76(3): 818 (1989): Calliandra filipes.
formosa
subsp. gracilis *(Griseb.) H.M. Hernández in Ann. Missouri Bot. Gard.*, 76(3): 847 (1989): Calliandra gracilis.
subsp. mollicula *(M. Martens & Galeotti) H.M. Hernández in Ann. Missouri Bot. Gard.*, 76(3): 851 (1989): Acacia mollicula.
subsp. rosei *(Wiggins) H.M. Hernández in Ann. Missouri Bot. Gard.*, 76(3): 848 (1989): Calliandra rosei.
subsp. salvadorensis *(Britton & Rose) H.M. Hernández in Ann. Missouri Bot. Gard.*, 76(3): 847 (1989): Anneslia salvadorensis.
subsp. schottii *(Torrey ex S. Watson) H.M. Hernández in Ann. Missouri Bot. Gard.*, 76(3): 852 (1989): Calliandra schottii.
microcephala *(Britton & Killip) H.M. Hernández in Ann. Missouri Bot. Gard.*, 76(3): 857 (1989): Calliandra microcephala.
nervosa *(Urban) H.M. Hernández in Ann. Missouri Bot. Gard.*, 76(3): 808 (1989): Pithecellobium nervosum.

ZAPOTECA :–
portoricensis
subsp. flavida *(Urban) H.M. Hernández in Ann. Missouri Bot. Gard.*, 76(3): 825 (1989): Calliandra flavida.
subsp. pubicarpa *H.M. Hernández in Ann. Missouri Bot. Gard.*, 76(3): 824 (1989)—Mexico.
scutellifera *(Benth.) H.M. Hernández in Ann. Missouri Bot. Gard.*, 76(3): 852 (1989): Calliandra scutellifera.
tehuana *H.M. Hernández in Ann. Missouri Bot. Gard.*, 76(3): 853 (1989)—Mexico.

ZEPHYRANTHES (Amaryllidac.).
yaviensis *P. Ravenna in Onira, Bot. Leafl.*, 1(8): 55 (1988)—Argentina.

ZEUXINE (Orchidac.).
uraiensis *S.S. Ying, Coloured Illustr. Fl. Taiwan*, 3: 620 (1988)—Taiwan.

ZEYHERELLA (Sapotac.).
rwandensis *(Troupin) L. Liben in Bull. Jard. Bot. Nation. Belg.*, 59(1–2): 258 (1989): Chrysophyllum rwandense.

ZINGIBER (Zingiberac.).
martinii *R.M. Smith in Notes Roy. Bot. Gard. Edinburgh*, 45(3): 418 (1988 publ. 1989)—Borneo (Sarawak).
monglaense *S.J. Chen & Z.Y. Chen in Guihaia*, 8(4): 315 (1988)—China.
nigrimaculatum *S.Q. Tong in Acta Phytotax. Sin.*, 27(4): 287 (1989)—China.
pseudopungens *R.M. Smith in Notes Roy. Bot. Gard. Edinburgh*, 45(3): 413 (1988 publ. 1989)—Borneo (Sabah).
puberulum
var. borneense *R.M. Smith in Notes Roy. Bot. Gard. Edinburgh*, 45(3): 419 (1988 publ. 1989)—Borneo (Sarawak).

ZINOWIEWIA (Celastrac.).
aymardii *J.A. Steyermark in Ann. Missouri Bot. Gard.*, 75(3): 1064 (1988)—Venezuela.

ZIZIPHORA (Labiat.).
vichodceviana *V. Tkatsch. ex M. Tulyaganova in Opred. Rast. Sred. Azii*, 9: 362 (1987)—U.S.S.R. (Middle Asia).

ZORNIA (Legumin.).
acauensis *M. Brandão & N.M.S. Costa in Prog. Resum. Soc. Bot. Brasil, XXXVI Congr. Nac. Bot., Curitiba:* 368 (1985), without latin descr. or type—Brazil.

ZORNIA :–
acuta *S.T. Reynolds & A.E. Holland in Austrobaileya*, 3(1): 18 (1989)—Australia (N. Terr., W. Australia).
dichotoma *E. Balátová-Tuláčková & P. Herrera Oliver in Folia Geobot. Phytotax.*, 23(4): 447 (1988)—Cuba.
disticha *S.T. Reynolds & A.E. Holland in Austrobaileya*, 3(1): 22 (1989)—Australia (N. Terr.), New Guinea (Papua New Guinea).
dyctiocarpa
var. filifolia *(Domin) S.T. Reynolds & A.E. Holland in Austrobaileya*, 3(1): 30 (1989): Z. filifolia.
floribunda *S.T. Reynolds & A.E. Holland in Austrobaileya*, 3(1): 26 (1989)—Australia (Queensland, New South Wales).
linearifoliolata *N.F. Mattos in Loefgrenia*, 90: 1 (1986)—Brazil.
maritima *S.T. Reynolds & A.E. Holland in Austrobaileya*, 3(1): 23 (1989)—Australia (Queensland).
muelleriana
subsp. congesta *S.T. Reynolds & A.E. Holland in Austrobaileya*, 3(1): 21 (1989)—Australia (W. Australia, N. Terr.).
muriculata
subsp. angustata *S.T. Reynolds & A.E. Holland in Austrobaileya*, 3(1): 25 (1989)—Australia (N. Terr., W. Australia, Queensland, New South Wales).
oligantha *S.T. Reynolds & A.E. Holland in Austrobaileya*, 3(1): 34 (1989)—Australia (N. Terr.).
pedunculata *S.T. Reynolds & A.E. Holland in Austrobaileya*, 3(1): 28 (1989)—Australia (Queensland).
prostrata *S.T. Reynolds & A.E. Holland in Austrobaileya*, 3(1): 32 (1989)—Australia (Queensland, W. Australia, N. Terr.).
var. macrantha *S.T. Reynolds & A.E. Holland in Austrobaileya*, 3(1): 33 (1989)—Australia (N. Terr.).
ramosa *S.T. Reynolds & A.E. Holland in Austrobaileya*, 3(1): 27 (1989)—Australia (Queensland).

ZYGOTRITONIA (Iridac.).
hysterantha *P. Goldblatt in Bull. Mus. Nation. Hist. Nat., B, Adansonia*, Ser. 4, 11(2): 208 (1989)—Central African Republic.

NEW AND OVERLOOKED GENERIC NAMES LISTED UNDER FAMILIES

A

ACANTHACEAE
SALPINCTIUM *T.J. Edwards*

AMARYLLIDACEAE
×CALICHARIS *A.W. Meerow*

ARACEAE
ANAPHYLLOPSIS *A. Hay*
UROSPATHELLA *G.S. Bunting*

B

BROMELIACEAE
BREWCARIA *L.B. Smith, J. Steyermark & Robinson*
CHEVALIERA *Gaudichaud ex J.G. Beer*
PLATYAECHMEA *(Baker) L.B. Smith & W.J. Kress*
PODAECHMEA *(Mez) L.B. Smith & W.J. Kress*
STEYERBROMELIA *L.B. Smith*

C

CACTACEAE
×SELENIPHYLCHIA *F. Süpplie*

CASUARINACEAE
QUADRANGULA *M.G. Baumann-Bodenheim*

COMPOSITAE
ACILEPIDOPSIS *H. Robinson*
CUATRECASANTHUS *H. Robinson*
CUATRECASASIELLA *H. Robinson*
CYRTOCYMURA *H. Robinson*
DIELITZIA *P.S. Short*
EIRMOCEPHALA *H. Robinson*
ERYMOPHYLLUM *Paul G. Wilson*
FELDSTONIA *P.S. Short*
FITZWILLIA *P.S. Short*
IRANECIO *B. Nordenstam*
JOSEANTHUS *H. Robinson*
LEMOORIA *P.S. Short*

COMPOSITAE :-
OBLIVIA *J.L. Strother*
PARAPRENANTHES *Chang ex C. Shih*
PSEUDOJACOBAEA *(Hook. f.) R. Mathur*
PTEROCYPSELA *C. Shih*
SONDOTTIA *P.S. Short*
TAPLINIA *N.S. Lander*
TUXTLA *J.L. Villaseñor & J.L. Strother*
ZANDERA *D.L. Schulz*

CRUCIFERAE
DACTYLOCARDAMUM *I.A. Al-Shehbaz*

CYCLANTHACEAE
CHORIGYNE *R. Eriksson*
DIANTHOVEUS *B.E. Hammel & G.J. Wilder*

CYPERACEAE
ACTINOSCIRPUS *(Ohwi) R.W. Haines & K.A. Lye*
KOYAMAEA *W.W. Thomas & G. Davidse*
MAXIMOVICZIA *A.P. Khokhryakov*
MAXIMOWICZIELLA *A.P. Khokhryakov*
OREOBOLOPSIS *T. Koyama & E.R. Guaglianone*

D

DIPSACACEAE
SCABIOSIOPSIS *K.H. Rechinger*

E

EPACRIDACEAE
DAENIKERANTHUS *M.G. Baumann-Bodenheim*
OREOTHAMNUS *M.G. Baumann-Bodenheim*

EUPHORBIACEAE
LOBANILIA *A. Radcliffe-Smith*

F

FAGACEAE
COLOMBOBALANUS *K.C. Nixon & W.L. Crepet*
FORMANODENDRON *K.C. Nixon & W.L. Crepet*

G

GENTIANACEAE
ROGERSONANTHUS *B. Maguire & B.M. Boom*
SIPAPOANTHA *B. Maguire & B.M. Boom*
TETRAPOLLINIA *B. Maguire & B.M. Boom*

GESNERIACEAE
×JANKAEBERLEA *J. Halda*
×JANKAENDRON *J. Halda*
×JANKAESSANDRA *J. Halda*
×RAMBERLEA *J. Halda*

GRAMINEAE
DAVIDSEA *T.R. Soderstrom & R.P. Ellis*
PARODIOLYRA *T.R. Soderstrom & F.O. Zuloaga*
PSEUDOXYTENANTHERA *T.R. Soderstrom & R.P. Ellis*
SPHAEROBAMBOS *S. Dransfield*

GUTTIFERAE
PARAMAMMEA *J.F. Leroy*

L

LABIATEAE
AETHYOPYS *(Benth.) F.M. Opiz*

LACANDONIACEAE
LACANDONIA *E. Martínez & C.H. Ramos*

LEGUMINOSAE
LEMURODENDRON *J.F. Villiers & P. Guinet*
OBOLINGA *R.C. Barneby*

M

MALVACEAE
ALLOSIDASTRUM *(Hochreutiner) A. Krapovickas, P.A. Fryxell & D.M. Bates*

MALVACEAE :–
NAYARIOPHYTON *T.K. Paul*

MELASTOMATACEAE
COMOLIOPSIS *J.J. Wurdack*
PENTOSSAEA *W.S. Judd*

MYRSINACEAE
EMBLEMANTHA *B.C. Stone*

MYRTACEAE
MICROSIDEROS *M.G. Baumann-Bodenheim*

O

ORCHIDACEAE
ANTERIORCHIS *E. Klein & D. Strack*
×BIFRENIELLA *hort.*
×BOLLOPETALUM *hort.*
×BRASSIODA *hort.*
CERATOCENTRON *K. Senghas*
×COOKARA *hort.*
×DEISELARA *hort.*
×FIALAARA *hort.*
×GEORGEBLACKARA *hort.*
×GOHARTIA *hort.*
×GOMETTIA *hort.*
×LOCKOCHILETTIA *hort.*
×NEOGLOSSUM *hort.*
NOTHODORITIS *Z.H. Tsi*
×NOTYLETTIA *hort.*
×ROTORARA *hort.*
SVENKOELTZIA *P. Burns-Balogh*

P

PALMAE
VOANIOALA *J. Dransfield*

PINACEAE
×HESPEROTSUGA *C.N. Page*
NOTHOTSUGA *H.H. Hu ex C.N. Page*

PODOCARPACEAE
AFROCARPUS *(Buchh. & E. Gray) C.N. Page*
METADACRYDIUM *M.G. Baumann-Bodenheim*
RETROPHYLLUM *C.N. Page*
SUNDACARPUS *(Buchh. & Gray) C.N. Page*

PODOSTEMACEAE
TERNIOPSIS *H.C. Chao*

POLYGONACEAE
ARISTOCAPSA *J.L. Reveal & C.B. Hardham*

POLYGONACEAE :–
DODECAHEMA *J.L. Reveal & C.B. Hardham*
SYSTENOTHECA *J.L. Reveal & C.B. Hardham*

R

RANUNCULACEAE
ARSENJEVIA *V.N. Starodubtsev*

RAPATEACEAE
MARAHUACAEA *B. Maguire*

RHAMNACEAE
SMYTHEA *B.C. Seeman*

RUBIACEAE
CHLOROCHORION *C. Puff & E. Robbrecht*
FURCATELLA *M.G. Baumann-Bodenheim*

S

SANTALACEAE
ELAPHANTHERA *N. Hallé*

SCROPHULARIACEAE
GLEKIA *O.M. Hilliard*
MAEVIELLA *R.A. Rossow*
MELANOSPERMUM *O.M. Hilliard*
NUTTALLANTHUS *D.A. Sutton*
SAIROCARPUS *D.A. Sutton*
TRIEENEA *O.M. Hilliard*

U

UMBELLIFERAE
ARAFOE *M.G. Pimenov & T.V. Lavrova*
AUTUMNALIA *M.G. Pimenov*
MACROSCIADIUM *V.N. Tikhomirov & T.V. Lavrova*
NEOPLATYTAENIA *A.M. Gel'dikhanov*
POLEMANNIOPSIS *B.L. Burtt*

W

WINTERACEAE
SARCODRIMYS *(Baill.) M.G. Baumann-Bodenheim*

Z

ZINGIBERACEAE
PYRGOPHYLLUM *(Gagnep.) T.L. Wu & Z.Y. Chen*
ROSCOEA *J. Smith*

FAMILY UNKNOWN

HAPTANTHUS *A. Goldberg & C. Nelson S.*

APPENDIX — FERNS & FERN ALLIES

NAMES AT GENERIC RANK AND BELOW

A

ACRORUMOHRA
dissecta *Ching ex Y.T. Hsieh in Bull. Bot. Res. North-East. Forest. Inst.*, 9(3): 47 (1989)—China.

ADIANTUM
isthmicum *B. Zimmer in Willdenowia*, 19(1): 215 (1989)—Panama.
lamrianum *A. Bidin & R. Jaman in Gard. Bull. Singapore*, 41(2): 45 (1988 publ. 1989)—Borneo (Sabah).

ALEURITOPTERIS
subvillosa
var. dilatata *(Brause) H.S. Kung in Fl. Sichuanica*, 6: 256 (1988): Cheilanthes yunnanensis var. dilatata Brause
var. tibetica *(Ching & S.K. Wu) H.S. Kung in Fl. Sichuanica*, 6: 258 (1988): Leptolepidium subvillosum var. tibeticum Ching & S.K. Wu

ALSOPHILA
subgen. Gymnosphaera *(Bl.) Q. Xia in Acta Phytotax. Sin.*, 27(1): 9 (1989): gen. Gymnosphaera Bl.
gigantea
var. polynervata *(Miou) Q. Xia in Acta Phytotax. Sin.*, 27(1): 11 (1989): Cyathea gigantea var. polynervata Miou
metteniana
var. subglabra *Ching & Q. Xia in Acta Phytotax. Sin.*, 27(1): 14 (1989)—China.

ASPLENIUM
ser. Falcata *Ching & S.H. Wu in Bull. Bot. Res. North-East. Forest. Inst.*, 9(2): 80 (1989).
sect. Holophyllus *Ching & S.H. Wu in Bull. Bot. Res. North-East. Forest. Inst.*, 9(2): 79 (1989).
ser. Trichomanoidea *Ching & S.H. Wu in Bull. Bot. Res. North-East. Forest. Inst.*, 9(2): 80 (1989).
ser. Unilateralia *Ching & S.H. Wu in Bull. Bot. Res. North-East. Forest. Inst.*, 9(2): 80 (1989).
ser. Variantia *Ching & S.H. Wu in Bull. Bot. Res. North-East. Forest. Inst.*, 9(2): 80 (1989).
ser. Wrightiana *Ching & S.H. Wu in Bull. Bot. Res. North-East. Forest. Inst.*, 9(2): 80 (1989).

ASPLENIUM :-
ambohitantelense *F. Rakotondrainibe in Bull. Mus. Nation. Hist. Nat., B, Adansonia*, Ser. 4, 10(4): 373 (1988 publ. 1989)—Madagascar.
bicentenniale *D.L. Jones in Austrobaileya*, 2(5): 477 (1988)—Australia (Queensland).
capitisyork *D.L. Jones in Austrobaileya*, 2(5): 477 (1988)—Australia (Queensland).
consimile *Ching ex S.H. Wu in Bull. Bot. Res. North-East. Forest. Inst.*, 9(2): 90 (1989)—China.
duplicatoserratum *Ching ex S.H. Wu in Bull. Bot. Res. North-East. Forest. Inst.*, 9(2): 86 (1989)—China.
fujianense *Ching ex S.H. Wu in Bull. Bot. Res. North-East. Forest. Inst.*, 9(2): 88 (1989)—China.
glanduliserrulatum *Ching ex S.H. Wu in Bull. Bot. Res. North-East. Forest. Inst.*, 9(2): 83 (1989), as 'glanduli-serrulatum'—China.
gulingense *Ching & S.H. Wu in Bull. Bot. Res. North-East. Forest. Inst.*, 9(2): 84 (1989)—China.
harmanii *D.L. Jones in Austrobaileya*, 2(5): 475 (1988)—Australia (Queensland, New South Wales).
×helii
nothosubsp. alberti-nieschalkii *W. Bennert, R.E.G. Pichi Sermolli, H. Rasbach, K. Rasbach & T. Reichstein in Bauhinia*, 9(1): 104 (1988)—Spain. (A. petrarchae subsp. petrarchae × trichomanes subsp. inexpectans).
lakshmananii *M.B. Viswanathan in Current Sci.*, 58(6): 318 (1989)—India.
longjinense *Ching & S.H. Wu in Bull. Bot. Res. North-East. Forest. Inst.*, 9(2): 89 (1989)—China.
neomutijugum *Ching ex S.H. Wu in Bull. Bot. Res. North-East. Forest. Inst.*, 9(2): 88 (1989)—China.
qiujiangense *Ching ex S.H. Wu in Bull. Bot. Res. North-East. Forest. Inst.*, 9(2): 90 (1989)—China.
sarelii
var. magnum *H.S. Kung in Fl. Sichuanica*, 6: 377 (1988): China.
sebungweense *J.E. Burrows in Bothalia*, 19(2): 173 (1989)—Zimbabwe, Zambia.
serratissimum *Ching ex S.H. Wu in Bull. Bot. Res. North-East. Forest. Inst.*, 9(2): 87 (1989)—China.

ASPLENIUM :-
subcrenatum *Ching ex S.H. Wu in Bull. Bot. Res. North-East. Forest. Inst.*, 9(2): 86 (1989)—China.
sublongum *Ching ex S.H. Wu in Bull. Bot. Res. North-East. Forest. Inst.*, 9(2): 82 (1989)—China.
subtoramanum *Ching ex S.H. Wu in Bull. Bot. Res. North-East. Forest. Inst.*, 9(2): 91 (1989)—China.
subtrapezoideum *Ching ex S.H. Wu in Bull. Bot. Res. North-East. Forest. Inst.*, 9(2): 84 (1989)—China.
taiwanense *Ching ex S.H. Wu in Bull. Bot. Res. North-East. Forest. Inst.*, 9(2): 85 (1989)—Taiwan.
tenuifolium
var. minor *Ching ex S.H. Wu in Bull. Bot. Res. North-East. Forest. Inst.*, 9(2): 92 (1989)—China.
trichomanes
nothosubsp. lucanum *P. Cubas, J.A. Rosselló & E. Pangua in Candollea*, 44(1): 182 (1989)—Balearic Is. (A. trichomanes subsp. inexpectans × trichomanes subsp. quadrivalens).
subsp. maderense *M. Gibby & J.D. Lovis in Fern Gaz. (U.K.)*, 13(5): 285 (1989)—Madeira.
unilaterale
var. decurrens *(Bedd.) H.S. Kung in Fl. Sichuanica*, 6: 357 (1988): Asplenium resectum var. decurrens Bedd.
xinyiense *Ching & S.H. Wu in Bull. Bot. Res. North-East. Forest. Inst.*, 9(2): 82 (1989)—China.

ATHYRIOPSIS
oldhami *(Hook. & Bak.) Y.C. Oh & C.S. Lee in Korean J. Pl. Taxon.*, 16(1): 13 (1986), without basionym ref.: Basionym not stated.

ATHYRIUM
×heterosporum *Y.T. Hsieh & Z.R. Wang in Acta Phytotax. Sin.*, 27(2): 156 (1989)—China.
scandicinum
f. bipinnata *F. Rakotondrainibe in Bull. Mus. Nation. Hist. Nat., B, Adansonia*, Ser. 4, 10(4): 375 (1988 publ. 1989)—Madagascar.
xichouense *Y.T. Hsieh & Z.R. Wang in Acta Phytotax. Sin.*, 27(2): 155 (1989)—China.

B

BLECHNUM
atropurpureum *A.R. Smith in Acta Bot. Venezuel.*, 14(3): 5 (1984)—Venezuela.
norfolkianum *(Heward) J.H. Maiden in Proc. Linn. Soc. New S. Wales*, 28(4): 732 (1903 publ. 1904): Lomaria norfolkiana Heward†

BOTRYPUS
decurrens *(Ching) Ching & H.S. Kung in Fl. Sichuanica*, 6: 96 (1988): Botrychium decurrens Ching

C

CALOCHLAENA (Maxon) M.D. Turner & R.A. White in Amer. Fern J., 78(3): 91 (1988): Culcita subgen. Calochlaena Maxon
dubia *(R. Br.) M.D. Turner & R.A. White in Amer. Fern J.*, 78(3): 92 (1988): Davallia dubia R. Br.
javanica *(Blume) M.D. Turner & R.A. White in Amer. Fern J.*, 78(3): 93 (1988): Dicksonia javanica Blume
novae-guineae *(Rosenstock) M.D. Turner & R.A. White in Amer. Fern J.*, 78(3): 93 (1988): Davallia novae-guineae Rosenstock
straminea *(Labill.) M.D. Turner & R.A. White in Amer. Fern J.*, 78(3): 92 (1988): Dicksonia straminea Labill.
villosa *(C. Chr.) M.D. Turner & R.A. White in Amer. Fern J.*, 78(3): 93 (1988): Culcita villosa C. Chr.

CERADENIA
auroseiomena *L.E. Bishop in Amer. Fern J.*, 79(1): 18 (1989)—Colombia.
comosa *L.E. Bishop in Amer. Fern J.*, 79(1): 16 (1989)—Bolivia.
dendrodoxa *L.E. Bishop in Amer. Fern J.*, 79(1): 15 (1989)—Peru, Ecuador.
maxoniana *L.E. Bishop in Amer. Fern J.*, 79(1): 24 (1989)—Colombia.
mirabilis *L.E. Bishop in Amer. Fern J.*, 79(1): 22 (1989)—Bolivia, Peru.
phloiocharis *L.E. Bishop in Amer. Fern J.*, 79(1): 19 (1989)—Panama, Costa Rica.
praeclara *L.E. Bishop in Amer. Fern J.*, 79(1): 16 (1989)—Peru.
terrestris *L.E. Bishop in Amer. Fern J.*, 79(1): 23 (1989)—Peru.

CETERACH
lolegnamense *M. Gibby & J.D. Lovis in Fern Gaz. (U.K.)*, 13(5): 287 (1989)—Madeira.

CHEILANTHES
buchtienii *(Rosenst.) R.M. Tryon in Fieldiana, Bot.*, n.s., 22: 34 (1989): Notholaena buchtienii Rosenst.
cantangensis *(Tryon) R.M. Tryon in Fieldiana, Bot.*, n.s., 22: 32 (1989): Notholaena cantangensis Tryon
cavernicola *D.L. Jones in Austrobaileya*, 2(5): 469 (1988)—Australia (N. Terr., W. Australia).

CHEILANTHES :-
dunlopii *D.L. Jones in Austrobaileya,* 2(5): 470 (1988)—Australia (N. Terr.).
pinnatifida *D.L. Jones in Austrobaileya,* 2(5): 472 (1988)—Australia (N. Terr., W. Australia).
praetermissa *D.L. Jones in Austrobaileya,* 2(5): 472 (1988)—Australia (N. Terr.).
pseudovellea *(H. Quirk & T. Chambers) D.L. Jones in Austrobaileya,* 2(5): 479 (1988): Cheilanthes sieberi subsp. pseudovellea H. Quirk & T. Chambers

CRYPTOGRAMMA
cascadensis *E.R. Alverson in Amer. Fern J.,* 79(3): 95 (1989)—U.S.A. (Washington, California, Idaho, Montana, Oregon), Canada (British Columbia).

CTENITIS
ochrorachis
var. violacea *F. Rakotondrainibe in Bull. Mus. Nation. Hist. Nat., B, Adansonia,* Ser. 4, 10(4): 375 (1988 publ. 1989)—Madagascar.
parishii *(Hook.) N.S. Jamir & R.R. Rao, Ferns of Nagaland:* 368 (1988): Nephrodium parishii Hook.

CTENOPTERIS
khasyana *(Hooker) E.B. Copeland in Philippine J. Sci,* 81(1): 45 (1952): Polypodium khasyanum.

CYATHEA
bullata *(Baker) F. Rakotondrainibe in Bull. Mus. Nation. Hist. Nat., B, Adansonia,* Ser. 4, 10(4): 376 (1988 publ. 1989): Alsophila bullata Baker

CYCLOSORUS
thwaitesii *(Hook.) B.A. Abeywickrama, Check List of Pteridophytes of Sri Lanka:* 9 (1978): Meniscium thwaitesii Hook.

CYRTOGONELLUM
falcilobum *Ching ex Y.T. Hsieh in Bull. Bot. Res. North-East. Forest. Inst.,* 9(3): 17 (1989)—China.
minimum *Y.T. Hsieh in Bull. Bot. Res. North-East. Forest. Inst.,* 9(3): 17 (1989)—China.
omeiense *Ching ex Y.T. Hsieh in Bull. Bot. Res. North-East. Forest. Inst.,* 9(3): 15 (1989)—China.
salicifolium *Ching & Y.T. Hsieh in Bull. Bot. Res. North-East. Forest. Inst.,* 9(3): 16 (1989)—China.
simile *Ching ex Y.T. Hsieh in Bull. Bot. Res. North-East. Forest. Inst.,* 9(3): 16 (1989)—China.

D

DAVALLIA
seramensis *M. Kato in J. Fac. Sci. Univ. Tokyo, Bot.,* 14(4): 223 (1989)—Moluccas.
sessilifolioides *M. Kato in J. Fac. Sci. Univ. Tokyo, Bot.,* 14(4): 227 (1989)—Moluccas.

DAVALLODES
seramense *M. Kato in J. Fac. Sci. Univ. Tokyo, Bot.,* 14(4): 222 (1989)—Moluccas.

DICRANOPTERIS
seramensis *M. Kato in Acta Phytotax. Geobot.,* 40(1–4): 82 (1989)—Moluccas.

DIPHASIASTRUM
complanatum
subsp. alpinum *(L.) A.C. Jermy in Fern Gaz. (U.K.),* 13(5): 260 (1989): Lycopodium alpinum L.
subsp. issleri *(Rouy) A.C. Jermy in Fern Gaz. (U.K.),* 13(5): 260 (1989): Lycopodium alpinum race issleri Rouy

DIPLAZIUM
zakamenense *(Tard.) F. Rakotondrainibe in Bull. Mus. Nation. Hist. Nat., B, Adansonia,* Ser. 4, 10(4): 376 (1988 publ. 1989): Athyrium zakamenense Tard.

DIPLOPTERYGIUM
maximum *(Ching) Ching & H.S. Kung in Fl. Sichuanica,* 6: 134 (1988): Hicriopteris maxima Ching

DRYOPTERIS
angustipalea *D. Darnaedi, M. Kato & K. Iwatsuki in J. Jap. Bot.,* 64(10): 308 (1989)—Moluccas.
austroindica *C.R. Fraser-Jenkins in Bull. Brit. Mus. (Nat. Hist.), Bot.,* 18(5): 370 (1989), as 'austro-indica'—India.
barbigera
subsp. komarovii *(Kossinsky) C.R. Fraser-Jenkins in Bull. Brit. Mus. (Nat. Hist.), Bot.,* 18(5): 384 (1989): Dryopteris komarovii Kossinsky
blanfordii
subsp. nigrosquamosa *(Ching) C.R. Fraser-Jenkins in Bull. Brit. Mus. (Nat. Hist.), Bot.,* 18(5): 388 (1989): Dryopteris nigrosquamosa Ching
bonatiana *(Brause) C.R. Fraser-Jenkins in Bull. Brit. Mus. (Nat. Hist.), Bot.,* 18(5): 363 (1989): Polystichum bonatianum Brause
caroli-hopei *C.R. Fraser-Jenkins in Bull. Brit. Mus. (Nat. Hist.), Bot.,* 18(5): 422 (1989)—India, Assam, Bhutan, Nepal, Xizang, China, ?Burma.

DRYOPTERIS :-

caudifolia *R.C. Ching & P.C. Chiu in Bot. Res., Inst. Bot. (Beijing),* 1987(2): 21 (1987)—China.

chiui *R.C. Ching in Bot. Res., Inst. Bot. (Beijing),* 1987(2): 32 (1987)—China.

conferta *R.C. Ching & P.C. Chiu in Bot. Res., Inst. Bot. (Beijing),* 1987(2): 17 (1987)—China.

consimilis *R.C. Ching & P.C. Chiu in Bot. Res., Inst. Bot. (Beijing),* 1987(2): 25 (1987)—China.

cuneatiloba *R.C. Ching & P.C. Chiu in Bot. Res., Inst. Bot. (Beijing),* 1987(2): 11 (1987)—China.

darjeelingensis *C.R. Fraser-Jenkins in Bull. Brit. Mus. (Nat. Hist.), Bot.,* 18(5): 335 (1989)—India, Sikkim.

deparioides

subsp. ambigua *(Sledge) C.R. Fraser-Jenkins in Bull. Brit. Mus. (Nat. Hist.), Bot.,* 18(5): 447 (1989): Dryopteris ambigua Sledge

subsp. gracillima *(Ching) C.R. Fraser-Jenkins in Bull. Brit. Mus. (Nat. Hist.), Bot.,* 18(5): 449 (1989): Dryopteris gracillima Ching

excelsior *R.C. Ching & P.C. Chiu in Bot. Res., Inst. Bot. (Beijing),* 1987(2): 8 (1987)—China.

exindusiiata *R.C. Ching & P.C. Chiu in Bot. Res., Inst. Bot. (Beijing),* 1987(2): 32 (1987)—China.

fengyangshanensis *R.C. Ching & P.C. Chiu in Bot. Res., Inst. Bot. (Beijing),* 1987(2): 33 (1987)—China.

×flemingii *C.R. Fraser-Jenkins in Bull. Brit. Mus. (Nat. Hist.), Bot.,* 18(5): 458 (1989)—India. (D. chrysocoma × juxtaposita).

fuyangensis *R.C. Ching & P.C. Chiu in Bot. Res., Inst. Bot. (Beijing),* 1987(2): 26 (1987)—China.

×ghatakii *C.R. Fraser-Jenkins in Bull. Brit. Mus. (Nat. Hist.), Bot.,* 18(5): 460 (1989)—India. (D. austroindica × cochleata).

grandiosa *R.C. Ching & P.C. Chiu in Bot. Res., Inst. Bot. (Beijing),* 1987(2): 8 (1987)—China.

himachalensis *C.R. Fraser-Jenkins in Bull. Brit. Mus. (Nat. Hist.), Bot.,* 18(5): 367 (1989)—India, China.

hirtipes

subsp. atrata *(Kunze) C.R. Fraser-Jenkins in Bull. Brit. Mus. (Nat. Hist.), Bot.,* 18(5): 333 (1989): Aspidium atratum Kunze

huanglungensis *R.C. Ching in Bot. Res., Inst. Bot. (Beijing),* 1987(2): 10 (1987)—China.

indonesiana *D. Darnaedi, M. Kato & K. Iwatsuki in J. Jap. Bot.,* 64(10): 304 (1989)—Moluccas, Java, Sumatra.

jiangshanensis *R.C. Ching & P.C. Chiu in Bot. Res., Inst. Bot. (Beijing),* 1987(2): 34 (1987)—China.

khullarii *C.R. Fraser-Jenkins in Bull. Brit. Mus. (Nat. Hist.), Bot.,* 18(5): 362 (1989)—India.

DRYOPTERIS :-

laodianensis *R.C. Ching & P.C. Chiu in Bot. Res., Inst. Bot. (Beijing),* 1987(2): 16 (1987)—China.

lepidocaulon *R.C. Ching & P.C. Chiu in Bot. Res., Inst. Bot. (Beijing),* 1987(2): 19 (1987)—China.

×liddarensis *C.R. Fraser-Jenkins in Bull. Brit. Mus. (Nat. Hist.), Bot.,* 18(5): 460 (1989)—India. (D. barbigera subsp. barbigera × serratodentata).

lingii *R.C. Ching in Bot. Res., Inst. Bot. (Beijing),* 1987(2): 23 (1987)—China.

×loyalii *C.R. Fraser-Jenkins in Bull. Brit. Mus. (Nat. Hist.), Bot.,* 18(5): 462 (1989)—India. (D. caroli-hopei × marginata).

lungjingensis *R.C. Ching & P.C. Chiu in Bot. Res., Inst. Bot. (Beijing),* 1987(2): 27 (1987)—China.

lungquanensis *R.C. Ching & P.C. Chiu in Bot. Res., Inst. Bot. (Beijing),* 1987(2): 1 (1987)—China.

lungshanensis *R.C. Ching in Bot. Res., Inst. Bot. (Beijing),* 1987(2): 5 (1987)—China.

lushanensis *R.C. Ching & P.C. Chiu in Bot. Res., Inst. Bot. (Beijing),* 1987(2): 20 (1987)—China.

×macdonellii *C.R. Fraser-Jenkins in Bull. Brit. Mus. (Nat. Hist.), Bot.,* 18(5): 464 (1989)—India. (D. filix-mas × ramosa).

macrochlamys *(Fée) C.R. Fraser-Jenkins in Bull. Brit. Mus. (Nat. Hist.), Bot.,* 18(5): 453 (1989): Aspidium macrochlamys Fée

madrasensis *C.R. Fraser-Jenkins in Bull. Brit. Mus. (Nat. Hist.), Bot.,* 18(5): 359 (1989)—India, Sri Lanka.

medialisora *R.C. Ching & P.C. Chiu in Bot. Res., Inst. Bot. (Beijing),* 1987(2): 6 (1987)—China.

neofuscipes *R.C. Ching & P.C. Chiu in Bot. Res., Inst. Bot. (Beijing),* 1987(2): 4 (1987)—China.

nudistipes *R.C. Ching & P.C. Chiu in Bot. Res., Inst. Bot. (Beijing),* 1987(2): 30 (1987)—China.

oblongipinnula *R.C. Ching & P.C. Chiu in Bot. Res., Inst. Bot. (Beijing),* 1987(2): 5 (1987)—China.

occidentalizhejiangensis *R.C. Ching & P.C. Chiu in Bot. Res., Inst. Bot. (Beijing),* 1987(2): 17 (1987), as 'occidentali-zhejiangensis'—China.

parafuscipes *R.C. Ching & P.C. Chiu in Bot. Res., Inst. Bot. (Beijing),* 1987(2): 4 (1987)—China.

paravaria *R.C. Ching & P.C. Chiu in Bot. Res., Inst. Bot. (Beijing),* 1987(2): 22 (1987)—China.

pseudofuscipes *R.C. Ching & P.C. Chiu in Bot. Res., Inst. Bot. (Beijing),* 1987(2): 3 (1987)—China.

qinyuanensis *R.C. Ching & P.C. Chiu in Bot. Res., Inst. Bot. (Beijing),* 1987(2): 7 (1987)—China.

DRYOPTERIS :–

rheophila *Mitsuta ex D. Darnaedi, M. Kato & K. Iwatsuki in J. Jap. Bot.*, 64(10): 302 (1989)—Thailand.
rufosquamosa *R.C. Ching & P.C. Chiu in Bot. Res., Inst. Bot. (Beijing)*, 1987(2): 6 (1987)—China.
scabripes *R.C. Ching in Bot. Res., Inst. Bot. (Beijing)*, 1987(2): 9 (1987)—China.
sessilipinna *R.C. Ching & P.C. Chiu in Bot. Res., Inst. Bot. (Beijing)*, 1987(2): 29 (1987)—China.
shandongensis *J.X. Li & F. Li in Acta Phytotax. Sin.*, 26(5): 406 (1988)—China.
shanghaiensis *R.C. Ching & P.C. Chiu in Bot. Res., Inst. Bot. (Beijing)*, 1987(2): 24 (1987)—China.
shanmenensis *R.C. Ching & P.C. Chiu in Bot. Res., Inst. Bot. (Beijing)*, 1987(2): 2 (1987)—China.
sledgei *C.R. Fraser-Jenkins in Bull. Brit. Mus. (Nat. Hist.), Bot.*, 18(5): 361 (1989)—India, Sri Lanka.
sri-lankensis *C.R. Fraser-Jenkins in Bull. Brit. Mus. (Nat. Hist.), Bot.*, 18(5): 451 (1989), nom. nov.: Lastrea deltoidea *Beddome*
subchampionii *R.C. Ching in Bot. Res., Inst. Bot. (Beijing)*, 1987(2): 14 (1987)—China.
subtenuicula *R.C. Ching & P.C. Chiu in Bot. Res., Inst. Bot. (Beijing)*, 1987(2): 29 (1987)—China.
tieanzuensis *R.C. Ching & P.C. Chiu in Bot. Res., Inst. Bot. (Beijing)*, 1987(2): 24 (1987)—China.
tsoongii *R.C. Ching in Bot. Res., Inst. Bot. (Beijing)*, 1987(2): 14 (1987)—China.
×vidyae *C.R. Fraser-Jenkins in Bull. Brit. Mus. (Nat. Hist.), Bot.*, 18(5): 466 (1989)—Sikkim. (D. sparsa × splendens).
wangii *R.C. Ching in Bot. Res., Inst. Bot. (Beijing)*, 1987(2): 13 (1987)—China.
×wechteriana *C.R. Fraser-Jenkins in Bull. Brit. Mus. (Nat. Hist.), Bot.*, 18(5): 466 (1989)—India. (D. chrysocoma × nigropaleacea).
woodsiisora
var. tenuissima *(Tagawa) Li & J.Z. Wang in Bull. Bot. Res. North-East. Forest. Inst.*, 9(4): 80 (1989): Dryopteris tenuissima Tagawa
wuyishanica *R.C. Ching & P.C. Chiu in Bot. Res., Inst. Bot. (Beijing)*, 1987(2): 34 (1987)—China.
wuyishanicola *R.C. Ching & P.C. Chiu in Bot. Res., Inst. Bot. (Beijing)*, 1987(2): 31 (1987)—China.
wuyuanensis *R.C. Ching in Bot. Res., Inst. Bot. (Beijing)*, 1987(2): 15 (1987)—China.
yushanensis *R.C. Ching & P.C. Chiu in Bot. Res., Inst. Bot. (Beijing)*, 1987(2): 28 (1987)—China.
zhenangensis *R.C. Ching & P.C. Chiu in Bot. Res., Inst. Bot. (Beijing)*, 1987(2): 11 (1987)—China.
zunyiensis *R.C. Ching in Bot. Res., Inst. Bot. (Beijing)*, 1987(2): 18 (1987)—China.

DRYOPTERIS :–

×zygoparentalis *C.R. Fraser-Jenkins in Bull. Brit. Mus. (Nat. Hist.), Bot.*, 18(5): 467 (1989)—India. (D. darjeelingensis × scottii).

E

EQUISETUM

×bowmanii *C.N. Page in Watsonia*, 17(3): 273 (1989)—Great Britain. (E. sylvaticum × telmateia).

G

GLEICHENIA

peltophora
var. seramensis *M. Kato in Acta Phytotax. Geobot.*, 40(1–4): 87 (1989)—Moluccas.

GRAMMITIS

rigescens *(Bory ex Willd.) J.E. Burrows in Bothalia*, 19(2): 171 (1989): Polypodium rigescens Bory ex Willd.

H

HUPERZIA

acifolia *(Rolleri) C. Rolleri & A.M. Deferrari in Not. Mus. La Plata, Bot.*, 21(100): 155 (1988), without exact basionym page: Lycopodium acifolium Rolleri
acuta *(Rolleri) C. Rolleri & A.M. Deferrari in Not. Mus. La Plata, Bot.*, 21(100): 155 (1988), without exact basionym page: Lycopodium acutum Rolleri
arcuata *B. Øllgaard in Fl. Ecuador*, 33(1): 15 (1988)—Ecuador, Colombia.
aristei *(Nessel) C. Rolleri & A.M. Deferrari in Not. Mus. La Plata, Bot.*, 21(100): 155 (1988), without exact basionym page: Urostachys aristei Nessel
austroecuadorica *B. Øllgaard in Fl. Ecuador*, 33(1): 31 (1988)—Ecuador.
balansae *(Herter) M.G. Baumann-Bodenheim, Syst. Fl. Neu-Caledonien*, 7: 167 (1989), without basionym ref., isonym; *see J. Holub in Folia Geobot. Phytotax.*, 20(1): 71 (1985): Lycopodium balansae Herter
boliviana *(Ros.) C. Rolleri & A.M. Deferrari in Not. Mus. La Plata, Bot.*, 21(100): 155 (1988): Lycopodium bolivianum Ros.

HUPERZIA :-

brauseana *(Herter) C. Rolleri & A.M. Deferrari in Not. Mus. La Plata, Bot.*, 21(100): 155 (1988): Lycopodium brauseanum Herter

campiana *B. Øllgaard in Fl. Ecuador,* 33(1): 109 (1988)—Ecuador.

columnaris *B. Øllgaard in Fl. Ecuador,* 33(1): 45 (1988)—Ecuador.

coriacea *(Rolleri) C. Rolleri & A.M. Deferrari in Not. Mus. La Plata, Bot.*, 21(100): 156 (1988), without exact basionym page: Lycopodium coriaceum Rolleri

crassa

var. gelida *B. Øllgaard in Fl. Ecuador,* 33(1): 56 (1988)—Ecuador.

var. manus-diaboli *B. Øllgaard in Fl. Ecuador,* 33(1): 56 (1988)—Ecuador.

densifolia *(Bk.) C. Rolleri & A.M. Deferrari in Not. Mus. La Plata, Bot.*, 21(100): 156 (1988), without basionym page: Lycopodium reflexum var. densifolius Bk.

espinosana *B. Øllgaard in Fl. Ecuador,* 33(1): 47 (1988)—Ecuador.

eversa *(Poiret) B. Øllgaard in Fl. Ecuador,* 33(1): 28 (1988): Lycopodium eversum Poiret

hastata *B. Øllgaard in Fl. Ecuador,* 33(1): 39 (1988)—Ecuador.

hoffmannii *(Maxon) C. Rolleri & A.M. Deferrari in Not. Mus. La Plata, Bot.*, 21(100): 156 (1988): Lycopodium hoffmannii Maxon

hypogaea *B. Øllgaard in Fl. Ecuador,* 33(1): 58 (1988)—Ecuador, Colombia, Peru.

leptodon *(Maxon ex Herter) C. Rolleri & A.M. Deferrari in Not. Mus. La Plata, Bot.*, 21(100): 156 (1988), without exact basionym page, basionym as 'Lycopodium': Urostachys leptodon Maxon ex Herter

linifolia

var. tenuifolia *(Nessel) B. Øllgaard in Fl. Ecuador,* 33(1): 90 (1988): Urostachys linifolius var. tenuifolius Nessel

llanganatensis *B. Øllgaard in Fl. Ecuador,* 33(1): 35 (1988)—Ecuador.

loxensis *B. Øllgaard in Fl. Ecuador,* 33(1): 17 (1988)—Ecuador.

mexiae *(Copeland) C. Rolleri & A.M. Deferrari in Not. Mus. La Plata, Bot.*, 21(100): 156 (1988): Lycopodium mexiae Copeland

nesselii *(Brause ex Nessel) C. Rolleri & A.M. Deferrari in Not. Mus. La Plata, Bot.*, 21(100): 156 (1988): Urostachys nesselii Brause ex Nessel

parvifolia *(Nessel) C. Rolleri & A.M. Deferrari in Not. Mus. La Plata, Bot.*, 21(100): 156 (1988): Urostachys parvifolius Nessel

pflanzii *(Nessel) C. Rolleri & A.M. Deferrari in Not. Mus. La Plata, Bot.*, 21(100): 156 (1988), without exact basionym page: Urostachys pflanzii Nessel

polyclada *(Sodiro) C. Rolleri & A.M. Deferrari in Not. Mus. La Plata, Bot.*, 21(100): 156 (1988): Lycopodium polycladum Sodiro

polydactyla *B. Øllgaard in Fl. Ecuador,* 33(1): 64 (1988)—Ecuador.

HUPERZIA :-

polylepidetorum *B. Øllgaard in Fl. Ecuador,* 33(1): 42 (1988)—Ecuador.

reflexa

var. minor *(Spring) B. Øllgaard in Fl. Ecuador,* 33(1): 26 (1988): Lycopodium reflexum var. minus Spring

sanctae-barbarae *(Rolleri) C. Rolleri & A.M. Deferrari in Not. Mus. La Plata, Bot.*, 21(100): 156 (1988), without exact basionym page: Lycopodium sanctae-barbarae Rolleri

scabrida *B. Øllgaard in Fl. Ecuador,* 33(1): 60 (1988)—Ecuador.

sellifolia *B. Øllgaard in Fl. Ecuador,* 33(1): 68 (1988)—Ecuador, Colombia.

serpentiformis *(Herter) C. Rolleri & A.M. Deferrari in Not. Mus. La Plata, Bot.*, 21(100): 156 (1988): Lycopodium serpentiforme Herter

socratis *(Herter) C. Rolleri & A.M. Deferrari in Not. Mus. La Plata, Bot.*, 21(100): 157 (1988), as 'socratea' & without exact basionym page: Urostachys socratis Herter

spongiosa *(Rolleri) C. Rolleri & A.M. Deferrari in Not. Mus. La Plata, Bot.*, 21(100): 157 (1988), without exact basionym page: Lycopodium spongiosum Rolleri

sydowiora *(Herter) C. Rolleri & A.M. Deferrari in Not. Mus. La Plata, Bot.*, 21(100): 157 (1988): Lycopodium sydowiorum Herter

talpiphila *B. Øllgaard in Fl. Ecuador,* 33(1): 44 (1988)—Ecuador.

trachyloma *(Herter) C. Rolleri & A.M. Deferrari in Not. Mus. La Plata, Bot.*, 21(100): 157 (1988): Urostachys trachyloma Herter

unguiculata *B. Øllgaard in Fl. Ecuador,* 33(1): 26 (1988)—Ecuador, Colombia, Peru.

HYMENOPHYLLUM

fucoides

var. calodictyon *(Bosch) R.G. Stolze in Fieldiana, Bot., n.s.,* 20: 56 (1989): Hymenophyllum calodictyon Bosch

var. chachapoyense *R.G. Stolze in Fieldiana, Bot., n.s.,* 20: 57 (1989)—Peru, Colombia, Ecuador.

lamellatum *R.G. Stolze in Fieldiana, Bot., n.s.,* 20: 58 (1989)—Peru.

maderense *M. Gibby & J.D. Lovis in Fern Gaz. (U.K.),* 13(5): 285 (1989)—Madeira.

myriocarpum

var. endiviifolium *(Desv.) R.G. Stolze in Fieldiana, Bot., n.s.,* 20: 63 (1989): Hymenophyllum endiviifolium Desv.

var. nigrescens *(Liebm.) R.G. Stolze in Fieldiana, Bot., n.s.,* 20: 63 (1989): Hymenophyllum nigrescens Liebm.

tarapotense *R.G. Stolze in Fieldiana, Bot., n.s.,* 20: 75 (1989)—Peru.

undulatum

var. fendlerianum *(J.W. Sturm) R.G. Stolze in Fieldiana, Bot., n.s.,* 20: 64 (1989): Hymenophyllum fendlerianum J.W. Sturm

I

ISOETES

coreana *Y.H. Chung & H.K. Choi in Korean J. Pl. Taxon.*, 16(1): 2 (1986)—Korea.

L

LINDSAEA

mesarum *K.U. Kramer in Ann. Missouri Bot. Gard.*, 76(2): 605 (1989)—Venezuela.

LYCOPODIELLA

descendens *B. Øllgaard in Fl. Ecuador,* 33(1): 143 (1988)—Ecuador, Peru.
steyermarkii *B. Øllgaard in Fl. Ecuador,* 33(1): 149 (1988)—Ecuador, Panama, Venezuela.

LYCOPODIOIDES

albocincta *(Ching ex H.S. Kung) H.S. Kung in Fl. Sichuanica,* 6: 61 (1988): Selaginella albo-cincta Ching ex H.S. Kung
chrysocaulos *(Hook. & Grev.) H.S. Kung in Fl. Sichuanica,* 6: 78 (1988): Lycopodium chrysocaulos Hook. & Grev.
compta *(Hand.-Mazz.) H.S. Kung in Fl. Sichuanica,* 6: 76 (1988): Selaginella compta Hand.-Mazz.
delicatula *(Desv.) H.S. Kung in Fl. Sichuanica,* 6: 67 (1988): Lycopodium delicatulum Desv.
doederleinii *(Hieron.) H.S. Kung in Fl. Sichuanica,* 6: 73 (1988): Selaginella doederleinii Hieron.
gebaueriana *(Hand.-Mazz.) H.S. Kung in Fl. Sichuanica,* 6: 65 (1988): Selaginella gebaueriana Hand.-Mazz.
labordei *(Hieron. ex Christ) H.S. Kung in Fl. Sichuanica,* 6: 77 (1988): Selaginella labordei Hieron. ex Christ
mairei *(Levl.) H.S. Kung in Fl. Sichuanica,* 6: 68 (1988): Selaginella mairei Levl.
moellendorffii *(Hieron.) H.S. Kung in Fl. Sichuanica,* 6: 75 (1988): Selaginella moellendorffii Hieron.
omeiensis *(Ching ex H.S. Kung) H.S. Kung in Fl. Sichuanica,* 6: 71 (1988): Selaginella omeiensis Ching ex H.S. Kung
prostrata *(H.S. Kung) H.S. Kung in Fl. Sichuanica,* 6: 76 (1988): Selaginella prostrata H.S. Kung
pulvinata *(Hook. & Grev.) H.S. Kung in Fl. Sichuanica,* 6: 64 (1988): Lycopodium pulvinatum Hook. & Grev.
remotifolia *(Spring) H.S. Kung in Fl. Sichuanica,* 6: 65 (1988): Selaginella remotifolia Spring
sanguinolenta
f. kantzensis *(H.S. Kung) H.S. Kung in Fl. Sichuanica,* 6: 61 (1988): Selaginella sanguinolenta f. kantzensis H.S. Kung
sichuanica *(H.S. Kung) H.S. Kung in Fl. Sichuanica,* 6: 71 (1988): Selaginella sichuanica H.S. Kung

LYCOPODIOIDES :-

tamariscina *(Beauv.) H.S. Kung in Fl. Sichuanica,* 6: 62 (1988): Stachygynandrum tamariscinum Beauv.

LYCOPODIUM

cernuum
f. ingens *B.C. Tan & D.B. Tolentino in Philippine J. Sci.*, 116(4): 438 (1987)—Philippine Is.
clavatum
subsp. contiguum *(Klotzsch) B. Øllgaard in Fl. Ecuador,* 33(1): 126 (1988): Lycopodium contiguum Klotzsch
hickeyi *W.H. Wagner, J.M. Beitel & R.C. Moran in Amer. Fern J.*, 79(3): 121 (1989), nom. nov.: Lycopodium obscurum var. isophyllum *R.J. Hickey*
selago
var. miniatum *(Spring) B.C. Tan & D.B. Tolentino in Philippine J. Sci.*, 116(4): 437 (1987): Lycopodium miniatum Spring
tauri *(Hert.) M. Kato in Acta Phytotax. Geobot.*, 39(4–6): 137 (1988): Urostachys tauri Hert.

M

MARSILEA

farinosa
subsp. arrecta *J.E. Burrows in Bothalia,* 19(2): 169 (1989)—South Africa (Transvaal), Botswana.
scalaripes *D.M. Johnson in Amer. Fern J.*, 78(2): 68 (1988)—Malaya.

MECODIUM

corrugatum
var. elongatum *(Christ) H.S. Kung in Fl. Sichuanica,* 6: 144 (1988): Hymenophyllum corrugatum var. elongatum Christ

MICROLEPIA

macrosora *A. Biswas in J. Econ. Taxon. Bot.*, 12(2): 322 (1988 publ. 1989)—India.
oblongifolia *A. Biswas in J. Econ. Taxon. Bot.*, 12(2): 321 (1988 publ. 1989)—India.
sikkimensis *A. Biswas in J. Econ. Taxon. Bot.*, 12(2): 322 (1988 publ. 1989)—Sikkim.
tamenlongensis *A. Biswas in J. Econ. Taxon. Bot.*, 12(2): 315 (1988 publ. 1989)—India.
viridula *A. Biswas in J. Econ. Taxon. Bot.*, 12(2): 315 (1988 publ. 1989)—India.

MOHRIA

caffrorum
var. ferruginea *J.E. & S.M. Burrows in Bothalia,* 19(2): 168 (1989)—South Africa (Natal, Transvaal, Orange Free State, Cape Province), Lesotho.

N

NEPHROLEPIS

arida *D.L. Jones in Austrobaileya*, 2(5): 474 (1988)—Australia (N. Terr., W. Australia).

O

OPHIOGLOSSUM

convexum *J.E. Burrows in Bothalia*, 19(2): 167 (1989)—South Africa (Transvaal), Zambia, Zimbabwe, Madagascar, Réunion.

P

PACHYPLEURIA

dissecta *(Alston) M. Kato in J. Fac. Sci. Univ. Tokyo, Bot.*, 14(4): 233 (1989): Humata dissecta Alston

obtusata *(v.A.v.R.) M. Kato in J. Fac. Sci. Univ. Tokyo, Bot.*, 14(4): 232 (1989): Humata obtusata v.A.v.R.

papuana *(Copel.) M. Kato in J. Fac. Sci. Univ. Tokyo, Bot.*, 14(4): 231 (1989): Humata papuana Copel.

werneri *(Copel.) M. Kato in J. Fac. Sci. Univ. Tokyo, Bot.*, 14(4): 232 (1989): Humata werneri Copel.

PALHINHAEA

cernua

var. curvata *(Sw.) N.C. Nair & S.R. Ghosh in J. Econ. Taxon. Bot.*, 12(1): 197 (1988): Lycopodium curvatum Sw.

var. pendula *(Hook.) N.C. Nair & S.R. Ghosh in J. Econ. Taxon. Bot.*, 12(1): 197 (1988): Lycopodium pendulum Hook.

f. sikkimensis *(Mull.) H.S. Kung in Fl. Sichuanica*, 6: 50 (1988): Lycopodium sikkimense Mull.

PELLAEA

glabella

var. missouriensis *G.J. Gastony in Amer. Fern J.*, 78(2): 64 (1988)—U.S.A. (Missouri).

occidentalis

subsp. simplex *(Butters) G.J. Gastony in Amer. Fern J.*, 78(2): 64 (1988): Pellaea glabella var. simplex Butters

PHEGOPTERIS

triangularis *J.P. Saint-Lager in A. Cariot, Etude Fl.*, 2: 964 (1889)—France.†

PHLEGMARIURUS

macrostachys *(Hook. ex Spring) N.C. Nair & S.R. Ghosh in J. Econ. Taxon. Bot.*, 12(1): 194 (1988): Lycopodium macrostachys Hook. ex Spring

PHYMATOPSIS

hastata

var. pygmaea *(Maxim.) Li & J.Z. Wang in Bull. Bot. Res. North-East. Forest. Inst.*, 9(4): 80 (1989): Polypodium hastatum f. pygmaeum Maxim.

POLYSTICHUM

auritum *(Fée) G. Yatskievych in Amer. Fern J.*, 79(1): 26 (1989): Phanerophlebia aurita Fée

bulbiferum *D.S. Barrington in Ann. Missouri Bot. Gard.*, 76(2): 371 (1989)—Ecuador.

chilense

var. dusenii *(C. Chr.) Looser ex R. Rodríguez R. in Gayana, Bot.*, 44(1–4): 48 (1987 publ. 1988): Polystichum multifidum var. dusenii C. Chr.

concinnum *Lellinger ex D.S. Barrington in Ann. Missouri Bot. Gard.*, 76(2): 367 (1989)—Costa Rica, Panama.

habaense *Ching & H.S. Kung in Bull. Bot. Res. North-East. Forest. Inst.*, 9(3): 4 (1989)—China.

lichiangense *(Wright) Ching ex H.S. Kung in Bull. Bot. Res. North-East. Forest. Inst.*, 9(3): 6 (1989): Nephrodium lichiangense Wright

melanostipes *Ching & H.S. Kung in Bull. Bot. Res. North-East. Forest. Inst.*, 9(3): 10 (1989)—China.

multifidum

var. pearcei *(Phil.) R. Rodríguez R. in Gayana, Bot.*, 44(1–4): 51 (1987 publ. 1988): Polystichum pearcei Phil.

nigrum *Ching & H.S. Kung in Bull. Bot. Res. North-East. Forest. Inst.*, 9(3): 8 (1989)—China.

obtusipinnum *Ching & H.S. Kung in Bull. Bot. Res. North-East. Forest. Inst.*, 9(3): 6 (1989)—China.

×pseudosemifertile *T. Nakaike & V.L. Gurung in Bull. Nation. Sci. Mus., B (Tokyo)*, 14(3): 81 (1988), as 'pseudo-semifertile'—Nepal. (P. lentum × semifertile).

salwinense *Ching & H.S. Kung in Bull. Bot. Res. North-East. Forest. Inst.*, 9(3): 7 (1989)—China.

subintegerrimum *(Hook. & Arn.) R. Rodríguez R. in Gayana, Bot.*, 44(1–4): 48 (1987 publ. 1988): Aspidium subintegerrimum Hook. & Arn.

subintegerrimum *(Hooker & Arnott) D.S. Barrington in Ann. Missouri Bot. Gard.*, 76(2): 373 (1989), Tisonym: Aspidium subintegerrimum Hooker & Arnott

talamancanum *D.S. Barrington in Ann. Missouri Bot. Gard.*, 76(2): 369 (1989)—Costa Rica, Panama.

POLYSTICHUM :-

×tare-bhirense *T. Nakaike & V.L. Gurung in Bull. Nation. Sci. Mus., B (Tokyo),* 14(3): 83 (1988)—Nepal. (P. piceopaleaceum × squarrosum).

PTERIDIUM

aquilinum

subsp. atlanticum *C.N. Page in Watsonia,* 17(4): 431 (1989)—Great Britain.

subsp. latiusculum *(Desv.) C.N. Page in Watsonia,* 17(4): 429 (1989): Pteris latiuscula Desv.

PTERIS

shimianensis *H.S. Kung in Fl. Sichuanica,* 6: 231 (1988), nom. nov.: Pteris aspericaulis var. cuspigera *Ching*

PTEROZONIUM

retroflexum *Mickel in Acta Bot. Venezuel.,* 14(3): 7 (1984)—Venezuela.

S

SACCOLOMA

elegans

subsp. chartaceum *G.B. Nair ex G. Cremers & K.U. Kramer in Bot. Helvetica,* 99(1): 47 (1989)—Guyana, Guatemala, Belize, Costa Rica, Panama, Cuba, Jamaica, Hispaniola (Dominican Republic), Venezuela, French Guiana, Colombia, Ecuador, Bolivia, Peru, Brazil.

SCEPTRIDIUM

officinale *(Ching) Ching & H.S. Kung in Fl. Sichuanica,* 6: 101 (1988): Botrychium officinale Ching

SCLEROGLOSSUM

koealagoguambaense *M.G. Baumann-Bodenheim, Syst. Fl. Neu-Caledonien,* 7: 167 (1989)—New Caldeonia.

SELAGINELLA

serratosquarrosa *N. Quansah in Fern Gaz. (U.K.),* 13(5): 282 (1989), as 'serrato-squarrosa'—Cameroun, Gabon.

tuberosa *B. McAlpin & D.B. Lellinger in Brenesia,* 24: 409 (1985 publ. 1986)—Costa Rica.

SOROLEPIDIUM

integrilobum *Ching ex Y.T. Hsieh in Bull. Bot. Res. North-East. Forest. Inst.,* 9(3): 48 (1989)—China.

T

THELYPTERIS

altitudinis *M.M. Ponce in Darwiniana,* 28(1–4): 345 (1987 publ. 1988)—Argentina.

arborea *(Brause) A.R. Smith in Acta Bot. Venezuel.,* 14(3): 7 (1984): Dryopteris arborea Brause

binervata *A.R. Smith in Acta Bot. Venezuel.,* 14(3): 6 (1984)—Venezuela.

brevisora *(Rosenst.) M.M. Ponce in Darwiniana,* 28(1–4): 349 (1987 publ. 1988): Dryopteris diplazioides var. brevisora Rosenst.

browniana *M.M. Ponce in Darwiniana,* 28(1–4): 365 (1987 publ. 1988)—Argentina.

patens

var. smithiana *M.M. Ponce in Darwiniana,* 28(1–4): 373 (1987 publ. 1988)—Argentina, Bolivia, Brazil.

rhachiflexuosa *R. Riba in Amer. Fern J.,* 79(3): 122 (1989)—Mexico.

TRICHIPTERIS

flava *R.M. Tryon in Fieldiana, Bot., n.s.,* 20: 125 (1989)—Peru.

serpens *R.M. Tryon in Fieldiana, Bot., n.s.,* 20: 126 (1989)—Peru.

TRICHOMANES

polyphlebius *V. Marcano in Ernstia,* 52: 5 (1989)—Venezuela.

V

VITTARIA

montana *V.S. Manickam in Indian Fern J.,* 5(1–2): 180 (1988)—India, Sri Lanka.

Z

ZYGOPHLEBIA L.E. Bishop in Amer. Fern J., 79(3): 107 (1989).

cornuta *(Lellinger) L.E. Bishop in Amer. Fern J.,* 79(3): 112 (1989): Grammitis cornuta Lellinger

dudleyi *L.E. Bishop in Amer. Fern J.,* 79(3): 113 (1989)—Peru.

eminens *(C. Morton) L.E. Bishop in Amer. Fern J.,* 79(3): 113 (1989): Grammitis eminens C. Morton

longipilosa *(C. Chr.) L.E. Bishop in Amer. Fern J.,* 79(3): 109 (1989): Polypodium longipilosa C. Chr.

mathewsii *(Kunze ex Mett.) L.E. Bishop in Amer. Fern J.,* 79(3): 108 (1989): Polypodium mathewsii Kunze ex Mett.

ZYGOPHLEBIA :-

sectifrons *(Kunze ex Mett.) L.E. Bishop in Amer. Fern J.*, 79(3): 110 (1989): Polypodium sectifrons Kunze ex Mett.

villosissima *(Hook.) L.E. Bishop in Amer. Fern J.*, 79(3): 117 (1989): Polypodium villosissimum Hook.

werffii *L.E. Bishop in Amer. Fern J.*, 79(3): 115 (1989)—Peru.